Computer Aided Design of 3D Printable Anatomically Shaped Medical Devices

Methodologies and Applications

Filip Górski
Poznan University of Technology
Institute of Materials Technology
Poznan, Poland

CRC Press
Taylor & Francis Group
Boca Raton London New York

CRC Press is an imprint of the
Taylor & Francis Group, an **informa** business
A SCIENCE PUBLISHERS BOOK

Cover credit: The cover illustration was generated using DALL-E (version 2), a generative AI tool by OpenAI, under the author's supervision, Filip Górski. The tool was employed to create visual concepts aligning with the book's themes of computer-aided design and additive manufacturing. The author has reviewed and refined the content of the generated illustration to ensure its accuracy and relevance. All responsibility for the final version of the illustration rests solely with the author.

First edition published 2025
by CRC Press
2385 NW Executive Center Drive, Suite 320, Boca Raton FL 33431

and by CRC Press
4 Park Square, Milton Park, Abingdon, Oxon, OX14 4RN

CRC Press is an imprint of Taylor & Francis Group, LLC

Library of Congress Cataloging-in-Publication Data (applied for)

ISBN: 978-1-032-47920-0 (hbk)
ISBN: 978-1-032-47922-4 (pbk)
ISBN: 978-1-003-38654-4 (ebk)

DOI: 10.1201/9781003386544

Typeset in Times New Roman
by Prime Publishing Services

Preface

Today, in the early decades of the 21st century, there are growing voices expressing concern that science is experiencing a global crisis. Anti-science movements actively challenge and undermine the achievements of dedicated scientists who strive to communicate their findings to the public, while scientists themselves often find their progress hindered by the demands of rigor, advancing their fields only incrementally. In such a climate, it is easy to forget that science, at its core, exists to serve people. It is the result of brilliant minds daring to explore the nature of things, shaping the wonders of our advanced technical civilization. Over the past 200 years, these advancements have transformed our lives in ways once unimaginable, and today, few of us can envision a world without them.

When I started my work in the field of technical sciences, I have always imagined that I would be able to create inventions that help people – just to make daily lives easier and better. In the last couple of years I have been able to pursue that dream. With a background in mechanical engineering and industrial computer science, I have spent over a decade at my alma mater, Poznan University of Technology, where my research has focused on harnessing the power of computer-aided design (CAD) and additive manufacturing (AM) for the development of patient-specific medical devices. My work represents an ongoing endeavor to make highly customized orthopedic, prosthetic, and surgical devices accessible, effective, and affordable. This book gathers the insights and methodologies I have developed with my research team and aims to illustrate both the technical pathways and collaborative approaches that drive meaningful progress in the realm of medical engineering, through the use of expertise from mechanical engineering, materials engineering and computer science, among others.

The work presented here is not solely the result of theoretical investigation. It reflects active collaboration of myself and my team with hospitals, doctors, and—perhaps most importantly—patients. Feedback from those who use these devices in everyday life informs every step of the development process, ensuring that the outcomes are not only technically sound but also relevant and practical. The chapters containing case study results on orthopedic devices and surgical supplies emphasize this perspective, detailing the challenges and rewards of designing for specific patient anatomies and the logistical demands of healthcare settings.

This book also recognizes that CAD and 3D printing for healthcare remain emerging fields, with challenges such as clinical validation, regulatory pathways, and long-term patient outcomes requiring ongoing attention. Advancements like automated design workflows and adaptive AM techniques can address these gaps, but they also necessitate an interdisciplinary approach that unites engineering with clinical insights. Through an attempt at a holistic view, I hope to provide readers of this book with both the technical grounding and strategic foresight to pursue new directions in medical device innovation.

My aim is that this book will serve as a resource for engineers, researchers, and clinicians alike—anyone driven by a commitment to enhance patient care through innovative, yet accessible, scientifically grounded solutions. From the smiles of young prosthesis recipients to the satisfaction of surgeons utilizing custom anatomical models, the impact of these technologies is clear and compelling. It is my hope that this book contributes to a broader movement in healthcare, one that values both precision engineering and the patient-centered mindset essential for a future where personalized medical solutions are within everyone's reach.

Acknowledgements

This book would not have been possible without the invaluable contributions of many individuals and organizations. I would like to extend my deepest gratitude to a number of people and organizations for their support and collaboration throughout this journey.

Co-Workers and Collaborators

I am profoundly grateful to my whole team at Faculty of Mechanical Engineering of Poznan University of Technology, as well as other co-workers and collaborators from many different institutions, whose expertise and insights have been instrumental in shaping the contents of this book.

Dr Wiesław Kuczko
Dr Radosław Wichniarek
Dr Magdalena Żukowska
Dr Przemysław Zawadzki
MSc Paweł Buń
MSc Justyna Rybarczyk
MSc Natalia Wierzbicka
MSc Radosław Paszkiewicz
MSc Jakub Gapsa
Prof. Adam Hamrol
Prof. Michał Rychlik
Dr Remigiusz Łabudzki
Dr Dominik Rybarczyk
Dr Krzysztof Łukaszewski
MSc Beata Czerkas
Dr Sabina Siwiec
Dr Jacek Banaszewski with the team
Dr Maciej Okła with the team
Dr Krzysztof Grandys with the team
Prof. Razvan Pacurar and Dr Ancuta Pacurar
Prof. Filippo Sanfilippo
and many others.

Students

Special thanks to my dedicated students, whose hard work and enthusiasm have greatly contributed to the research and writing of this book and many other publications. They realized many interesting clinical cases, studies and engineering work under my supervision working as part of the whole research team, contributing to the greater picture which is presented in this book.

- Olga Komorowska
- Daria Boberska
- Emilia Smolarek
- Klaudia Jańczak
- Piotr Dorna
- Monika Lulkiewicz
- Marta Burdzińska
- Michalina Jankowska
- Paula Andrzejewska
- Anna Pasierbek
- Sandra Słupińska
- Katarzyna Turostowska
- Aleksandra Grohs
- Aleksandra Kupaj
- Renata Jezińska
- Weronika Czerniejewska
- Natalia Sahaj
- Beata Wysocka
- Izabela Wesołowska
- and many others.

Research Projects

This book is the culmination of several research projects, which have provided the foundational data and insights:

- EMERALD European Network for 3D Printing of Biomimetic Mechatronic Systems (2022–2023)
- *AutoMedPrint* – Automation of the design and rapid production of individualized orthopedic and prosthetic products based on data from anthropometric measurements (2018–2021)
- *BRIGHT* - Boosting the scientific excellence and innovation capacity of 3D printing methods in pandemic period (2021–2023)

Funding Sources

I am also deeply appreciative of the funding sources and agencies that have made this research possible:

- Polish National Center for Research and Development
- Polish National Science Center
- Polish Ministry of Science and Higher Education
- European Commission
- EEA and Norway Grants

Personal Acknowledgements

I would like to thank my family for their full support and encouragement throughout this project. Your patience and understanding have been a very important part in helping me complete this work.

I would also like to extend my gratitude to all the participants and subjects, who contributed to the research studies.

Lastly, I thank you, the reader, for your interest in this work. I hope that this book will provide valuable insights and contribute to the ongoing dialogue in the scientific community.

AI Use Disclaimer

During the creation of this book, DALL-E (version 2), a generative AI tool by OpenAI, was utilized to develop conceptual illustrations. The tool was employed solely for generating preliminary visual ideas based on the author's detailed instructions. The final illustration was curated and refined by the author to ensure its integrity, thematic relevance, and accuracy. The use of this tool complied with Taylor & Francis guidelines for the responsible application of Generative AI technologies.

Contents

CHAPTER 1
Introduction

Invention and rapid development of three-dimensional printing, also known as additive manufacturing, brings a new era in various fields, including medicine, by offering unprecedented flexibility, speed, and customization capabilities. Among its many applications, the design and production of anatomically shaped medical devices is one of the most potentially useful in the context of personalized healthcare and life quality improvement. This book, "Computer Aided Design of 3D Printable Anatomically Shaped Medical Devices: Methodologies and Applications," aims to somehow bridge the gap, observed on the basis of tangible, proven possibilities and perceived reality, between the potential of 3D printing technology and its application in creating innovative and highly personalized, yet still accessible medical devices.

To dive into the complex realms of medical design and manufacturing, a thorough understanding of various disciplines is necessary, including—first and foremost—biomedical engineering, but also mechanical engineering, material science, and, crucially, the human anatomy. The target audience of this book, as thought of by the author, is formed by a wide spectrum of professionals and enthusiasts with mixed specialties: from biomedical and mechanical engineers to reconstructive surgery practitioners, orthopedists, physiotherapists, and MedTech startup owners. The book draws its appeal from a detailed examination of practical use cases, pioneering methodologies, and compelling accounts of how digital design and 3D printing are revolutionizing modern medicine.

At its core, this publication asserts that low-cost 3D printing technologies represent a largely untapped opportunity in the medical sector. Although transformative for over three decades, the capabilities of 3D printing to produce cost-effective, anatomically accurate medical devices have not been fully leveraged. Challenges abound, particularly in the design of 3D printable items that can faithfully mimic complex human anatomies. This book aims to address these challenges by introducing robust, efficient Computer-Aided Design (CAD) methodologies for creating devices that are anatomically conformant and can be economically manufactured through 3D printing.

The shift from traditional to additive manufacturing marks a significant paradigm change in medical device creation. Unlike conventional methods that typically remove material, additive manufacturing adds it layer by layer, enabling more intricate designs that were once either impossible or economically impractical. This is not just a technical shift but a philosophical one as well, advocating for bespoke, patient-specific solutions over generic ones, thereby fitting individual human anatomies with extraordinary precision. However, for effective implementation of additive manufacturing, a digital representation of the intended product is essential, necessitating a new suite of computer tools which is often lacking in current literature.

The book explores the methodologies, applications, and transformative potential of CAD and 3D printing in the medical domain. The subsequent chapters offer readers a detailed guide through the complex processes of designing and manufacturing anatomically shaped devices, for use by patients, doctors, engineering specialists and learners. Whether an engineer, a medical professional, a student, or simply an enthusiast of 3D printing technology, this book was written to inspire, educate, and encourage everyone to explore the vast possibilities of this particular branch of personalized medicine and convince the readers of its usefulness.

Central to the creation of these personalized devices is the science of Computer Aided Design (CAD), perceived by some as an art in the context of shaping organic geometries. This book delves into both the basics and more advanced problems of designing anatomical models, emphasizing the connection between use of available engineering software and deep anatomical understanding. A significant part of the book is devoted to the use of intelligent CAD models and automated design systems that leverage automation to streamline the design process, making it faster, more accurate, and adaptable to the unique needs of each patient.

The design process is relying heavily on anthropometric data acquisition. The methods presented in the book range through 3D scanning and photogrammetry to medical imaging techniques such as Computed Tomography (CT) and Magnetic Resonance Imaging (MRI). The transformation of this data into properly constructed 3D printable models is a journey through advanced features of available CAD software and methodologies developed through years of research and real-world applications. Each step in this process, from data capture to model refinement, is described in a way to ensure that the final product is not only an accurate representation of the patient's anatomy, but also practical for use in medical treatment or rehabilitation.

Furthermore, this book navigates through the spectrum of additive manufacturing technologies, quite wide despite relatively short time that passed since the first inventions in that field. The book is focused on the most popular technologies: polymer extrusion, photopolymerization technologies, sintering (powder-based) processes, and vacuum casting. Each technology is examined for its suitability in medical device fabrication, with a keen eye on the possibilities and limitations inherent in each method. This overview ensures that readers are able to assess and decide about selection of appropriate manufacturing techniques for their designs.

The book was created on the basis of a serious real-life experience, i.e., numerous, actual case studies and stories from patients and doctors. These narratives provide a tangible context for the technical discussions, highlighting how the intelligent application of CAD models and 3D printing technology can lead to remarkable outcomes in patient care.

The first part of the case study exploration is focused on the personalized products made by easily available, inexpensive 3D printing technologies, aimed to use directly by the patients, especially the little ones – small children and teenagers. Their rapidly changing bodies, when met with a necessity of aiding either through a prosthetic replacement, or an orthotic supply, require these customized products to grow alongside them, posing serious challenges for mass and cheap production. This chapter of the book is dedicated to the selected orthotics and prosthetics and focused mostly on the results of AutoMedPrint project realized by the author and his team, with great success, but many remaining challenges. Examples of personalized, cheap, 3D printed orthotic devices for treatment of injuries and long-term conditions are presented, along with full technical description of how these products came to life and what issues were resolved along the way. These narratives highlight not only the technical challenges and solutions but also the human stories behind each device. From orthotic devices that restore mobility and independence to prosthetics that offer new hope and functionality, each case study is a testament to the power of personalized medical solutions.

The second part of the case study description, and also the penultimate chapter of the book, is focused on what todays 3D printing and CAD can do for doctors, practicing personalized surgery, especially aimed at curing cancer or bringing back important functionalities and looks of human body through reconstructive operations. Different approaches are being presented, with case studies based on cooperation with medical specialists of various hospitals, with differentiated needs and preferences, working on real and difficult patient cases. The ability to 3D print anatomically accurate models for surgical planning and training, as well as customized tools and implants for use in surgery, represents a significant advancement in surgical practice. Apart from digital design and 3D printing, use of interactive immersive technologies, such as mixed and virtual reality, is mentioned and presented, to complete state of the art description of what modern digital technology can do.

As we look to the future, the role of CAD and 3D printing in advancing personalized medicine is both promising and profound. The methodologies and applications presented in this book are at the forefront of this revolution, offering a glimpse into a future where medical devices are tailored not just to the medical condition, but to the individual patient, on demand. This approach has the potential to improve outcomes, reduce recovery times, and enhance the quality of life for patients worldwide, similarly helping doctors increase their work satisfaction and reduce their workload.

In the author's opinion, this book is not just a technical manual and presentation of some scientific results. It is a vision for the hopefully probable future of medicine. It is an invitation to engineers, medical professionals, researchers, and innovators to explore the potential of 3D printing and computer-aided design in creating solutions that are as unique as the individuals they are designed to help. Analyzing what has been done so far and how quickly these technologies evolve, it is clear that the convergence of technology, medicine, and personalization will play a pivotal role in shaping the future of healthcare.

This future, while promising, is not without its challenges. The book shows and mentions some of the problems in widespread adoption, including the need for streamlined design processes, improved materials, and regulatory considerations, as well as some human constraints. However, it also offers a roadmap for overcoming these obstacles, drawing on the author's extensive experience and years of research in the field.

To end on a high note: we are not just designing devices; we are reshaping lives. We are not just engineering solutions; we are crafting hope. And in every filament, every layer, and every printed body part, we are building a future where the care we provide is as unique as the people that we are helping. This is the essence of this book —a guide to the future of customized, tailored medicine, where technology meets humanity in the service of health and well-being.

CHAPTER 2

CAD and 3D Printing in Medicine

Basic Concepts

2.1 Computer-Aided Design basics

2.1.1 Design processes and their role in modern engineering

Human beings – *homo sapiens*, are predators by nature. We rely on visual communication and sight is the single most important sense that we use to perceive our surroundings. That is why, since the beginning of the species, humans have communicated using drawings, not least because they are mostly independent on the language used in daily life by their recipients. This is translated to engineering activities performed over the course of progress of our technical civilization. Every non-trivial technical device must be first designed – that is, conceptualized and drawn – before obtaining physical form of prototype of ready product.

By the Cambridge Dictionary, design, in a general sense, is the art of making plans or drawings for something, or the way in which something is planned and made ("Design" 2024). It is usually associated with creating a graphical abstract of an object, that does not yet physically exist at the moment of performing the design – to easily convey the idea and its characteristic features from one person to another, in a more efficient manner than words (text – written or oral).

Design engineering is a set of activities performed by engineers, aimed at preparing a detailed, technical description of a new product (or its part), containing its visual representation – more or less abstract – accompanied by various text descriptions at the very core of the process (Kroes and Kroes 2012). What we call a project, or design, or layout of a new technical object or a system of objects, ultimately is destined to facilitate its coming into physical existence, usually by various manufacturing processes. As such, design engineering can be basically considered as one of first stages in the production process, preceded only by the concept of the thing that will be produced. The concept, in turn, is a result of answering to a specific need, or problem, expressed by a sufficiently large group of people. Therefore, tracing back, design is a mean of problem solving and fulfilling needs of people – that is what it was through almost the whole history of humanity and its development as technical civilization.

An essence of the engineering design processes is creation of models. A model is a representation of various aspect of an entity that exists or can theoretically come into existence (Wartofsky 2012). Usually, during the design processes, a number of different models are created – there can be mathematical models (Ilhami and Masruroh 2020), geometrical models (Canciglieri et al. 2010), simulation models (Hussaini and Lahrman 2012) and others. The act of creating models is called "modeling" and often is the most time-consuming part of the whole design process. In the computer era, models are mostly digital, with geometrical models functioning as full, visual, three-dimensional representations of features of a product-to-be so-called 3D models. Figure 2.1 shows examples of various models that can be used in the design processes.

The design engineering processes (and their results) nowadays, in the era of information and digital world, have more and more functions to fulfill. The original function of engineering design – to store features of a yet non-existing technical object in an abstract form able to be exchanged between different specialists – remains of course at the very core. However, many other functions are also present. As we increase our ability to create digital, virtual worlds and environments resembling the real world in many aspects, the virtual representation of our future product gets more and more sophisticated and usable, long before it takes any physical form. This product, fully formed and usable to a certain extent, but existing only in digital

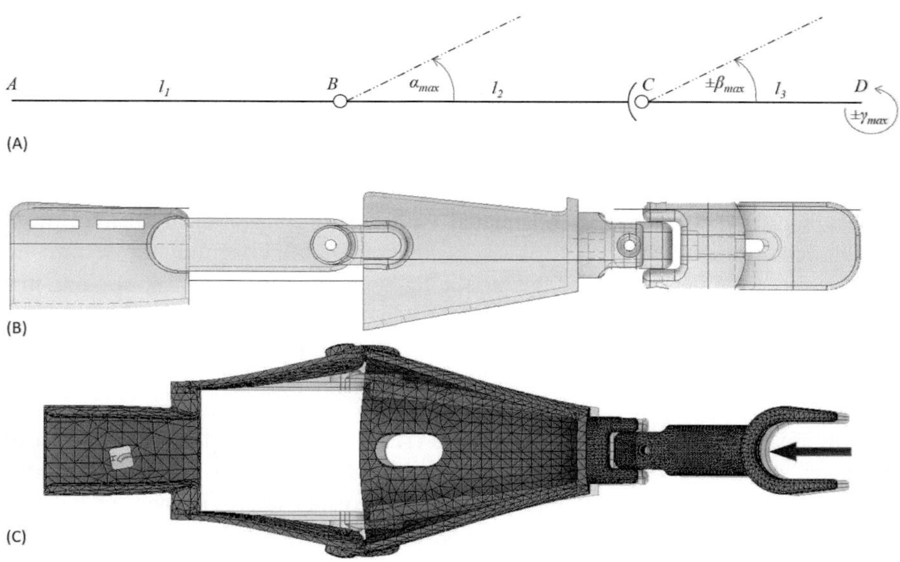

Figure 2.1 Models in the design processes, on example of hand prosthesis: (A) mathematical model; (B) geometrical model; (C) simulation model.

Figure 2.2 Testing a virtual product (A) along its physical prototype (B), on example of a hand prosthesis.

space, is often called a virtual product (Hirschel 2003). The act of its creation and putting into digital use is known in literature as virtual prototyping (Wang 2002). A concept of digital twins is also known and widely used to describe the general idea, as well as its companion notions of digital shadow and digital model (Grieves 2023), entities functioning at various levels of connection between a real, physical object (or system) and its digital counterpart. Modern virtual technologies, such as Virtual Reality or Mixed Reality (VR/MR) can be used in the process of creating and testing of virtual products (Arrighi and Mougenot 2019), as demonstrated in Fig. 2.2.

However, virtual prototyping is only the initial phase of use of models created at the design stage. Product representations can be then used in multitude of other systems, to proceed with their production. In a modern production company, design does not function as an island activity, but rather is integrated digitally with all the rest of processes happening in the company, to allow smooth implementation of ready design into production and then use by the clients, as soon as possible. This integrated digital approach is usually introduced in largest companies (Steier et al. 2022) and its details will not be covered in this book.

The design processes consist in various activities undertaken by design engineers, aimed at giving a product its shape and features. Depending on the type of product and level of intellectual challenge related to its designing, different types of design can be distinguished (Schaub 2004):

1. *Routine design* – when the task is to design a product with similar functions or shape to a product that already exists or is available somewhere. The challenges here are usually minor (even if the design is realized from the scratch) and might consist in adjusting the material, technology, or size of the product, or introducing of minor new functions or elements into already known design. An example – in the realm of medical products – might be adjusting a known design of a prosthetic device to a new group of patients.
2. *Innovative design* – when the task is to design a product fulfilling the known need in a new way, not previously known. Some major challenges might be needed to overcome here, including geometrical, material or processing constraints,

or unusual requirements of product recipients. Usually an entirely new design, partially comparable to previously available solutions, is created as a result, and it can be considered as a product innovation (one of main innovation types (Reguia 2014)). In the realm of medical products, it might be building a new class of implants, for patients having specific condition, not previously treated that way.

3. *Creative design* – the most difficult and rare one – when the need and problem is new, or previously not tackled, and an entirely new class of solutions must be conceived to solve the problem. Here, no previous solutions exist, or they are bad and should not serve as inspiration. As the name suggests, a true dose of creativity is needed to approach the problem with a fresh set of ideas, and create new systems, with new geometries and functions, spawning not only singular innovations, but rather whole groups of them. In case of medical products, invention of the first hip implant (Markatos et al. 2016) or first bionic prosthesis (Clement et al. 2011) might be called an example of creative design. In the consumer world, the birth of the iPhone was one of the better known examples of creative design, when the need was not even formulated (Tyhaar 2021) – one might say that is was rather created.

According to various studies, most of the time is spent on routine design (Schaub 2004, Rigger et al. 2018), however it is the least productive and most repetitive time and there is a strong trend to reduce it whenever possible (da Silva Vieira et al. 2010). Creative design is usually very rare and most design engineers – even highly specialized ones – might not encounter the necessity of it through their whole careers. Innovative design is frequently required to be implemented at company level, especially in highly competitive branches of industry, such as automotive or medical industries, as its motivation is frequently striving towards being at least slightly better (in terms of various numbers characterizing the products and their manufacturing processes) than the current competitors. Creative design, when implemented with success, results in products characterized by significantly better characteristics than existing solutions and as such can spawn ranges of entirely new products and business models (Schaub 2004, Li et al. 2007) – but how, when, and why it happens is not within a scope of this book.

The role of the design engineer in a modern company constantly changes and evolves. However, the key component is to be a stable foundation, on which the further stages of the production processes are placed. The design engineer constantly delivers projects that must be improved in comparison with their predecessors, or contain new contents that answers to current production demand. Depending on the industry branch, type of product and the market, the requirements and activities might be drastically different. Especially in industries where customization (or even mass customization, which will be explained later) plays a vital role, design engineers are under a huge pressure to deliver constantly new results of high quality, in a short time. Fortunately, modern computer technologies can aid these processes (e.g., by automating or facilitating them), which is covered in the further parts of this book.

2.1.2 CAD place in product lifecycle and production companies

Design – regardless of whether it is computer-aided or not – is only a part of the first stage of development of a new or improved technical solution, as mentioned in the previous chapter. General framework of emergence, use and discontinuation of a given type of technical object has a name of product lifecycle, although there are many different definitions for this notion (Corallo et al. 2013). In general, two main types of lifecycles might be distinguished:

1. *Product lifecycle* – pertaining to the whole product as an abstract class, ranging from emergence of a need and concept, till the discontinuation of such need, and as such – production of instances of this class.
2. *Single instance lifecycle* – pertaining to a single instance of a given product, starting from its conceptualization (ordering by a client), ending with cease of use, often accompanied by recycling procedures.

Distinction between these two cycles has a profound meaning when referring to customized, individualized products. In these, each instance usually requires repeating almost the whole cycle, including the design process or its fragment. Typical, general product lifecycle divided into phases and stages is presented in Fig. 2.3, while more detailed lifecycle of a single instance of customized product is shown in Fig. 2.4.

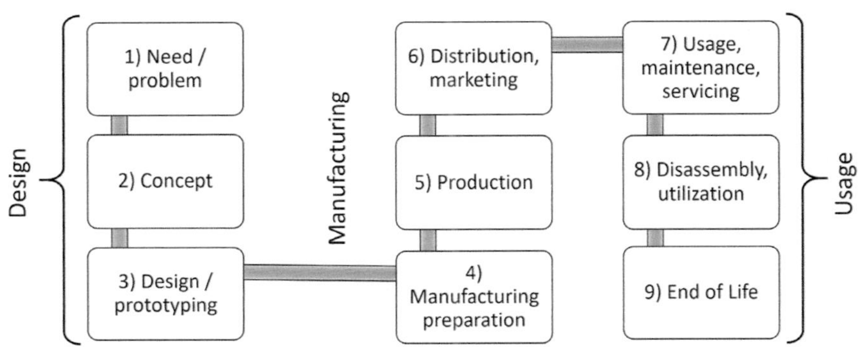

Figure 2.3 Product class lifecycle.

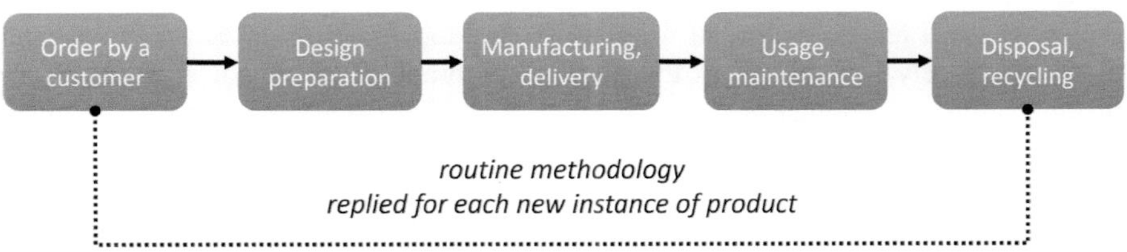

Figure 2.4 Product instance lifecycle.

In both cases of the cycles, three basic stages can be distinguished:

1. Design stage, where product does not yet have a physical form and cannot be touched or interacted with in the real (non-digital, non-abstract) world. In other words, there are models, drawings, and virtual prototypes, but no materials have been processed yet.
2. Manufacturing stage, at which the product becomes physical, as it comes from the digital realm into factory floor level, but it is kept inside the company factory and not yet made available and distributed to its final recipients.
3. Use stage, where the product is advertised and shipped to its recipients – clients – and comes into use in real conditions. This stage also incorporates all the servicing and maintenance activities, as well as decommissioning of the product and reverting its components back to raw material stage (hopefully through the means of recycling).

If the product is a typical example of a serial production, there should be high disproportion between time spent on its design and the rest of the stages. As the product is repetitive, its design is relatively short and after it is done, engineers may focus on improvements or entirely new solutions. Any innovations focused on the reduction of the time to market should be as such placed in the manufacturing processes – what has been done since the start of the industrial era, through industrial revolutions (currently at the stage of so-called Industry 4.0 (Culot et al. 2020), with a key concept of smart factory (Hozdić 2015)).

However, in the case of customized, individualized products, the story is a bit different. The time spent on various lifecycle stages is more evenly distributed between design – repeated every time a new product instance is ordered by a customer – and manufacturing, that additionally needs to flexibly adjust to the constantly changing shapes and features of created products. Here, innovations implemented in the design stage, or holistically at the whole lifecycle, could yield much better results than the improvement of manufacturing processes themselves.

In the era of the fourth industrial revolution it is worth considering, as the producer market is being replaced with a consumer market, and mass production is – will be – replaced by mass customization, with each piece of a given product being a bit different than the other piece (Fogliatto et al. 2012). The mass customization – described in more details in one of the subsequent chapters – is a concept well known since at least a couple of decades (Tseng and Jiao 2001), recently more and more intensively introduced, in the automotive branch (Kabasakal et al. 2017) or in other consumer branches (Kayğın Sel 2013). From the perspective of medical products aimed to use with specific patients and their condition, this approach seems perfectly natural. However, it is much more challenging than the standard, routine design, and production of the same products in mass quantities – that is why in the medical production modern serial manufacturing techniques have not yet been fully introduced in cases of many products – only singular publications can be found for, e.g., prosthetic sockets (Xu and Qin 2023) or leg orthoses (Pallari et al. 2010), not being a common practice.

As a result of putting the design in the lifecycle, CAD – or widely putting, CAx systems must be considered just one part of the whole ecosystem in modern production companies, where the product's life is overseen and controlled, with attempts at integration of all knowledge about the product inside the digital environment of a given company. These attempts are a domain of Product Lifecycle Management (PLM) systems (Saaksvuori and Immonen 2008), which integrate documentation, models, applications, knowledge, and workflows related to various products in a single computer system, accessible by company employees at different levels, depending on their functions and tasks, seriously influencing performance of product development (Tai 2017). However, it is important to notice that as Small and Medium Enterprise (SME) level, PLM concept is largely underdeveloped (Duigou et al. 2011), being mostly a domain of large companies. It also pertains to medical production, where PLM concept implementation encounters more difficulties than in automotive or other technologically advanced branches (Martínez Gómez et al. 2019).

2.1.3 Computer-Aided Design and CAx systems

In shortest terms, Computer-Aided Design (CAD) is the use of computer technology to help engineers prepare a design of a new or improved product (Encarnacao et al. 2012). In technical terms, Computer-Aided Design (CAD) systems are computer software and hardware tools used to create, modify, analyze, and optimize 2D and 3D designs and drawings. CAD systems have revolutionized the design and engineering fields, making the creation and optimization of complex designs more efficient and accurate. They continue to evolve and play a pivotal role in various industries. CAD systems are widely used in architecture, mechanical engineering, production engineering, and medicine. The hardware side of these systems usually pertains to computers of sufficiently high computing power and specialized graphics cards, as well as proper input devices, with no upper limit of required performance, especially considering working on advanced shapes or complex products. The software part are usually dedicated programs, with various tools for creation of geometrical models, both in two- and three-dimensional space.

The key feature of CAD systems is visualization – real-time rendering of how the product shape looks like at the current stage of its (virtual) development. In the basic form, visualization is required to work with the subsequent design stages – to observe the changes and new features introduced by the engineer working on the product. The visualization in CAD systems is often accompanied with a hierarchical description – so-called model tree – showing the features, elements and often history of modeling of a given object. More detailed examples and ways of conduct are presented in Chapter 4.

The CAD systems, such as SolidWorks, Inventor or CATIA are to be distinguished from the 3D modeling software. Both enable creation of three-dimensional geometrical models of various objects. However, 3D modeling software is a slightly broader term, covering also the solutions that are not used in engineering processes, but, e.g., for entertainment, education and so on – examples of well-known non-CAD solutions, used widely for decades now, involve 3D Studio Max, Maya, Rhinoceros or Blender.

Traditionally, CAD software is also associated with parametric modeling – a notion of defining and maintaining relationships (constraints) between various geometric features, both in 2D (sketches) and 3D (modeling operations), often ordered by time of creation, in the already mentioned model tree. More on parametric modeling and its key concepts is presented in further chapters. However, non-parametric approaches are also utilized in CAD systems, including working on meshes, or directly introducing certain features, without basing on any kinds of parameters. The non-parametric approach is often called direct modeling (Fandáková et al. 2019) and in its pure form is rather rarely met. Frequently, in a software offering such possibility, direct modeling is intertwined with parametrically generated features (Zou et al. 2023), thus creating a notion of synchronous modeling.

The parametric modeling is a standard way of conduct in many industries, where machine parts take form of more simple geometries and predictive model trees, such as in automotive industry. Non-parametric approach, on the other hand allows easier editing of geometrical features and introducing new, complex ones, but usually history and parameters of these changes are not kept. As such, the non-parametric approach has advantages when dealing with organic, intricate geometries and rarely repeatable structures. Non-parametric modeling is a less predictive process, which can be compared to a more artistic approach ("digital sculpting"), while parametric approach is highly technical, with each process and parameter staying under full control of the model creator – at least in theory.

As shown in further parts of this book, both approaches found their profound applications in medical engineering, depending very much on the source of data used for design and a class of a problem to be solved in a given design task. Medical design also more rarely than industrial design relies on typical, standard universal CAD solutions, frequently delving into open-source software with specific sets of function or specialized, dedicated programs aimed at designing of a single, predetermined type of products (e.g., MediACE3D software used for orthotic design (Khandagale and Pise 2022)).

Current development trends in CAD software include introduction of more free and less rigid approach to creating model trees. Hybrid solutions and freeform modeling are widely available, as well as more free exchange of files created in various platforms. This is also related to CAD going into the cloud – more and more cloud solutions for storing and working on CAD projects are available (Wu et al. 2015). Also, as mentioned before, integration between CAD and modern virtual (XR) technologies is more and more common, enabling to visualize designed objects in virtual reality or Augmented Reality (AR) for immersive design experiences, potentially helping with perception of a new product and making it better before actually starting production.

The newest trend is the use of generative Artificial Intelligence (AI) together with CAD, for easier creation of new models at the level of innovative design. AI was previously used together with CAD on more than one occasion (Hunde and Woldeyohannes 2022), but generative solutions are relatively new and their potential is not yet fully known (Ali et al. 2024). Potentially, it could revolutionize the field and allow easy automation of even very difficult design tasks, as is already happening considering use of generative AI in the field of graphics design (Tomić et al. 2023). However, in terms of 3D modeling such wide possibilities are yet to come. General automation of design (not always including AI) is also a trend, implemented widely by the author in previous research and implementation work (Górski et al. 2020a).

As mentioned earlier, CAD is a basic member of a larger family of computer systems aimed at aiding human specialists in development and maintaining of products through their full lifecycle. There are also other classes of systems belonging to that group, often applied together with CAD, over the course of development of a product. While CAD is the basic system class, allowing to obtain a digital representation (a model) of a given part or a product, the other CAx systems enable an engineer to perform further operations, once the basic shape is set. The most important classes of these systems are Computer-Aided Manufacturing (CAM) and Computer-Aided Engineering (CAE).

Computer-Aided Manufacturing is a very wide class of systems that are used for preparation of manufacturing processes, on the basis of a product representation obtained in the CAD modeling phase. In the most frequent use case, their operation boils down to preparing programs for Computer Numerical Control (CNC) machines for subtractive manufacturing processes, such as milling, laser cutting, lathing and others. Such CAM systems are very often accompanying CAD systems, forming single, integrated environments (an example is CATIA with its various modules). There are also plenty of other CAM systems, both dedicated (separate software) and integrated, for various types of manufacturing processes (Elanchezhian and Sundar 2007). A separate class of CAM software are slicers – programs, which are used to prepare and execute programs for additive manufacturing devices (3D printers). When discussing slicers, a notion of "CAM" is rarely used – however they belong to exactly this class of software. Slicers are an important part of additive manufacturing of medical products and as such will be described in one of the next chapters.

The second class of beyond-CAD CAx systems, very frequently used also for medical, anatomically adjusted products, are CAE systems. These are mostly used for aiding all sorts of engineering computations and simulations involving the actually designed geometry. The most fundamental and most popularly used class of applications here are the static strength calculations, implemented using finite element analysis (Lacroix et al. 2009), allowing to determine deformations and stresses in a constrained and loaded object or assembly. However, CAE software can be used for much more than just simple static computations – the so-called "multiphysics" software can be used to simulate loading in dynamic conditions, flow of liquid and gas in and around the object, change of temperature fields, electrical currents and much more. The CAE software is not a part of this book, however there are plenty of textbooks on the topic (Raphael and Smith 2003) that, albeit aimed at generic products, can be successfully applied for anatomical, biomedical products as well.

Apart from CAD, CAM and CAE, there are a number of other systems belonging to the CAx spectrum and also outside it, used together for product development and management. These include, but are not limited to (Lee 1999):

- Computer-Aided Process Planning (CAPP),
- Computer-Aided Quality (CAQ),
- Material Requirements Planning (MRP).
- Manufacturing Resource Planning (MRP II),
- Enterprise Resource Planning (ERP),
- Product Data Management (PDM),
- Product Lifecycle Management (PLM),
- Virtual Prototyping (VP).

Depending on the type, class, scale, used processes and requirements, these systems may or may not be used in development of anatomical medical products. However, systems usually associated with mass production and large companies, such as PLM, ERP or CAPP are rarely used for individualized production of customized medical devices. This situation, however, can change in the future.

2.2 Design for medicine and biomedical engineering

2.2.1 Medical products – concepts and key characteristics

A medical product, as broadly conceptualized, encompasses a diverse array of items and substances designed with the primary intent of interacting with biological systems – e.g., humans or animals – to achieve a wide range of health-related goals. These goals may include, but are not limited to, diagnosing diseases, preventing health issues, monitoring physiological conditions, treating, or alleviating diseases, injuries, or handicaps, and even modifying physiological processes or anatomical structures. The scope of medical products is vast, covering everything from direct interventions that sustain life and control conception to more auxiliary roles, such as disinfecting medical devices or facilitating *in vitro* examinations of biological specimens (Aronson et al. 2020, Tobin and Walsh 2023).

What really defines medical products is their application. They can be used directly, such as in the case of surgical instruments, implants, or pharmaceuticals that interact with the body's systems, or indirectly, as in the software used for the analysis of medical images or the sterilization agents used on surgical equipment. This broad application spectrum underscores the fundamental criteria for something to be classified as a medical product: its manufacture and use must be oriented toward health-related objectives, emphasizing safety, effectiveness, reliability, and compatibility with its intended purpose. By the very essence of classifying these products – abstract or physical – we may distinguish between software and hardware (Ahmadi 2023), digital products (software, models, simulations) and physical objects – implants, prosthetics, orthotics, surgical supplies, and many others.

Diving deeper into the classification, medical products can be segmented into several main types, including but not limited to pharmaceuticals (drugs and biologicals), medical devices (ranging from simple tools to complex machinery), biologics (such as vaccines and blood products), and diagnostic products (including *in vitro* diagnostic reagents and systems). Each category follows stringent regulatory standards that ensure they meet the required safety and efficacy benchmarks before they can be introduced to the market (Tobin and Walsh 2023, Chen et al. 2018b).

Basic definitions of various types of medical products can be found in widely accessible review papers (Tettey et al. 2024, Aronson et al. 2020). Pharmaceuticals, for example, are chemical or biological substances that interact with the body's systems to treat, cure, prevent disease, or affect the structure or any function of the body. Medical devices, on the other hand, can range from simple items like bandages or tongue depressors to sophisticated machines like Magnetic Resonance Imaging (MRI) scanners, often designed to aid in diagnosis, treatment, or prevention of diseases without interacting chemically with the body. Biologics include a wide range of products derived from living organisms, such as vaccines, which are used to invoke an immune response to protect against disease (Woo 2014). Diagnostic products encompass a variety of tools and reagents used in laboratories to examine specimens from the human body, aiding in the diagnosis of diseases and conditions (World Health Organization 2017).

On the other hand, items or substances not intended for any direct or indirect health-related use, or those that do not meet the criteria of safety, effectiveness, reliability, and compatibility for their claimed health-related purposes, do not fall under the category of medical products. Common knowledge examples involve nutritional supplements (not aimed at preventing or curing diseases), personal care products like cosmetics, used primarily for beautification or alteration of appearance (without claiming health benefits). The distinction between what constitutes a medical product and what does not can sometimes blur, especially with products at the intersection of health and consumer goods. However, the key differentiator remains the product's intended use in the diagnosis, prevention, monitoring, treatment, or alleviation of diseases or health conditions, supported by a foundation of scientific evidence and regulatory approval ensuring their safety and efficacy for those purposes (Marcus et al. 2016).

Medical products, with their direct application in diagnosis, treatment, prevention, and the general maintenance of health, serve as the tangible instruments through which the principles of medicine are applied, directly impacting patient care and health outcomes. However, a process of conceptualizing these products and developing their practical application, later delving into the healthcare system integration is not the same as in the case of developing consumer goods or engineering devices. In other words, an interdisciplinary approach is necessary, and as such we often use a notion of medical engineering, or biomedical engineering.

Medical engineering, interchangeably, disputably referred to as biomedical engineering (Enderle and Bronzino 2011), represents a dynamic and interdisciplinary field, being a mix of various branches of traditional engineering, medicine, and biology. This innovative domain utilizes methodologies and creative problem-solving mechanisms inherent in engineering disciplines, such as mechanical engineering, computer science and material engineering, applying them directly to the complex challenges of medicine and biology with the overarching aim of enhancing healthcare delivery. The essence of medical engineering lies in its foundational premise, which is to connect and use the potential of engineering achievements with the nuanced, intricate understanding of biological systems and medical protocols, thereby advancing the prevention, diagnosis, monitoring, and treatment of diseases and construct new, unique devices at the touch between technology and nature (Sree et al. 2023).

At the core of this integration is the development and refinement of technologies and methodologies that can improve the quality of patient care. This includes the creation of advanced diagnostic devices, innovative therapeutic strategies, and sophisticated monitoring systems, alongside the design of cutting-edge medical equipment and healthcare information systems. By translating engineering solutions into viable healthcare applications, medical engineering, and technology (shortly med-tech) plays a pivotal role in the evolution of medical practices and treatments, making healthcare more accessible, efficient, and effective (Chaturvedi and Srinivas 2021).

Biomedical engineering is a parachute term, consisting of several specialized areas such as biomechanics, biomaterials, medical imaging, and bioinformatics, among others. Biomechanics, for example, focuses on the mechanical aspects of biological systems, aiding in the design of prosthetics and orthotic devices that mimic or enhance natural movements, as well as monitoring movement and dynamics of living beings to enhance their capabilities and prevent or treat illnesses or injuries. Biomaterials, on the other hand, is a branch – a discipline in itself – that delves into the development of novel materials compatible with biological systems on various levels (externally and internally) for use in implants, prosthetics, tissue engineering, and drug delivery systems (Park 2012). Medical imaging combines advanced engineering techniques with applied physics and computer vision to enhance the visualization of the internal structures of the body, facilitating improved diagnosis and treatment planning (Suetens 2017). Bioinformatics and computational biology represent the convergence of biology, computer science, and information technology, aiming to understand biological data, particularly in the context of genomics and molecular biology, through the development of software and algorithms (Peter et al. 2019).

The interplay between engineering innovations and medical science not only fosters the development of technologies that improve clinical outcomes but also contributes to the basic understanding of biology and disease mechanisms. For instance, tissue engineering and regenerative medicine, sub-disciplines within biomedical engineering, are predicated on the principles of biology, chemistry, and materials science to develop biological substitutes that restore, maintain, or improve tissue function or a whole organ (Berthiaume et al. 2011).

Contrary to traditional engineering disciplines that primarily focus on the design and optimization of systems and materials for a broad range of industries, medical engineering's ultimate focus is mostly put on improving health outcomes. It means that the whole set of additional, formal requirements set in, involving a unique set of ethical considerations, regulatory compliance requirements, and a need for clinical validation to ensure that the engineered solutions are safe, effective, and beneficial to patients. An additional distinction, defining "clinical engineering" as a branch of its own, has been known for some time already (Bauld 1991), along with a job position of a clinical engineer – still not a widespread view, but as technological inventions start quickly spreading in healthcare, it is more and more justified to simply have an engineer in the patient treatment process all the time, to work alongside medical practitioners and aid them with technical knowledge.

Through its interdisciplinary approach, biomedical engineering embodies the convergence of engineering progress with medical insights, holding the promise of revolutionary advancements in healthcare. By bridging the gap between engineering and medicine, it paves the way for novel diagnostic tools, therapeutic strategies, and enhanced biomedical understanding, underpinning the future of medical practice and healthcare delivery.

Ultimately, though, a result of work of biomedical engineers – not without input from medical specialists and more traditional engineers hailing from selected branches – boils down to creation of certain products, as mentioned before, aimed at different use scenarios. Considering this view, medical products are categorized based on their application and risk factor. The most common classification, particularly for medical devices, ranges from Class I to Class III (Bevanda 2017):

1. Class I: Low risk (e.g., bandages, examination gloves, and hand-held surgical instruments), usually subject to the least regulatory control.
2. Class II: Intermediate risk (e.g., infusion pumps, surgical drapes), requiring more stringent regulatory controls to ensure safety and effectiveness.
3. Class III: High risk (e.g., implantable pacemakers, heart valves), subject to the highest level of regulatory control due to their significant risk of illness or injury.

The essential requirements for medical products revolve around safety, efficacy, quality, and regulatory compliance (Tettey et al. 2024, Marcus et al. 2016). The most important ones include:

- *biocompatibility* – ensuring that products do not provoke an adverse reaction from the body – types of these reactions are well-known and described in available literature (Black 2005); the biocompatibility may pertain to external (devices in contact with skin or mucosal) or internal use (implants, in contact with blood, muscles, intestines, etc.),
- *sterility* – related to biocompatibility, but often treated separately (Fu et al. 2021) – products must be free from harmful microorganisms,
- *stability* – products must maintain their intended physical, chemical, and microbiological properties, both when not used (in storage) and when operated outside or within patient's body (Mathieu et al. 2014),
- *traceability* – each product must be traceable back to its manufacturing source, with full documentations of realized processes available (Klein and Stolk 2018),

- *regulatory compliance* – products must meet specific standards and regulations set by governing bodies, such as the FDA (U.S. Food and Drug Administration) (Lauer et al. 2017) or EMA (European Medicines Agency) (Bevanda 2017).

Comparing the industrial and consumer products – cars, food, furniture, machine parts, cellphones, bicycles, etc. – with the medical products, it is immediately apparent and noteworthy that in the case of the latter, the structure of stakeholders, interested in the product one way or another, is quite different. It is more straightforward in the case of typical products, both from the technical and economical point of view – there is demand and there is supply, and the communication (feedback loop) is bidirectional – between the client and the producer, who – often in the case of customized products – work together on an idea (vision) of the product, making it better for both parties.

In the case of medical products, the situation is quite different. First of all, there are three parties involved – the clients now are replaced by doctors and patients, with varied interests and subconscious feelings about the product. The main user – patient – usually does not really want the product, in other words – they would prefer being healthy and not needing it. They also are not competent enough to decide about the product type, shape, and functions – this is a role of doctors, who – on the other hand – will not use the products for themselves, but for the health and wellbeing of their patients. The producers might work with both groups, depending on the product – some products are strictly for the doctors (patients do not even know of their usage – surgical tools are a good example) and some strictly for the patients (doctors prescribe them but never see patient directly using them). This creates very interesting dynamics of interactions, especially in the case when these products are individualized anatomically, tailored to a given human being either by shape or by function. At the same time, it poses high challenges in development of such products. This difference is illustrated in Fig. 2.5.

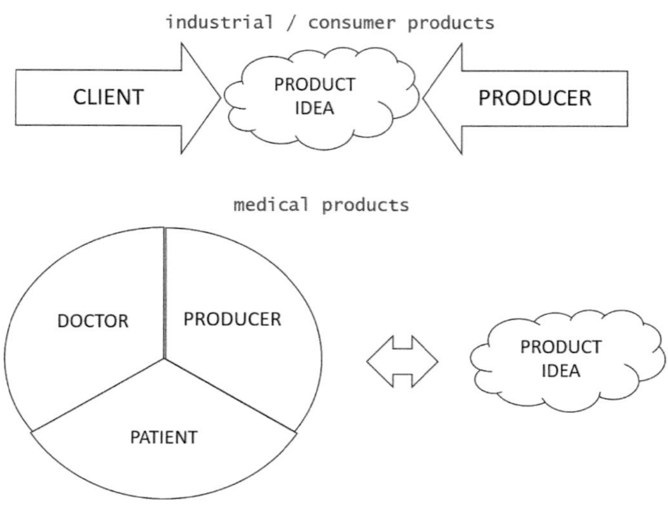

Figure 2.5 Participants of development process of typical and medical products.

The production of medical products involves several critical processes, each tailored to meet the stringent requirements of medical regulations and ensure patient safety (Ciurana 2014). These processes include, but are not limited to:

1. *Design and development* – this initial phase involves conceptualizing the product, including its intended use, safety features, and compliance with medical regulations. It often includes prototyping (frequently with use of 3D printing technologies) and extensive, lengthy testing of manufactured prototypes and first batches of ready products.
2. *Material selection* – choosing appropriate materials that are, first of all, biocompatible, then – durable, appropriately lightweight, suitable for the intended use of the product and possibly – ecologically friendly.
3. *Manufacturing* – this process involves the actual production, which usually occurs in a controlled environment, first to prevent contamination with unwanted microorganisms and toxins, second – to be able to repeat the process, obtaining products of high quality. Techniques vary widely from traditional machining to advanced methods like 3D printing, depending on the product.
4. *Quality control and assurance* – rigorous testing and inspection, conducted throughout the whole production process, to ensure that every product meets the required, aforementioned standards.
5. *Sterilization* – many medical products, especially in classes of II and higher undergo sterilization to eliminate any microorganisms, sometimes right before use.
6. *Packaging and labeling* – medical products are packaged in a manner that maintains their structural integrity and provides essential information, including usage instructions, warnings, and expiration dates, if applicable.

Apart from the products themselves, the manufacturing process in medical engineering is also highly regulated, to ensure that every product released to the market is safe, effective, and of the highest quality. This involves continuous oversight from regulatory bodies and in many cases, post-market surveillance to monitor the product's performance in the real world.

Considering all the above-mentioned considerations, it is quite difficult to introduce a new medical product into the market and that is why sometimes it takes a very long time and substantial money to do so (Kaule et al. 2020, Carol 2022). It is even more troublesome in the case of highly personalized products, where not only the shape, but the entire process might be changed from patient to patient, leading to issues with standardization and normalization of certain procedures (Jin et al. 2022). This, unfortunately, is frequently the case of anatomically individualized products, manufactured for specific patients, produced by various ways of 3D printing. Introducing such products to the market is even more troublesome, also considering that the time required to undergo all the formal procedures may well exceed the lifetime of a product that is needed in a specific moment. That is why new, improved regulations must be considered if the prospect of rapid, decentralized manufacturing of medical products would be taken seriously – and it should, in the author's opinion.

However, this book does not focus on the regulatory and formal side of things related to anatomically customized devices – so that the reader should look in other, available sources, such as standards, regulations and review publications on the matter, as the ones already mentioned (Tobin and Walsh 2023, Tettey et al. 2024, Marcus et al. 2016, Lauer et al. 2017, Kaule et al. 2020).

2.2.2 Customized medical products

In the realm of medical product design and manufacturing, the concepts of customization, individualization, and personalization represent crucial approaches to enhancing patient care by tailoring medical products to meet individual needs (Zema et al. 2017, Gubin et al. 2016). These concepts, though related, differ significantly in their application and implications for product design. By examining these distinctions, along with Pine's classification of customization (Pine 1999) and the challenges of designing customized medical products, we can gain a comprehensive understanding of the nuanced field of medical product customization. These definitions are also crucial for understanding the further parts of the book, as we will frequently refer to them and use them in many contexts, assuming that the reader is familiar with the basic nuances.

Customization refers to a process involving the customer in making conscious binary decisions about the characteristics of the final product. In medical products, customization might involve choosing between options for material, color, or type of attachment in devices like wrist braces, thereby allowing the product to meet specific functional requirements or preferences.

Individualization focuses on the shape and functionality of the product, making it suitable for use by specific groups or individuals without direct input from the customer. This approach is applied in medical products through anatomical shaping and material choice that cater to the patient's specific physical requirements, such as in custom-fit prosthetics or endoprostheses, where the product's design is predestined for a particular user's anatomy. Individualization by anatomical shaping lies at the heart of this book and is a foundation of its core concepts. There is also functional individualization, which means adjusting the product functionally (not geometrically) to meet needs of a specific client – or patient, in the case of medical products.

Personalization changes only those attributes of a product that affect how it is perceived by individual customers (patients, recipients), often concerning visual or auxiliary characteristics. For medical products, personalization can include adding names, stickers, or other decorations to daily used appliances, such as wheelchairs or other orthopedic devices, making the product feel more unique to the individual without altering its functional design.

These definitions were coined by the author, for the sake of this book and realized research. They are based on more traditional approaches to customization, most notably original work by B.J. Pine II (1999) with recent updates (Pine 2018).

B.J. Pine's classification presents four forms of customization, each varying by the degree of variability in the product and the level of customer involvement:

1. Transparent Customization involves minimal variability and customer involvement, aiming to offer personalized solutions without explicitly involving the customer in the design process. In medical products, it is akin with the individualization, either anatomical or functional.
2. Collaborative Customization entails a high degree of both product variability and customer involvement, where products are designed "step by step" with the customer to meet their unique needs, hence akin with the basic "customization" term introduced earlier.
3. Adaptive Customization allows for significant product variability but requires less customer involvement, as products are designed to be just modified from the standard base, not configured step by step – that is the second flavor of customization with the client.
4. Cosmetic Customization involves low product variability and customer involvement, focusing on surface-level modifications that do not alter the core functionality of the product – it can be mapped to personalization, according to previous definitions.

Figure 2.6 shows how Pine's customizations can be mapped to the notions of individualization, customization and personalization used in this book. Examples of how various medical products can be customized, individualized, and personalized are shown in Table 2.1.

The customization level of medical devices can be broadly categorized into four classes, reflecting the degree of variability and the nature of customization possible.

- Class I includes basic medical tools and equipment, where a finite number of variants can be pre-designed and stored.
- Class II encompasses a wider range of highly customizable products or product arrangements – like hospital rooms or prosthetic modules, which have a vast number of possible variants but lack full anatomic individualization.

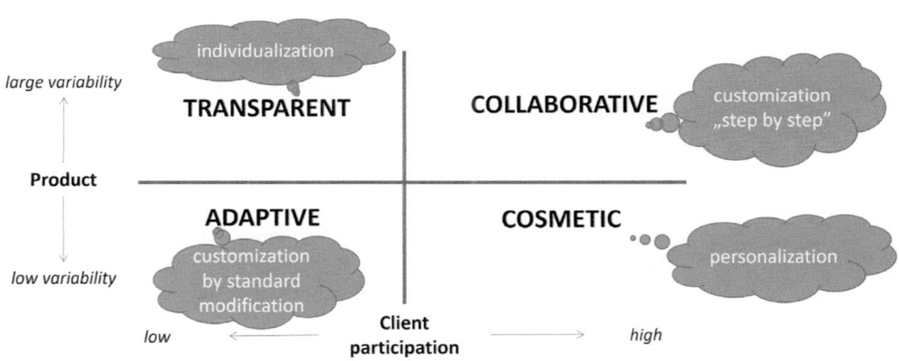

Figure 2.6 Translating Pine's customization (transparent-collaborative-adaptive-cosmetic) into notions introduced in this book.

Table 2.1 Examples of customization of medical products.

Process / Product	Customization	Individualization	Personalization
wrist hand orthosis	• material • color • mounting type	• shape (anatomical) • thickness	• n/a
joint endoprosthesis	• n/a	• shape (anatomical) • material • mounting type	• n/a
wheelchair	• functions • equipment • appearance	• version • size • shape	• markings • stickers and other decorative elements

- Class III and Class IV products involve infinite variability, with Class III allowing for full anatomical individualization (like implants and orthotics) and Class IV extending to bionic prostheses and artificial organs, requiring data storage of constructed variants for further customization.

How these classes correspond with each other in the scope of the introduced concepts of customization and individualization is shown in Fig. 2.7.

Designing customized medical products involves several challenges, primarily due to the need for incorporating individual anatomical data into the design process efficiently and accurately. The first of these challenges lies within incorporating patient anatomy into CAD designs. This requires a reliable source of anatomical data and knowledge about anatomical structures to characterize the design. Efficient design process also poses certain problems, as the goal is to design new products without repeating the entire process for each patient, thereby streamlining the creation of customized solutions. It is also crucial for the successful implementation of customization in medical products to possibly achieve a uniform design process across different medical cases, while meeting regulatory requirements and norms.

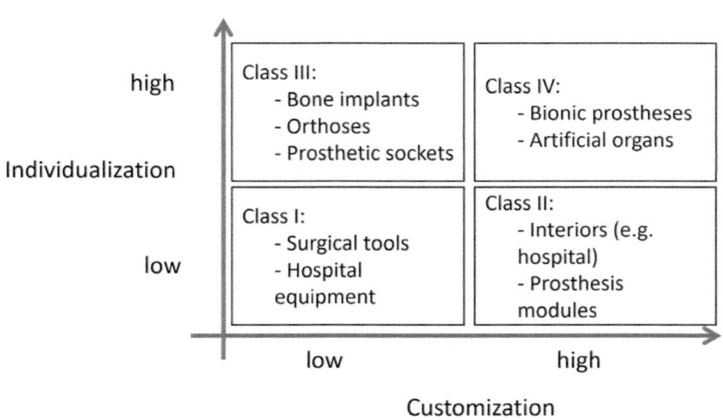

Figure 2.7 Classification of products by customization and individualization levels.

The intricacies of customization, individualization, and personalization in medical products not only demonstrate the potential for improved patient care through tailored solutions but also highlight the challenges that designers and manufacturers face. Addressing these challenges requires a multidisciplinary approach, blending medical knowledge with advanced manufacturing and design techniques to create products that meet the unique needs of individuals while adhering to the highest standards of safety and effectiveness.

2.2.3 Anatomical and bionic design – concepts and principles

The intertwining concepts of anatomical design, bionics, and biomimetics represent an intersection between biology, medicine, and engineering, offering profound implications for the field of medical engineering and the development of medical products. Each of these concepts, while related through their inspiration from the biological world, serves distinct purposes and emphasizes different aspects of the natural and artificial interface.

In the scope of this book, the most important concept is anatomical design. It is grounded in the detailed understanding of human or animal anatomy, applying this knowledge to create devices, structures, or products that align with biological forms (Rimini et al. 2022). The essence of anatomical design in medical engineering is its focus on compatibility with the human body's structure and function. This approach ensures that medical devices, ergonomic tools, and wearable technologies not only fit the natural contours and movements of the body but also enhance the user's comfort and the device's functionality. For instance, anatomically designed prosthetic limbs aim to replicate the movement and load distribution of natural limbs, reducing discomfort and improving the user's mobility and quality of life (Paxton et al. 2022).

The other, frequently used (sometimes even overused) concept is bionics. Although often used interchangeably with biomimetics (Iouguina et al. 2014), bionics traditionally emphasizes the integration of biological principles with electronic systems, focusing on enhancing or restoring human abilities (Nachtigal and Wisser 2014). In the context of medical engineering, bionics plays a crucial role in developing prosthetics, artificial organs, and sensory aids that mimic the function of their natural counterparts. For example, bionic limbs that incorporate muscle sensors and electronic actuators can offer users a range of movements and functionalities closely mirroring those of natural limbs, significantly advancing the field of rehabilitation and assistive technology (Clement et al. 2011). There are also a large number of other devices marked as bionic, such as heart valves (Ye et al. 2014), prosthetic eyes (Nowik et al. 2020) and many others.

The third concept, frequently used in modern engineering and also – in the author's opinion – least connected to the medical engineering is biomimetics. Biomimetics, or biomimicry, extends beyond the direct imitation of biological forms seen in the anatomical design and the functional mimicry of bionics. It seeks inspiration from the entirety of biological processes and systems to solve complex engineering and design challenges (Iouguina et al. 2014). Biomimetic principles have led to innovations in materials science, architecture, and robotics by emulating the efficiency, resilience, and adaptability of natural systems (Bar-Cohen 2006). In medical engineering, biomimetics can inspire the development of new materials that mimic the self-healing properties of the skin (Diesendruck et al. 2015), or surgical robots that replicate the precise and fluid movements of a human hand, offering new avenues for medical treatment and patient care (Song et al. 2022).

While the concepts of anatomical design, bionics, and biomimetics are interconnected through their biological inspirations, they differ in focus and application. Anatomical design concentrates on the physical compatibility with biological structures, bionics integrates biological functions with electronic systems, and biomimetics draws comprehensive inspiration from biological processes for innovative solutions. However, their connection lies in the shared goal of enhancing human life by bridging the gap between biological understanding and technological advancement. The differences are illustrated in Fig. 2.8, showing a prosthetic socket (a simple device adjusted to the patient's anatomy), a bionic hand orthosis (anatomically designed with electronic sensors) and a biomimetic leg orthosis (consisting of different parts, built in a way mimicking a human leg behavior).

In medical engineering, these concepts collectively contribute to the development of medical products that are not only functionally effective but also intuitive to use and harmonious with the human body. Anatomical design ensures ergonomic and efficient devices, bionics brings about functional restoration and enhancement, and biomimetics introduces innovative materials and systems inspired by nature's ingenuity. The synergy of these concepts in medical product development promises a future where medical technology seamlessly integrates with the human body, improving health outcomes and enhancing the quality of life for individuals around the globe.

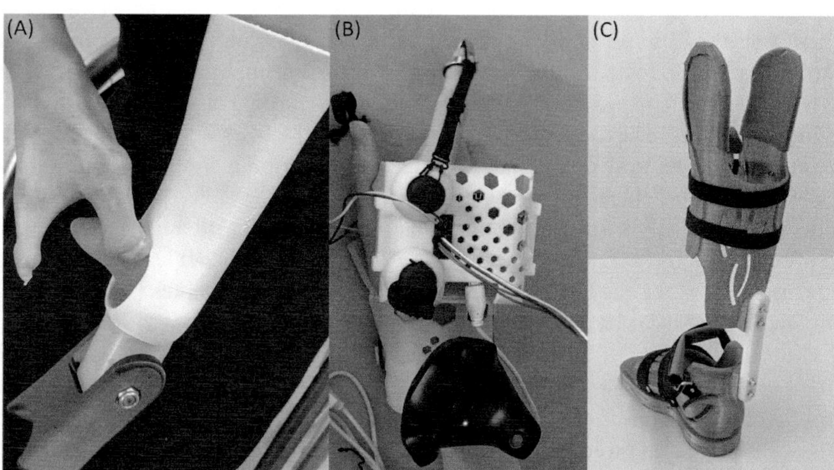

Figure 2.8 Concepts of: anatomical design (A), bionic design (B) and biomimetic design (C) on example of orthopedic devices.

Biomimetic and bionic designs have led to significant advancements across multiple fields:

- *medicine* – most obvious from the context of this book, these applications include prosthetics that mimic the movement and functionality of natural limbs, artificial organs that smoothly replicate biological functions, and surgical instruments designed after animal appendages for minimally invasive procedures (e.g., as described by Nakajima and Schwarz (2014));
- *robotics* – using bionics and biomimetic inspirations allow to build robots with locomotion inspired by animals (e.g., a fish, as described by Yu et al. (2004)), improving maneuverability and adaptability in complex environments;
- *materials science* – biomimetic approach enables inventing materials or manufacturing processes leading to nature-inspired, sometimes very useful structures, e.g., self-cleaning surfaces inspired by lotus leaves (Collins and Safiuddin 2022) or "strong lightweight" structures inspired by bone architecture;
- *architecture* – buildings designed with natural ventilation systems inspired by termite mounds (Turner and Soar 2008) or self-cooling mechanisms inspired by certain plant leaves (Rupp and Gruber 2022).

Bionic products must adhere to several critical requirements to ensure their effectiveness and safety. Some of these were already mentioned in previous subchapters – these are biocompatibility, functionality and efficiency, durability (especially when referring to products operating within the human body, in frequently corrosive environments or under heavy loads) and sustainability. Developing and building bionic products must also be accompanied with appropriate ethical considerations, as often it requires acquiring private, anatomical information that should not always be made available for the general public. Moreover, such products frequently work inside a living being's body, sometimes disturbing certain processes, not related to their main functionality. That is why serious and careful thoughts must be given when designing and introducing these products (Sharma et al. 2024).

The design of a bionic or biomimetic product involves a complex, interdisciplinary process that typically follows these steps:

1. *Problem identification* – the first step involves identifying a specific problem or challenge that could benefit from a solution inspired by nature.
2. *Biological research* – this involves studying biological systems, organisms, or processes. Researchers look for principles, strategies, and mechanisms that can be abstracted and applied to the design challenge.
3. *Concept development* – using insights gained from biological research, designers and engineers develop concepts that mimic the natural solutions identified. This phase may involve brainstorming, sketching, and preliminary modeling.
4. *Prototyping* – early prototypes are created to test the feasibility of the concepts. This iterative process involves refining the design based on trial and error, testing, and feedback. Additive manufacturing (3D printing) is more than frequently used for that purpose.

5. *Testing and optimization* – prototypes undergo rigorous testing to evaluate performance, safety, and reliability. The design is optimized based on test results, user feedback, and compliance with relevant standards and regulations, especially if a result is potentially a medical product, with the possibility of application in treatment of people with certain illnesses or injuries.
6. *Production and implementation* – after the design and prototyping phase is finalized and the devised product meets all the requirements, it moves into production. Post-production, the focus shifts to implementation, monitoring, and further optimization based on real-world performance.
7. *Ethical and regulatory review* – last but not least, this process underlines the whole lifecycle, as ethical considerations and regulatory compliance are critically evaluated to ensure that the product adheres to societal norms and legal standards at each level of its development.

The development of biomimetic and bionic products represents the cutting edge of technology and design, requiring a deep understanding of both biological principles and engineering challenges. By learning from nature, designers and engineers can create innovative solutions that improve human health, enhance performance, and solve complex problems in sustainable and ethically responsible ways. This book contains descriptions of these development processes, leading to the creation of products that might be labeled as bionic or biomimetic. Therefore, all the above-mentioned considerations were adhered to when creating methodologies described in this book and implementing them in research and medical practice.

2.3 Additive manufacturing in medicine

2.3.1 Additive manufacturing among other manufacturing technologies

Additive manufacturing is a group of technologies used for manufacturing of physical elements of various materials, using a principle of layered material deposition in a continued manner, starting from an empty working area, and ending with a part having finished geometry. Characteristic and distinctive features of additive manufacturing include a complete digitality of process preparation (the produced part needs to have 3D digital representation beforehand), division into layers for manufacturing and lack of dedicated tooling for specific parts, thus making possible to instantly switch between produced assortments. These qualities are common for all the additive manufacturing processes, however different from each other would they be.

Additive Manufacturing Technologies (AMTs) in the recent decades have found their place among the other manufacturing technologies, as the third fundamental group – near the subtractive processes (machining) and volumetric processes (foundry, plastic molding). The subtractive processes – material removal to get a desired shape – can be considered as the oldest technique, known even to prehistoric societies. The volumetric processes were invented in ancient times, as they require considerable force and/or temperature applied to a given volume of material. The additive technologies require digital models – so they naturally were implemented together with development of 3D modeling tools and computers. Characteristic

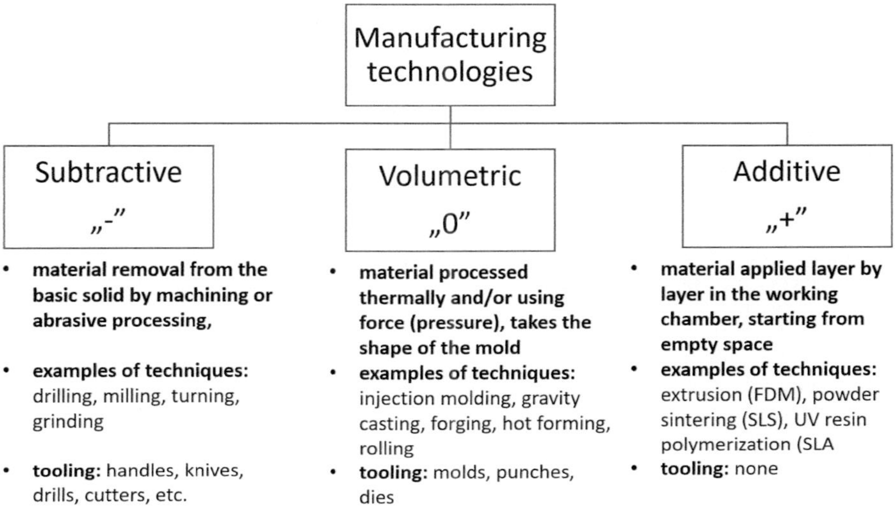

Figure 2.9 Additive manufacturing as compared to other groups of technologies.

features of all groups in comparison with the other groups are shown in the scheme in Fig. 2.9.

As opposed to the other two groups, additive manufacturing has been invented relatively recently. The first processes were discovered in the late 20th century. In 1984, Charles Hull created the stereolithography – selective liquid polymer cured by UV laser, as the first additive manufacturing technology (Hull 2015), in 1986 it was patented. In late 80s, Scott Crump invented the Fused Deposition Modeling – plastic extrusion through heated nozzle (Crump 1991), and likewise patented it in 1989. Both inventions gave rise to the first companies operating in the additive manufacturing market – 3D Systems Inc. started by Hull and his associates and Stratasys started by Crump and his wife (Brooke 2013). Today both companies are huge business players, owning most of the share of the additive manufacturing market.

Invention of additive manufacturing was relatively late mostly due to intrinsic digital nature of the process. The additive manufacturing processes lack their "conventional", "analog" counterparts, what makes the whole group distinct among other known manufacturing processes, which have been known for centuries, if not millennia. For instance, machining (subtractive processing) was most likely started in prehistoric times – in the stone age, for wood- and stonework, for creating tools, clothes, and elements of first buildings. Foundry and plastic processing are a later invention, as most of these processes usually require increased temperature – hence they were created shortly after humanity discovered fire and learnt how to control state of matter of some low-melting point metals (such as copper and tin), starting the bronze age. Over the next millennia of technical advancement of human civilization, these technologies were improved and refined, adding new process variations and new materials every now and then. However, it was not until the digital age as the third group of technologies was discovered and put into highly efficient course of progress.

Digitization of manufacturing is a trend that has been known for several decades now (Lorenz et al. 2020). At the current time of technical civilization development, it seems unimaginable to realize certain processes without computers – but it has not always been this way. The conventional manufacturing processes have, for the most time in their history, been dependent on manual skills of a craftsman performing them. Mechanization, automation, and digitization (marking subsequent industrial revolutions) enabled first to lower the required qualifications of production workers and then – to partially or totally remove human operators from the processes, handing the actual work to digitally controlled machines and robots (Growiec 2022). As such, traditionally manually performed processes such as turning, drilling, or forging were (and still are) gradually "converted" into mechanized and automated elements of a fully digitized manufacturing process.

In the case of the additive manufacturing, the situation was – and is – slightly different. There is no "analog" group of additive manufacturing processes – they need to be performed only digitally and mechanically and they also do not function outside their digital layer. That is why they are gaining huge popularity only now, in the recent decade or two, as the digital revolution in industry progresses – in the form of the fourth industrial revolution, also known as Industry 4.0 (Culot et al. 2020). Additive manufacturing – 3D printing – is routinely included as one of key technologies of the Industry 4.0 (Prashar et al. 2023), alongside collaborative robots, Augmented Reality and Industrial Internet of Things. As most of the products nowadays have their digital representation in form of CAD models, requirement of having a 3D model is no longer a hindrance preventing from use of the additive manufacturing.

Additive manufacturing is largely and popularly known under the name of 3D printing – coming from the early name of one of technologies, based on powders joined by liquid adhesive dosed through an inkjet printer head, today known as binder jetting (Mostafaei et al. 2021). Nowadays, "3D printing" is an umbrella term, covering all the processes, sometimes having nothing common with the actual printing. That is why in professional literature it is preferred to use the term of "additive manufacturing" and that is a convention that will be applied also in this book.

Each of available additive manufacturing processes involves the following steps, illustrated in Fig. 2.10:

1. Preparation of digital 3D model of a given part or product.
2. Applying basic principles of division into layers to the product (slicing), selecting and tuning parameters of the process.
3. Process simulation and verification, generation of NC (Numerical Control) code for a given machine.
4. Starting the prepared program on a machine, layer deposition.
5. Removing the finished part from the machine, post processing.
6. Assembly and use.

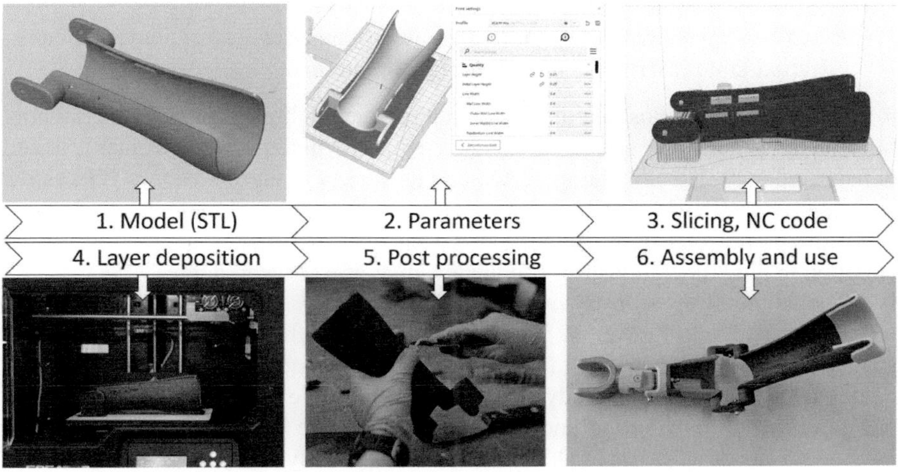

Figure 2.10 Steps of creation a product using additive manufacturing.

2.3.2 Applications, strengths, and weaknesses of additive manufacturing

Additive manufacturing has emerged as a pivotal technology across various sectors. This technology's ability to construct complex geometries layer by layer from digital 3D models offers distinct advantages over traditional manufacturing methods and it translates to a very wide scope of applications. In general, there is no particular branch that can be named as "best suited" for the additive manufacturing – the whole AM is a versatile group of technologies akin to subtractive or volumetric manufacturing, so it would be too much of generalization to say that. However, a rule of thumb is that whenever we are looking to make intricately, uniquely shaped parts, custom-designed products produced in very small batches, we need a quickly made prototype of a device that will be improved in many subsequent iterations or we need to maintain a control of the internal geometric structure of the material easily – 3D printing is the way to go.

In general, by the viewpoint of this book, the applications of additive manufacturing can be categorized into three primary sectors: engineering and industry, consumer applications, and medical and biomedical.

In the engineering context, traditionally there are three main application schemes (Pham and Dimov 2012):

- *Rapid Prototyping* – for building prototypes quickly and cheaper than with use of traditional methods; for visual, functional, and utility prototypes, most often using purely additive techniques, sometimes with help of other techniques (casting, foam milling, etc.) – the term is so widespread, that it is (by the author's experience) often used synonymously to the Additive Manufacturing or 3D printing itself;
- *Rapid Manufacturing* – quick production of single or small series of ready products, often with atypical shapes;

- *Rapid Tooling* – quick production of tooling for other processes (casting models, injection mold inserts, etc.) for short production runs.

Considering the most popular industrial production branches, rapid prototyping (or AM in general) is usually associated with the automotive, aeronautical and defense industry. The utilization of AM and its variety in these branches is largely a success story, with the constantly growing market and increase in application range. From the point of view of the customized medical production, it is worth paying attention to how these industries use the leverage of AM technologies.

The automotive industry is constantly driven by the need for innovation, efficiency, and customization, while maintaining high safety and performance standards. 3D printing and rapid prototyping are increasingly crucial in this context, offering significant advantages across multiple aspects of vehicle design and production. 3D printing and rapid prototyping are extensively utilized in the automotive sector (Tuazon et al. 2022), impacting nearly every stage of the vehicle development process. These technologies are applied in the creation of complex components like engine parts, transmissions, and interior cabin fixtures, as well as prototyping of aesthetically pleasing elements, such as body panels and lighting fixtures.

In the automotive industry, rapid prototyping is particularly valuable for its ability to quickly produce parts that meet precise specifications. It supports the design and testing phases by enabling quick iterations and refinements, significantly reducing the time from concept to production (Campbell et al. 2012). This capability not only accelerates the development cycle but also aids in customizing vehicles to consumer preferences without substantial cost increases.

Moreover, rapid prototyping contributes to reducing vehicle weight, a critical factor for enhancing fuel efficiency and reducing emissions. By allowing for the design and manufacture of parts that use less material without compromising strength or safety, it supports the industry's shift towards sustainability. Additionally, it improves supply chain efficiency by enabling on-demand production, reducing the need for large inventories and minimizing waste (Attaran 2017).

The aerospace sector is governed by regulatory standards and high expectations for quality and performance. Additive manufacturing helps in this regard – these technologies are instrumental across various design stages and functionalities, including the development of electronics, lighting systems, both interior and exterior components, and mechanical parts. They facilitate the aerospace industry's needs from initial design concepts to specific manufacturing parts and ongoing maintenance operations (Uriondo et al. 2015).

What really distinguishes additive manufacturing from the other technologies, however, is that it enabled a whole new branch of previously unknown applications: consumer 3D printing, or rather 3D printing of consumer goods (see Fig. 2.11 for concept illustration), frequently by the consumers themselves, in their houses, garages, small companies, schools and many other places. As 3D printing machines – especially extrusion-based ones – are quite simple to setup and operate, a new era in manufacturing is slowly becoming a reality, where production machines are legitimately used to create new products outside factories. This is of course

Figure 2.11 Industrial and consumer use of additive manufacturing (concept).

affecting many industries (especially trade companies (Leering 2017)), but also in a positive way – having 3D printers in schools, for example, may help printing basic medical supplies needed by children (Manero et al. 2019), let alone gadgets and tools that facilitate the learning processes, while use of 3D printers in hospitals and other healthcare facilities may greatly shorten the time between manufacturing of a customized medical product and its use directly by a doctor or a patient. Not to mention use of 3D printers at home, to enhance creativity and engineering interests in children by making customized toys or gadgets, or produce replacements for missing or damaged parts of household appliances.

The consumer uses of 3D printing go far beyond fabrication of technical objects out of synthetic materials – there are 3D printers that can use organic, edible materials, such as sugar, chocolate, and others, to create interestingly shaped food (Liu and Ciftci 2021). Some of them also allow printing of fabrics, so that new clothing items can be brought to life (Spahiu et al. 2020). At this moment, this is more of a hobby than a serious threat to certain manufacturing industries, but at some point its potential will surely rise in a way enabling a true revolution in manufacturing, leading to dispersed, "cloud" production.

Numerous examples of this potential can be found back in the COVID-19 pandemic time – we can find examples of groups of people printing medical equipment at their homes, in large quantities. The author of this book was directly involved in launching one of such activities himself ("#drukujdlalekarza", in English "#printfordoctor") in early 2020, with remarkable success (Study in Poland 2020), proving that it is fully possible to organize a community of more than a hundred people, that did not know each other beforehand, have never met in the real world, yet functioned like a very efficient production company, offering customizable medical products. That was – and still is – possible only thanks to unique capabilities of additive manufacturing in itself.

Considering these applications, several strengths of additive manufacturing that make it an attractive choice across different fields can be distinguished. The three main ones that come to mind, by the author's experience, are:

- *design complexity* – AM allows for the creation of complex, often "organic" shapes that would be challenging or impossible to achieve with traditional manufacturing;
- *flexibility in production* – there is a significant advantage in the ability to change the product assortment continuously at no additional cost related to lack of need for specialized production tooling;
- *speed and cost* – the speed of setting up manufacturing processes and achieving finished products is remarkable, coupled with relatively low operational costs and minimal qualification requirements for operators – compared with conventional, traditional technologies.

Despite its numerous benefits, additive manufacturing also faces many challenges. The three main groups of disadvantages of this technology group – treated holistically – would be:

- *physical and technical limitations* – issues such as the need for supports (in some processes), post-processing finishing (more or less cumbersome, depending on the process and material), and limitations in the size of the build chamber restrict the scope of AM,
- *material and machinery costs* – high costs of machinery and materials for precise production can be a barrier in the case of more advanced systems – the state of technology is interesting, as a variety of prices of commercial solutions is vast, from hundreds to millions (using USD as a reference),
- *performance issues* – there are many concerns regarding the accuracy, durability, and layering time of products made through AM (Iftekar et al. 2023), which may not always meet industrial standards.

As the additive manufacturing technology has multiple applications, as mentioned above, including industrial, medical, and consumer applications, it is crucial to evaluate its strengths and weaknesses in a detailed manner also within the contexts of industrial, medical and consumer applications separately.

Additive manufacturing allows for the creation of complex geometries that are either impossible or too costly to achieve with traditional manufacturing methods. This capability is particularly beneficial for industries such as aerospace and automotive, where complex, lightweight structures can significantly enhance performance and fuel efficiency. In industrial applications, it is also crucial to utilize the Rapid Prototyping potential, as AM accelerates the design process by allowing quick production of prototypes. Engineers can iterate designs faster and more cheaply than traditional methods (Campbell et al. 2012). Moreover, each part can be uniquely customized without additional tooling cost, which is ideal for manufacturing custom tools or parts in small volumes (more on the mass customization in the next subchapter).

However, there are also many weaknesses, mostly due to the nature of the technology itself, based on different foundations than in the case of subtractive or volumetric manufacturing. The range of materials suitable for additive manufacturing is limited compared to conventional methods. Materials often have different mechanical and thermal properties, which can be a disadvantage for applications requiring high-performance specifications (Tsioukas et al. 2020, Iftekar et al. 2023). The problem can be also noted in the case of surface quality. Parts produced by AM typically have a rougher surface finish and may be less uniform in strength and density, which can be a critical drawback in many industrial applications (Hartcher-O'Brien et al. 2019). The dimensional and shape accuracy also is not that high as in other, comparable technologies relying on similar materials (Górski et al. 2020a, Iftekar et al. 2023).

Also, from the economical point of view, while AM is cost-effective for small production runs, the economics of scale achieved by traditional manufacturing methods make them more viable for large-volume production – so not everything should be even considered to be produced by this technology.

In terms of medical applications, the main strength of AM lies in its capabilities of production of atypical, anatomical, "organic" shapes. AM is ideal for producing customized medical products like prosthetics, implants, and orthotics, which need to match the unique anatomical features of individual patients (Górski et al. 2020). That is, among other things, one of the crucial aspects that lies at the foundation of this book. Advances in AM technology have also enabled the use of biocompatible materials that can be safely implanted in the human body (Li et al. 2020). AM is also at the forefront of bioprinting (Leberfinger et al. 2019), including the development of tissue engineering where organs and tissues can be printed based on the patient's cells – but in practical terms, e.g., printing useful "body parts", it is rather in the far future, as the current scope of technologies is quite limited in its implant-production capabilities (Semba et al. 2020).

However, as noted previously, medical devices face significant regulatory scrutiny to ensure they are safe and effective and AM processes are no different here. Compliance can be complex and costly, diminishing benefits obtained by the advantages of the AM technological parameters (Carl and Hochmann 2022). Also, in terms of qualities of obtained products and materials of which they are made, there are many things still to be done. While there are advancements, the range of bio-compatible materials is still limited, restricting the types of products that can be manufactured (Iftekar et al. 2023, Guttridge et al. 2022). From the economical point of view, the initial investment for medical-grade 3D printers and the ongoing costs of materials and maintenance can be prohibitively high, especially when considering implants and other devices above class I (in contact with human internal tissues).

Considering the consumer applications, AM is beneficial as it allows for the mass customization of products from jewelry to furniture, catering to individual preferences without significant cost increases. Whenever used at home or in a factory, AM also generates less waste material compared to traditional subtractive manufacturing, aligning with growing consumer demand for sustainable products (Javaid et al. 2021). As previously mentioned, using AM production of parts on demand can be achieved, simplifying supply chains by reducing the need for inventory and improving responsiveness to market changes (Zhang et al. 2019).

Similarly, as in the case of medical devices, there are concerns with materials and their properties, lack of economic justification for large-scale production and problems with quality consistency, especially when considering low-cost machines made for home or hobby uses. Products made from 3D printing can and will often suffer from issues such as lower durability and strength, which may not meet all consumer expectations.

Summarizing and taking the industry state into consideration, it can be stated that as for now – well into the 21st century – additive manufacturing is positioned as a revolutionary force in modern manufacturing, with profound implications across industrial, consumer, and medical fields. While it offers unmatched flexibility and customization, the technology must overcome significant material, cost, and performance hurdles to fully realize its potential, especially in the case of medical products. The future of AM will likely see continued advancements that address these challenges while expanding its applications further. In the further chapters of this book, some successful endeavors at utilization of advantages of this group of technologies for paradigm shift in medical engineering and production is shown.

2.3.3 Additive manufacturing for medicine and biomedical engineering

Medicine and biomedical engineering are and have always been the key areas of application of the additive manufacturing technology (Tsioukas et al. 2020, Li et al. 2020), regardless of a process or material. Obviously, "medicine" and "biomedical engineering" are very general terms, covering a very broad range of possible applications. However, additive manufacturing seems to have found its way into almost all branches of medicine – in some of them it is more needed than in the others, but the potential of 3D printers in healthcare is, in general, indisputable by most authors (Hornick 2017). This is especially true now in the time which might be called a 3D printing "open era", after patents for basic technologies have been lifted in early 2010s and a multitude of FDM and SLA machines in all price ranges flooded the market, lowering the entry level for possible applications.

Additive manufacturing is seemingly a perfectly crafted concept from the viewpoint of personalized medicine and engineering of custom medical devices, for a given narrow group of patients or doctors, or even for a specific person at a given moment of selected medical procedure. This is mostly due to its lack of dedicated tooling – any 3D printing machine can be used to produce entirely different geometry after a currently realized process is finished. This is a huge difference between the two basic groups of technologies and additive manufacturing – having a 3D model of a shape that we need is usually everything that is required and if a particular model is at our disposal, manufacturing can be started in no time – programming usually takes only a couple of minutes for a skilled technician. It is also possible to create fully personalized devices, such as implants or prostheses – as there is (almost) no difference from the manufacturing point of view, if a 3D printer creates two identical implants or two implants for two different patients.

Another defining feature of AMTs showing high compatibility with requirements of personalized medical devices is that it is well suited for manufacturing of complex, organic shapes. It is not that AMTs are better in complex shapes than in the simple ones – it is rather that it makes a very slight or no difference, due to the layered nature of the process, as opposed, for instance, to machining, where it is a fundamental difference in viability, costs, and time when geometrical complexity increases. In medicine, or more specifically, in manufacturing of anatomically adjusted shapes, it is therefore more advisable to utilize additive manufacturing than conventional technologies, as the effect could be obtained more rapidly and effectively, at least in certain branches (Choonara et al. 2016).

The main areas of medical use of additive manufacturing technologies have been specified below:

1. Manufacturing of orthopedic and prosthetic supplies and their elements (Banga et al. 2021).

   Additive manufacturing is particularly advantageous in the production of orthopedic and prosthetic supplies due to its ability to customize items to the unique anatomical requirements of patients. Processes like Selective Laser Sintering (SLS) and Fused Deposition Modeling (FDM) allow for the creation of durable, lightweight components that fit perfectly with the user's body. This customization enhances comfort and functionality, improving patient outcomes by facilitating better mobility and reducing the risk of complications associated with ill-fitting prosthetics (Jin et al. 2015).

2. Manufacturing of customized implants, artificial organs, and their parts (Zadpoor and Malda 2017).

   Using technologies such as selective laser sintering or melting, as well as many others, manufacturers can produce implants that are tailored to the specific contours of a patient's anatomy, such as cranial plates or joint replacements, directly out of metallic or composite biomaterial (Munsch 2017). This precision contributes to improved integration with the body and longer-lasting outcomes. For artificial organs, bioprinting offers the potential to use cells and biocompatible materials to create complex tissue structures that mimic natural organs, promising significant advances in transplant medicine. There is also possibility of building resorbable customizable implants as scaffolds for bone structures – for that purpose, even FDM can be used, combined with special filaments, inducing growth of the bone structure in the place of implantation (Bhargav et al. 2018).

3. Preparing anatomical models for surgery, education, simulation, and diagnostics (Żukowska et al. 2023).

   The use of additive manufacturing to create detailed and accurate anatomical models is a breakthrough in medical education and surgical planning. By printing patient-specific replicas of organs being studied or operated on – of hard material (Żukowska et al. 2023) as well as of softer ones (Salmi 2016), surgeons can plan and simulate surgeries in a risk-free environment, increasing the likelihood of successful outcomes. Additionally, these models serve as tools for medical training and patient education, helping to explain complex medical conditions and treatments. Examples involve bone structures in the maxillofacial area (Thomas et al. 2014), kidneys (Bernhard et al. 2016) and many others.

4. Prototyping new therapeutic devices (Hofmann et al. 2016).

   Rapid prototyping enables the swift creation and iteration of designs for new therapeutic devices and their parts, such as massage devices, wheelchairs (Ahmad et al. 2022) and therapeutic robots. This accelerates the development process, allowing for rapid testing and modification based on clinical and user feedback. Additive manufacturing can produce these prototypes with a high degree of complexity and material specificity, which is critical for devices that must meet strict medical standards. This iterative cycle ensures that the final products are highly effective and safe for patient use.

5. Creating scaffolds for tissue engineering and bioprinting (Boopathi and Kumar 2024).

   In tissue engineering, 3D printing is used to create scaffolds that support the growth of biological tissues. These scaffolds are designed to mimic the extracellular matrix, providing a three-dimensional structure where cells can proliferate and differentiate into functional tissues (Qu 2020). The precision of additive manufacturing allows for the creation of pores within scaffolds, critical for cell migration, nutrient diffusion, and waste removal (Kumar et al. 2016). Bioprinting further extends these capabilities by incorporating live cells and growth factors into the printed structures, facilitating the development of functional biological tissues (Leberfinger et al. 2019).

6. Manufacturing customized surgical tools and medical instruments (Culmone et al. 2019).

   Additive manufacturing allows for the design and production of surgical tools and medical instruments tailored to specific surgeries or patient anatomies (Javaid and Haleem 2018). This customization can significantly enhance the efficiency and safety of surgical procedures. Tools such as surgical guides, with direct cutting and drilling during surgery, are printed to match patient-specific anatomical data, ensuring precision and reducing surgery times (Javaid and Haleem 2018). This not only improves surgical outcomes but also minimizes the risk of infection and speeds recovery. There might be also a possibility to produce custom multi-use tools personalized for advanced surgeons, with requirements stemming from their many years of experience.

7. Manufacturing models and tools usable in psychotherapy.

   In psychotherapy, 3D printing can be used to create customized therapeutic tools and models that assist in treatment. For example, objects used in therapy for stimulating recollection in memory loss, that help patients visualize and remember certain facts, can be custom-made (Garlinghouse et al. 2018). There are also reported cases of use of 3D printed models in improving psychological constructs in pregnancy (Cote et al. 2022). Such a personalized approach can help in creating a more interactive and responsive therapeutic environment.

Notable examples of medical applications of additive manufacturing – realized by the author together with his team of researchers and students – are presented in the Chapters 6 and 7 as case studies. Their technical foundations will be deeply explored in the further chapters of this book.

2.3.4 Recent trends and advancements

Additive manufacturing and 3D printing technology continue to evolve rapidly. Although the physical phenomena governing the basic processes remain mostly unchanged since they were first invented (more on that in Chapter 5), the processes, machinery, materials, and their combinations constantly improve, are mixed with each other and new applications are constantly pushed forward by the scientific and industrial community of AM aficionados.

The first group of trends is focused on materials. The materials have always been the highlight of AM technology, often laying at the foundations of the processes – vide photopolymer resins used in SLA, for example. The developments of a past decade have greatly expanded the range of materials that can be used in 3D printing. As scientists and engineers learnt to properly operate the 3D printers and the processes got normalized, the development of materials dedicated solely to 3D printing has become a solidified direction. Traditionally, 3D printing was limited to plastics and certain metals, the material palette now includes high-performance polymers, advanced composites, ceramics, and even glass. For instance, researchers have developed techniques to 3D print silica glass, which could revolutionize optical applications such as fiber optics and lenses (Zhang et al. 2022). Moreover, the emergence of metal matrix composites and conductive materials is paving the way for more durable and functional components suitable for electrical and thermal applications (Kumar et al. 2022). In view of medical applications, there is also a considerable effort put on various biocompatible materials (Subramani et al. 2024), as well as dynamic materials that can change their shape and function inside a human body, reacting to certain stimuli – in a process often dubbed as "4D Printing" (Zarek et al. 2017).

The second group of development trends is focused on improving processes and their performance indices. One of the critical challenges historically facing 3D printing has been the speed of production, mostly due to comparison with rapid speed of volumetric technologies, such as injection molding. Recent innovations aim to address this by enhancing the print heads (Hassan and Selvaganapathy 2022) and optimizing layering techniques (Abeykoon et al. 2020). For example, HP's Multi Jet Fusion technology (O'Connor et al. 2018) and Carbon's Digital Light Synthesis™ (Redmann et al. 2020) reduce printing times while improving the mechanical properties of the printed parts. These advances not only enhance productivity but also make AM more competitive with traditional manufacturing methods, albeit not in all applications (as discussed in previous subchapters).

The process improvement trend has led to a relatively new invention, which is called hybrid manufacturing (Strong et al. 2018). Hybrid manufacturing combines additive and subtractive manufacturing processes, often within a single machine. These systems integrate the layer-by-layer additive process with CNC milling capabilities to achieve parts with highly refined features and superior surface finishes (Barragan et al. 2023). This dual approach ensures greater dimensional accuracy and material integrity, addressing one of the significant limitations associated with standalone 3D printing systems. There are also other variants of hybrid manufacturing – such as processes of pouring chemically hardened resins inside hollow FDM structures mid-printing (Górski et al. 2018) or adding laser engraving to an FDM machine (Patel et al. 2019). A side-trend of hybrid manufacturing – usually based on large, multiaxial CNC devices – is building 3D printers operating in more than traditional three spatial axes. There are new proposals of 5-axial machines and more sophisticated approaches to use robotic arms with attached extrusion or sintering heads for greater flexibility (Molotla et al. 2023).

There are also entirely new processes being developed, based on phenomena previously unknown or unused in the realm of additive manufacturing. One example might be 3D printing with the use of ultrasonography – researchers found a way to the change shape and state of a material by stimulation through ultrasonic waves. This might help with some medical applications – a biomaterial can be injected first into the patient's body and then assume the full shape and properties after ultrasound stimulation through the skin and other tissues (Kuang et al. 2023). Another example of an entirely new process, also relying on specialized materials, is a wide group of experimental technologies known as 4D printing – where time, or dynamic change of the shape, becomes a factor. A special type of metamaterials (mostly metals) can be printed in one shape and then assume another shape when acted on with a proper force or other agent (Zarek et al. 2017, Zhou et al. 2023).

Finally, there are developments in well-known processes, adapting them to entirely new classes of applications. Construction industry is one example – the known extrusion printing was adapted (or rather thought anew, as one might say) to print large-scale structures and buildings. Technologies like contour crafting (Akhnough 2021) allow for the construction of building components or entire structures directly on-site, reducing labor costs and material waste (Rouhana et al. 2014). These technologies have been used to print everything from residential homes to bridges, demonstrating the scalability and utility of AM in heavy industries.

In the medical field, 3D bioprinting technology has seen some advancements. Utilizing known processes of inkjet printing, extrusion or stereolithography with a special class of materials – bioinks – studies on bioprinting have successfully demonstrated the potential to print organs and tissues that mimic the biological properties of their natural counterparts. For example, researchers at Tel Aviv University have 3D printed a heart using human cells (Noor et al. 2019) and at the University of Adam Mickiewicz in Poznan there are advanced studies on printing human meniscus (Semba et al. 2023). While still in the experimental stage, these developments hold immense potential for regenerative medicine, potentially addressing the critical shortage of donor organs, although not in the upcoming decade (in the author's opinion), due to scalability and biological stability issues (Murphy et al. 2020).

Additive manufacturing, in itself a rather fresh and new technology, is undergoing rapid transformation, characterized by technological innovations that broaden its applications and enhance its viability. As these technologies continue to mature, they promise to unlock unprecedented capabilities across multiple industries, making the future of 3D printing bright. That holds true especially in the medical domain, which is a focal point of this book.

2.4 Mass customization in medicine

2.4.1 Production types and place of mass customization in modern industry

Mass Customization (MC) represents a paradigm shift from traditional mass production to a more flexible, customer-centered approach. It arose as a response to the limitations of the mass production era, famously encapsulated by Henry Ford's statement that customers could have a car in any color "so long as it is black." This one-size-fits-all mentality no longer satisfies the diverse preferences of today's consumers, who demand products tailored to their specific needs and desires (Fogliatto et al. 2012, Kabasakal et al. 2017).

In this view, mass customization can be understood as a strategic approach that blends aspects of mass production and custom tailoring to deliver wide-ranging products tailored to individual preferences at a cost comparable to standard mass-produced items (Coletti et al. 2011). This strategy is particularly crucial in today's diverse market where customization offers a competitive edge. Understanding the different types of production can provide a clearer context for the unique position of mass customization.

1. *Piece Production (On Order)*: This type involves crafting products individually or in very small quantities, typically on a per-order basis. It is highly customized but also the most expensive production type, often used for items like tailored suits or custom jewelry.
2. *Small Series Production*: This approach is a compromise between single-piece production and mass production. It involves manufacturing small batches of products, allowing for some level of customization or variation between batches. It is less costly than piece production and is often employed for limited-edition items or specialized industrial equipment.
3. *Mass Production*: Here, products are manufactured in large volumes with standardized designs. This method emphasizes economies of scale, reducing the cost per unit by spreading the fixed production costs over a large number of items. However, it offers little to no customization, typified by the early 20th-century automotive industry.

Mass customization intersects these traditional methods by combining the low unit costs of mass production with the personalization aspects of piece production (Górski et al. 2016) (Fig. 2.12). It enables companies to produce highly customizable products in large volumes, thereby meeting individual customer preferences without sacrificing efficiency or significantly increasing costs. The transition to mass customization is fueled by advancements in manufacturing technologies and the rise

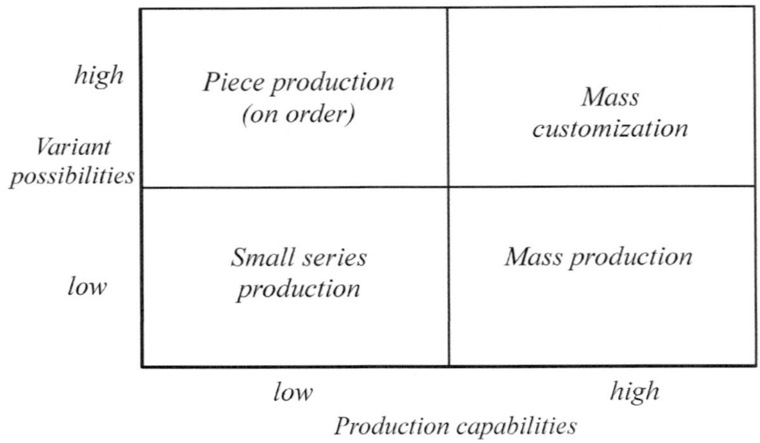

Figure 2.12 Production types and mass customization (Górski et al. 2016).

of Industry 4.0, which includes smart factories, cloud computing, and the Internet of Things (IoT) (Zawadzki and Żywicki 2016). These technologies facilitate the integration of high customization with high-volume production, allowing companies to produce configurable products on a large scale efficiently. This approach not only meets individual customer needs but also maintains competitive pricing and rapid production times.

In the modern industrial landscape, mass customization is becoming increasingly necessary across various sectors, including automotive, electronics, and fashion. The methodology enhances competitiveness by combining the economies of scale associated with mass production with the personalization typically associated with bespoke creation. By doing so, companies can attract a broader customer base and increase customer loyalty through products that more closely align with individual preferences.

The typical design process for configurable products often involves several steps, from initial customer input to final manufacturing, with multiple arrangements and departments involved. This process can be cumbersome and slow, often requiring significant manual input from sales teams, designers, and manufacturers, making it difficult to introduce the mass customization concept in business practice and making it profitable, rather than risky strategy of production (Górski et al. 2016).

The perfect design process for mass customization involves a more streamlined approach, involving Automated Design Systems (ADS), that allow customers to interact directly with the product configuration tools, bypassing the need for traditional sales and design roles. This direct interaction helps in capturing precise customer requirements. It also contains integration with manufacturing – once the design is finalized through ADS, the specifications are automatically sent to the manufacturing floor, where process planning is already optimized for quick setup and production (Lin and Hsu 2008, Kowalski et al. 2021). The comparison between these two processes is shown in Fig. 2.13.

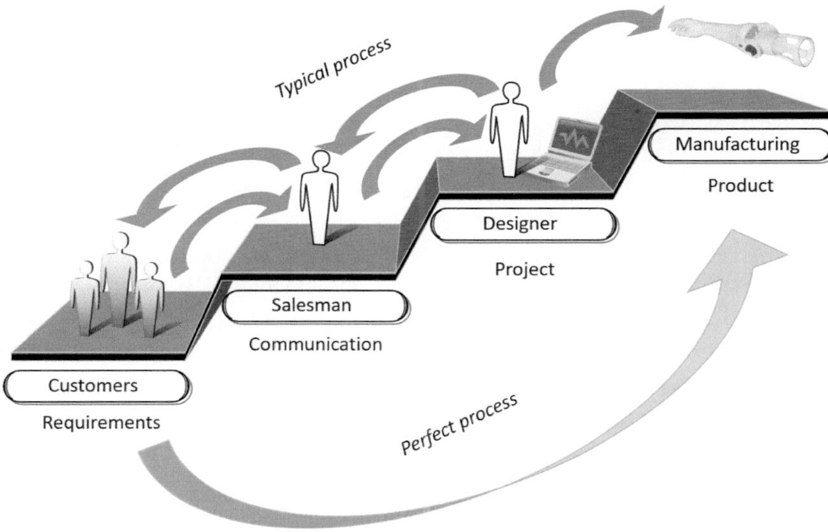

Figure 2.13 Typical (multiple human interaction) versus perfect (automated design systems) design process in development of customized products.

Implementing this perfect design process requires several key conditions:

- customer engagement tools, i.e., effective configurators that are intuitive and can guide customers through the customization process without overwhelming them (Zawadzki 2018),
- intelligent CAD models, created by proper knowledge acquisition and implementation (Górski et al. 2020b, Zawadzki 2018, Skarka 2007), broadly described in Chapter 4,
- advanced IT infrastructure – robust software and hardware that can support complex algorithms and maintain large databases,
- seamless integration of systems – customer interfaces, intelligent models and factory floor devices must be interconnected and communicate efficiently,
- staff training and adaptation – employees must be trained accordingly to manage these advanced systems and how to intervene not to disturb the automated design process.

Knowledge-Based Engineering (KBE) plays a critical role in facilitating mass customization (Zawadzki 2018, Górski et al. 2016, Stokes 2001). KBE systems utilize detailed knowledge about design and manufacturing processes, formalized by experts and embedded into automated systems that assist in or fully take over routine tasks. This not only speeds up the design process but also ensures consistency and adherence to technical standards.

Capabilities of KBE include:

1. *Reduction of design and production time* – as demonstrated in various industries as early as two decades ago, mostly in automotive, aeronautical and machine production (Stokes 2001, Kulon et al. 2006, Chapman and Pinfold 2001), KBE can reduce the time required for designing and planning from weeks to mere hours or minutes.
2. *Standardization and error reduction* – by automating routine tasks (partially or completely), as well as decisions, properly implemented KBE minimizes human error and standardizes output across different production cycles.
3. *Scalability* – the key component to the "mass" in mass customization, KBE systems can be scaled up to handle increased complexity in products and larger volumes, facilitating easier expansion into new markets or product lines.

In practical terms, implementing KBE in mass customization allows companies to maintain a high level of product variability and personalization while keeping production times and costs the same to those of standard mass production. This synergy of customization and efficiency is what makes mass customization a formidable strategy in modern manufacturing landscapes (Qi et al. 2020).

2.4.2 Mass customization: applications, enablers, and advantages

The application of mass customization extends across multiple industries, providing significant advantages over traditional manufacturing approaches. One key area is the ability to efficiently meet customer demands for personalization without sacrificing production speed or increasing costs prohibitively. This is achieved through modular design, advanced manufacturing processes like 3D printing, and flexible manufacturing systems that can quickly switch between different product configurations.

For example, in the automotive industry, customers can specify various features of their vehicles, including color, interior materials, and technological add-ons, which are then produced using a mass customization framework. This method ensures that each customer receives a product that feels personal and unique while keeping the cost the same to that of standard mass-produced models.

The automotive sector has been one of the pioneers in adopting mass customization, driven by consumer demand for vehicles that reflect personal style and functionality. Companies like Toyota and Audi utilize modular design strategies (Kaspar et al. 2020) that allow customers to select from a wide range of options for interiors, exteriors, and technological features. This modular approach means that while the basic chassis and engine components remain standardized, elements such as upholstery, paint colors, and accessory packages can be customized, allowing for thousands of unique combinations.

Another well-known branch to successfully implement mass customization is consumer electronics. Companies like Dell have successfully implemented mass customization to cater to diverse consumer needs. Dell's approach allows customers to select components such as memory, processor type, and software packages when ordering a computer (Li et al. 2022). This capability not only enhances customer satisfaction but also reduces inventory costs by assembling products on demand rather than stocking multiple variations.

From the less technical point of view, another industry that particularly benefits from mass customization is fashion and clothing, given the high variability in consumer tastes and body measurements. Companies like Nike offer customers the chance to design their own sneakers by choosing colors, materials, and even embroidering personal messages (Standaert 2022). This approach not only enhances customer engagement but also allows Nike to command a premium price for customized products.

The MC strategy is also becoming a tool in the food industry – key players in fast food market, such as McDonalds and KFC have introduced product configurators with personalization possibilities as part of their customer relationships strategy (Ban et al. 2024), while still maintaining a repetitive, quick process of food preparation, which is very similar in basic assumptions to industrial production.

Mass customization is underpinned by several key technologies. When considering usage of artificial materials, especially polymers, additive manufacturing or 3D printing is pivotal, particularly in industries such as dental, medical devices, and jewelry where highly personalized products are the norms. For example, Invisalign uses 3D printing to create custom orthodontic aligners based on individual patient scans (Valls 2023), reducing the need for frequent dentist visits and manual adjustments.

Advanced robotics and automation are also key to the successful implementation of MC. Robotics, especially cooperating robots, enable flexible manufacturing systems that can quickly switch between different product configurations (Makris 2021), what otherwise would be quite demanding logistically. In the automotive industry, robots perform tasks ranging from welding to painting, adaptable to different vehicle models without significant downtime. Robotics and 3D printing are often highlighted as key technologies in modern smart factories, enabling more flexible processes (Malik et al. 2022, Goel and Gupta 2020).

Computer-aided design (CAD) and computer-aided manufacturing (CAM) systems are also a must, as the digital approach with custom programming possibilities allows for quick translation of customer preferences into detailed product designs that can be directly fed into manufacturing equipment, reducing the time from design to production.

The economic benefits of mass customization are substantial for the companies that decide to implement this particular strategy, of course after careful consideration and justification. Offering customized products generally leads to higher customer satisfaction (Saniuk et al. 2020), which can translate into repeat business and positive word-of-mouth. Customized offerings with rapid manufacturing times can differentiate a company in a crowded market where many products and services often appear similar. Keeping that in mind, companies can often charge a premium for customized products, improving profit margins. Mass customization also has many common points with lean manufacturing assumptions (Tissir et al. 2020) – by manufacturing products to order, inventory costs can be reduced and the risk of overproduction is then minimized (Zhang et al. 2022). Finally, mass customization allows companies to be more responsive to market changes and customer trends (Siiskonen et al. 2020), adapting quickly without significant additional costs.

It is also worth remembering, that mass customization is not a golden path for each industry. There are a number of industries that should not consider this approach, as it would not bring them any benefits. These include companies that are involved in continuous production – such as basic food ingredients (e.g., sugar, oil, etc.), paper, petrol, etc. Another, less obvious production type is the whole group of non-customized products, produced in smaller or larger series, not requiring any particular customization – such as books, standardized machine parts (nuts, bolts, rivets, etc.), pharmaceuticals (although personalized medicine is also becoming a tool (Siiskonen et al. 2020)) and some part of low-key consumer electronics and home appliances. It is very important to remember, that even if mass customization and additive manufacturing are impressive and interesting technologies, not everything must be customized or 3D printed to be better – MC has its niches where it works perfectly and its potential is still underdeveloped – so these directions – such as production of personalized medical devices – must be pursued firsthand in the author's opinion.

Mass customization is not without its challenges, including the need for significant upfront investment in technology and training, as well as the complexity of managing supply chains that must be responsive to individual orders (Zawadzki and Żywicki 2016). However, as technology continues to advance and consumer expectations for personalization grow, the adoption of mass customization strategies is likely to expand, offering companies a powerful tool to differentiate themselves in a competitive market.

2.4.3 Premises for applying mass customization in medicine

The medical industry stands to benefit immensely from the principles of mass customization, particularly in the design of anatomically customized or bionic devices. The unique anatomical variations among individuals make mass customization an ideal approach in medical device production, such as in prosthetics, orthotics, and implantable devices. Some notions introduced in this book – notably in Chapters 4 (design) and 6 (case studies in orthopedics) show that this potential is viable and in future might become popular and widespread in general medicine. The integration of mass customization in medicine is supported by several key premises:

1. Individual patient needs – no two patients are alike; customized medical products can significantly improve patient outcomes by accommodating individual anatomical and physiological differences.
2. Advancements in medical imaging and 3D printing – technologies such as 3D scanning and printing have revolutionized the production of custom medical devices in recent years. These technologies allow for precise measurements and the creation of devices that perfectly fit the patient's body.
3. Knowledge-Based Engineering (KBE) and its successes in non-medical industry – this approach, proven in many industrial cases already, uses detailed knowledge

and rules to automate the design and manufacturing processes, thereby reducing the time and cost associated with custom devices.

The potential applications of mass customization in medicine are profound. For instance, in orthopedics, patient-specific implants and orthoses can be designed using CAD data derived from patient scans, ensuring optimal fit and functionality. Similarly, in cardiology, customized stents designed via mass customization techniques can better conform to a patient's unique vascular anatomy, improving the efficacy of interventions.

However, a couple of problems remain, due to current state of technology in scope of production of anatomically personalized medical devices. The main problem is that in the case of many products, the medical industry has not been keen on taking the technological leap that most other branches did a long time ago, that is from manual, piece production into serialized production with use of machines. Products such as prosthetic sockets, orthopedic corsets, wrist hand orthoses or ankle foot orthoses are still mostly produced entirely manually, by forming a material first around the limb of the patient (creating a negative mold) and then around its reproduced version (Abdalsadah et al. 2021). This process has the main advantage in that it works – patients are able to obtain the supplies they need. Also, it is not particularly expensive to set up, but otherwise it is inefficient, not repeatable, results are not stored digitally and the success is largely dependent of the manual skills of a technician.

The jump to the modern, digital process is gradually happening, with some companies embarking on this digital journey. An example of mapping between the traditional process and the digital one is presented in Fig. 2.14. It assumes use of digital measurement techniques – such as 3D scanning, or Magnetic Resonance Imaging (MRI), digital processing of anatomical data and then the use of CAD and additive manufacturing to fabricate the device, or part of it that has contact with the human body and must be geometrically adjusted to it. Here, we can use Knowledge Engineering (KE) techniques to extract the knowledge of technicians, doctors, physiotherapists, and patients and obtain a working process. However, the problem of time and skills remain – the manual "sculpting" is now replaced with a digital one and instead of a manually skilled technician, we now need an engineer, preferably skilled in surface modeling and knowledgeable in medical devices and human anatomy (i.e., a biomedical engineer). The time and cost do not change much, but the quality, repetitiveness and functionality of products may be increased.

The next step that occurs even more rarely requires use of KBE to automate the routine parts of design of anatomically shaped medical products. The problem is that however repetitive the process may seem, it is never the same each iteration, as each patient is different and requires adjustments to their anatomy. That is why, although possible, automating generation of anatomically shaped devices to be produced by 3D printing is an extremely difficult task. The example of how the digital process can be mapped to the automated one with use of KBE is shown in Fig. 2.15. This process has been successfully realized by the author together with a team of researchers in form of AutoMedPrint system, which is described in greater detail in Chapter 6. The

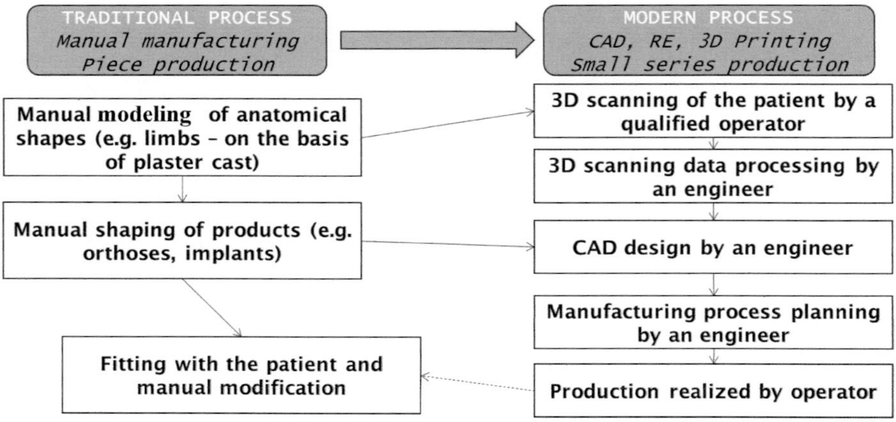

Figure 2.14 Production of anatomically adjusted medical devices – traditional versus modern process.

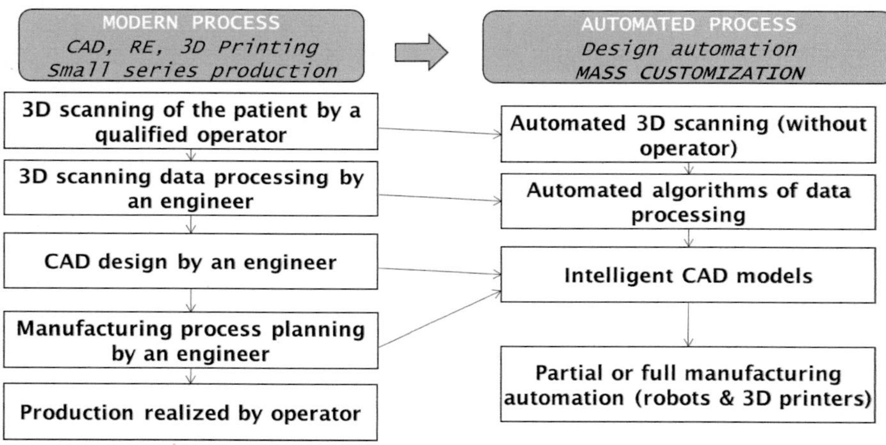

Figure 2.15 Production of anatomically adjusted medical devices – manual digital versus automated process.

steps of creating intelligent models and parts of automated design systems are laid out in Chapter 4.

In terms of automating the medical devices that represent the geometry of the human internal anatomy – implants, artificial organs, surgical supplies and many others – the steps of the process are generally the same, however there are a couple of problems related mostly to proper interpretation and extraction of 3D data using medical imaging techniques such as computed tomography (CT) and MRI. The problems of dealing with such data, both in the case of internal and external geometry, is detailed in Chapter 3. It is generally assumed in the scientific world that automating design of anatomically adjusted implants and other internal devices based on medical imaging, it is required to use some AI techniques, such as machine learning and neural networks (Li et al. 2021). Work on this matter is progressing, with some notable advancements (Li et al. 2023).

In conclusion, mass customization in the medical field not only promises greater patient satisfaction and better clinical outcomes but also represents a shift towards more personalized healthcare, where treatments and devices are tailored to meet the specific needs of individual patients. This approach aligns with broader trends towards personalization in medicine, including in pharmacology and treatment planning, making it a pivotal area of development in the coming years. However, many problems remain to be solved, especially when trying to replace skills and tacit knowledge of experts (such as radiologists, prosthetic and orthotic technicians, and medical design engineers), before mass customized implants, orthoses and prostheses can be achieved at a reasonable price in a short time, meeting the strict quality expectations and standards set by the medical industry.

Chapter 3

Medical Imaging and Anatomical Measurement

3.1 Overview of available technologies

3.1.1 Basic concepts and definitions

The capability of rapid manufacturing anatomically shaped medical devices by means of additive manufacturing can be achieved solely if a given device is possible to be represented as a digital, 3D model in a specific format. In this chapter, a comprehensive overlook of the problem of obtaining basic data for this representation will be presented. In simple terms, to be able to 3D print a device that is geometrically adjusted to fit a given individual, based on such an individual's anatomy, a measurement process must occur first and geometrical constraints of the anatomy must be acquired. When the data is successfully gathered, further design operations may be realized, ultimately leading to creation of a parametric (or non-parametric), complex 3D model representing a shape of a given device.

The chapter focuses on the principles and techniques of metrology as they apply to measuring the human body's interior and exterior. Basic metrology concepts are defined and contextualized for biological measurements, where variability and movement present unique challenges. The chapter explores methodologies including MRI and CT scans for internal measurements, alongside photogrammetry and 3D scanning for external measurements, discussing their precision, applications, and limitations, as well as data processing methods.

This book is not focusing on exploring basic engineering subjects, such as metrology – it is assumed that the reader has some basic understanding – if not, popular handbooks could be used for basic concepts (Webster and Eren 2018, Hudson 2022). The key concepts that will be used include, among others:

- *measurement* – the process of obtaining the magnitude of a quantity, in particular – spatial dimensions of a physical object,
- *measuring tool* – an instrument or device used to make measurements, analog (e.g., caliper) or digital (e.g., 3D scanner),

- *measurement resolution* – the smallest change in a quantity being measured that causes a perceptible change in the corresponding indication,
- *accuracy* – the degree to which the result of a measurement conforms to the true value,
- *precision* – obtaining the same results for repeated measurement under unchanged conditions (a degree of),
- *calibration* – the process of verifying the performance of a measuring tool by comparing it with a standard,
- *measurement uncertainty* – the doubt that exists regarding the value of a measurement, accounting for potential errors and the limitations of the measurement process.

Usually, in modern processes, measurement of the human body relies on non-contact techniques, meaning that the measuring tool does not touch or in other way directly interact with the measured object. Obviously, analog measurement methods may be also applied in certain cases (e.g., lack of access to proper measuring tools) – these involve use of various manual measuring tools (caliper, tailor's measure, tape measure, rulers, etc.). However, such processes are tedious, unprecise, rarely accurate and can be used efficiently only for a very limited range of products (e.g., simple mechanical prosthetic devices). As such, non-contact techniques, bringing digitally stored results, are used for majority of cases (Rumbo-Rodríguez et al. 2021).

In traditional engineering, use of non-contact measurement, such as 3D scanning, in correlation with subsequent design processes is a notion often named as "reverse engineering". In a general context, reverse engineering can be defined as analysis of a system to identify its components and the relationships between them in order to create its representation in a different form or at a different level of abstraction. In the design engineering context, it is related to measurement and then re-creation of measured geometry as a detailed 3D model, using parametric or non-parametric CAD methodologies (these will be described in further chapters). There are various definitions for this concept (Kumar et al. 2013, Raja and Fernandes 2007). The main idea assumes, that a physical object is measured, then the measurement data is processed, up until a faithful digital representation of an object is acquired that can be further used for different purposes – direct production (e.g., of machine parts with missing documentation (Raja and Fernandes 2007)), improvement or refining the original design concepts or visualization (e.g., using virtual reality (Robiony et al. 2007)).

In the case of medical use, it can be discussed (rather fruitlessly, in the author's opinion) whether using a term "reverse engineering" is correct as regards the human – a living being. However, the applied assumptions are generally the same as for the inanimate objects – measurement comes first, then the data is processed and a 3D model is obtained, which can be used for further design stages. Depending on the specific, anatomically-customized device, the original measurement of a human body area can be used directly (e.g., in case of prosthetic limbs), partially directly (e.g., in case of implants or mid-surgery aids) or indirectly (as in the case of orthotic devices). As such, "anatomical" (or "medical") reverse engineering can be defined as digitization of the patient's anatomical features to create an individualized model (prostheses, implants, artificial organs tailored to given patient).

As opposed to reverse engineering of machine parts (or more widely speaking, inanimate, rigid objects), anatomical reverse engineering processes encounter many problems related to the fact that the measured object belongs to a living, breathing, usually conscious being (moving aside considerations of measurement of body parts belonging to the deceased). The greatest challenges in measuring human interior anatomy are:

- *image resolution and clarity* – medical imaging techniques face limitations in resolution, affecting the ability to discern fine details,
- *patient movement* – even slight, involuntary movements can blur images, leading to inaccuracies,
- *tissue differentiation* – differentiating between various types of tissues, especially those with similar densities, can be challenging – that is why use of artificial intelligence is explored (Kim et al. 2018), otherwise it requires lots of work, especially regarding soft tissues,
- *artifacts* – ranging from metal implants to physiological processes, many things and phenomena can distort images, rendering parts of them useless for design.

Measuring external anatomy (e.g., limbs) has a bit different set of challenges to overcome – the basic ones are as following:

- *surface complexity* – human skin and contours, as well as hair and other features present complex surfaces for 3D scanning,
- *movement and stability* – keeping the subject still is crucial for accurate measurement, but difficult – even the patient sometimes is not aware of their own movements,
- *environmental factors* – lighting, reflections and shadows can affect the accuracy of external scans,
- *data processing* – converting raw data into usable 3D models requires sophisticated algorithms, which can introduce errors, requiring mostly a manual job in realizing the processing.

It is also very important to realize, that requirements toward the measurement results – especially regarding resolution and accuracy – are different for reverse engineering processes than in the standard use case. In the case of medical imaging, the standard use is for diagnostics – as such, high resolution of slices is not required. In the case of external optical measuring techniques, the issue is actually different – these processes, when using universal devices (all-purpose 3D scanners) are often too accurate as per the human anatomy (which is actually not a bad thing). However, they are not at all adjusted to the possibility of a shape changing mid-measure, which might happen if the patient moves, as typical 3D scanning is much more prone to errors than any medical imaging technique.

These issues, challenges, limitations, and possibilities of overcoming them are discussed more widely in appropriate subsequent subchapters.

3.1.2 Division of technologies

The interconnected use of 3D scanning and medical imaging techniques has greatly improved the possibilities in the realm of individualized design of anatomically shaped medical and non-medical products. These non-contact measurements can be used to scan the human body from the inside and outside to map the areas of interest, allowing to create, among other things, highly customized implants, orthoses, and prostheses.

The basic, non-destructive (generally harmless to the measured object) techniques available in the industry are presented in the schematic in Fig. 3.1. The schematic outlines the classification of popular digital 3D measurement techniques (analog, traditional methods, mentioned in the previous chapter, are omitted). Contact methods include contact scanners and Coordinate Measuring Machines (CMMs). Remote methods involve, in general, interaction with use of electromagnetic waves, with subcategories such as passive techniques including photogrammetry, and active techniques involving lasers and structured light scanners, which are further broken down into different types. Medical non-contact techniques include X-rays and Magnetic Resonance Imaging (MRI), with X-ray Computed Tomography (CT). The schematic also indicates sound-based method – 3D ultrasound scanners under the contactless category.

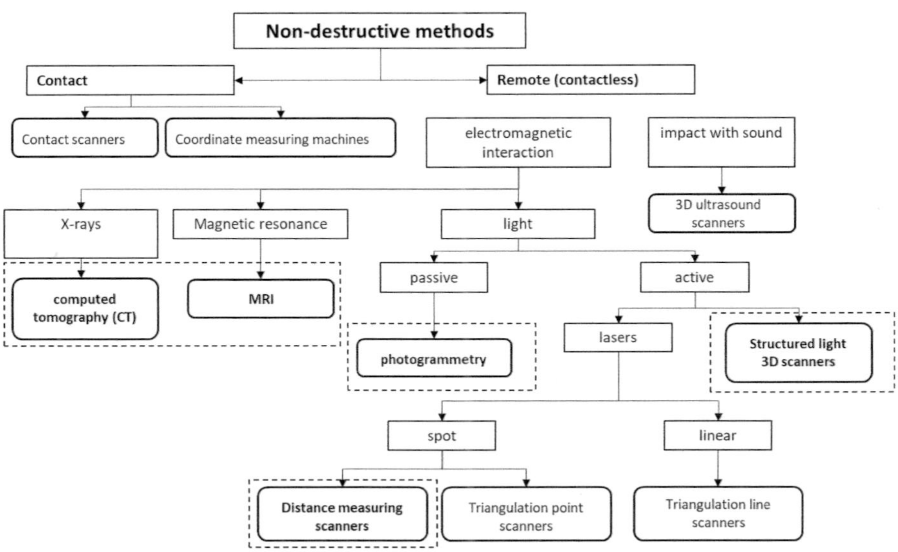

Figure 3.1 Basic division of non-destructive measurement technologies, highlighting techniques generally applicable in design of anatomically customized devices.

Of these measuring techniques, the most important ones to use in individualized design of medical, anatomically customized devices are the following ones:

- Computed Tomography (CT),
- Magnetic Resonance Imaging (MRI),
- optical scanning with structural light (commonly known as 3D scanning),
- photogrammetry,
- non-linear laser scanning (the distance measurement variant).

It is important to notice, that the list of techniques, as well as their usefulness in anatomical reverse engineering, could be a matter of discussion and new techniques could be easily added to the list. However, at the moment, these techniques are of most common use, mostly due to proper combination of obtained accuracy, tediousness of the measuring process, accessibility (price) and patient comfort, expressed in the time spent for single measurement and potentially harmful effects on measured subject's health.

It must be remembered, that the primary application of all these techniques is diagnostics – the medical imaging techniques are used predominantly to gather information about state of the internal organs of human body, e.g., to find dangerous tumors or to verify if the applied treatment brought tangible results. Similarly, 3D scanning, photogrammetry and laser scanning were originally used mostly for quality control of produced objects, or for archiving of existing state of objects (e.g., buildings post-construction). Use of all these techniques to gather data for design purposes is a relatively new concept, and they are rarely, if ever, mentioned together in available literature on the subject. It has been proven in numerous studies that CT, MRI and 3D scanning can be all used efficiently for design of customized implants and prostheses (Paxton et al. 2022).

Other technologies, such as ultrasound – profoundly used for various diagnostic and therapeutic purposes – could potentially also be used for the design. However, by the author's experience, they are not as accurate and it is difficult to gather some reliable data, as such their use in literature in that context is not mentioned. The same goes with contact methods – it is not technically impossible to make a measurement of a human limb using a CMM – it is simply impractical. Industrial-grade measuring devices also have one other quality that is important in the context of anthropometric measurement – their resolution, accuracy and precision are usually redundant (within fraction of a millimeter) for the measurement of human body parts, that can spontaneously change size even during one day, due to various factors (weather, physical activity, physiological processes). As such, industrial grade devices can be always considered for measuring the human body from that viewpoint – although usually their accessibility is low, due to the high price.

The next two subchapters (3.2–3.3) contain more detailed descriptions of the selected measurement technologies. In the further chapters, it is shown how the gathered data can be processed and used for design of specific medical devices.

3.1.3 Applications

The main uses of 3D imaging and anthropometric measurement techniques – with some known for more than two decades – can be divided into the following groups:

1. *Diagnostics* – measurements aimed at either verifying if current state of certain organs or tissues in human body stays within range of healthy (screening tests) or, when certain health issues have already been suspected, for detailed imaging of the source of a problem and its details (e.g., cancerous tumors location, type, and size), as well as decision-making toward possible treatment methods. Examples involve CT for bone cancer diagnosis (Ladd and Roth 2017), MRI in brain tumors detection (Natarajan et al. 2012) or 3D scanning to check the postural defects (Haleem and Javaid 2019).
2. *Surgical planning* – data obtained in the process of anthropometric measurement are used to conceptualize and plan the medical treatment requiring surgical intervention. This helps in practicing complex surgeries, reducing operating times, and improving surgical precision (Wang et al. 2015). In cases of trauma or congenital abnormalities, 3D imaging is critical for planning and executing reconstructive surgeries, allowing for the restoration of form and function (Markiewicz and Bell 2011, Nassar et al. 2022), with some examples shown in Fig. 3.2.
3. *Therapy* – mid-treatment measurement for control of current progress in patients' state of health. Not as detailed as the diagnostic measurement, aimed mostly at controlling certain dimensions (broadly speaking, lengths, diameters, angles, etc. – both in internal and external organs). Examples involve 3D scanning in physical therapy to monitor movement angles (Kurzydło et al. 2014), or repeated MRI to check the effects of chemotherapy (Rieber et al. 2002).
4. *Design of anatomical models for 3D printing* – the essence and main focus of this book. The design can involve creating pre-surgical models (Żukowska et al. 2024a, Żukowska et al. 2022), mid-surgery templates (Banaszewski et al. 2018),

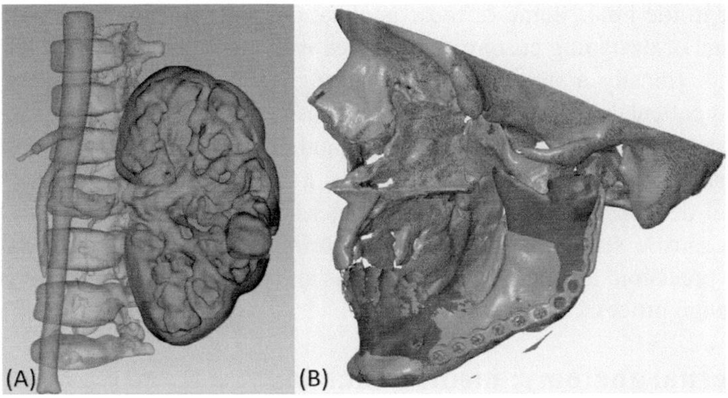

Figure 3.2 Examples of use of anthropometric measurement techniques in surgery planning, (A) kidney with cancerous tumor (imaging courtesy of Dr Krzysztof Grandys), CT scan; (B) lower jaw endoprosthesis – a model for surgical planning (imaging courtesy of BEng Emilia Smolarek).

educational models for doctors (Żukowska et al. 2024b) or products for direct use by patients – orthotics (Górski et al. 2020a,b), prosthetics (Górski et al. 2023a) or implants (Vitković et al. 2023a), as well as scaffolds for tissue engineering (Vitković et al. 2023b). It may also involve design not for the purpose of actual production, but rather for virtual prototyping, e.g., for educational purposes or for virtual planning of surgical processes (Górski et al. 2024b). Artificial intelligence can be applied here as well, to automate certain stages of the process (Wodziński et al. 2022).

5. *Quality control* – 3D scanning is routinely used for quality check (dimensional and shape accuracy, as well as precision) of produced parts (Haleem and Javaid 2019). The same use can be associated with medical imaging, when applied to inanimate objects (e.g., produced implants), this can be done using industrial computed tomography (Toth et al. 2015).
6. *Biomechanical analysis* – 3D models obtained via scanning can be utilized to understand the biomechanics of different body parts. This is particularly useful in designing orthopedic solutions and studying the impact of certain diseases like osteoarthritis (Cykowska et al. 2022), but also could be used in sports (Ortega and Olmedo 2017).
7. *Research and studies* – from case analyses, to statistical cohort studies, imagery taken by MRI, CT or 3D scanning could be used in order to advance scientific progress in measurement techniques themselves, interpretation of their results and their possible application range.
8. *Entertainment* – in the filmmaking industry or game industry, advanced measurement techniques can be used to accurately represent the human body behavior in certain Computer-Generated Imagery (CGI).
9. *Hobby and personal use* – in the wake of widespread cheap 3D printing and game making, it is not uncommon to use the medical imaging or 3D scanning data for home usage, to create hobby models, such as figurines or toys.

Later in the book, some of these applications will be further explored, mostly in the range of designing customized devices with anatomical shapes, either for use by a doctor (mostly a surgeon) for use during a patient's treatment, or for direct use by the patient, in form of prosthesis, orthosis, implant or another device. These applications are most demanding – in the authors' opinion – as the measurement and its results are not an aim in itself, but just a basis for a more complex process of anatomical design. Not only must be the measurement process realized according to the diagnostic standards, but also must bring results with a proper resolution, accuracy, precision, and clarity, to be used as an unambiguous source of constraints for the design process.

3.2 Internal anatomy: medical imaging

3.2.1 Computed Tomography

Computed Tomography (CT) is a diagnostic imaging method that uses a combination of X-rays and computer technology to produce cross-sectional images (often called

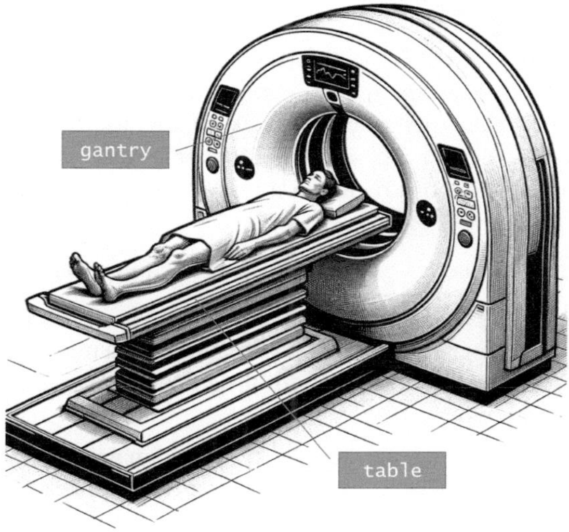

Figure 3.3 Computed Tomography scanner – medical.

slices), both horizontally and vertically, of the body. A CT scan shows detailed images of any part of the body, including bones, muscles, fat, and organs (John Hopkins Medicine 2024). CT scans are more detailed than standard X-rays (Storrs 2013). The measurement involves registering many tomographic images (cross-sections) of the examined object. The single image is a planar shot is realized in various positions of the object. Based on these images, 3D reconstruction is possible to realize (as described in further chapters). The technology enables measuring the geometric features of external surfaces and the internal structure of the object.

In CT, the X-ray beam moves in a circle around the body. This allows many different views of the same organ or structure. The X-ray information is sent to a computer that interprets the X-ray data and displays it in a two-dimensional form on a monitor. Typical CT device used for medical (living beings) measurement has a form of a tube, also known as the gantry, inside which a patient is lying still, on a table (Fig. 3.3). The gantry is the ring-shaped part of the CT Scanner, housing many of the components necessary to produce and detect x-rays. The components are mounted on the rotating scan frame. The diameter of the opening or aperture varies in size, as does the gantry as a whole (John Hopkins Medicine 2024, Harmonay 2020).

During the measurement, the patient first lies on a table that slides into the center of the CT scanner. An X-ray tube rotates around the patient, sending out narrowly focused beams of X-rays. Detectors on the opposite side of the patient pick up the X-rays and measure the quantity that makes it through. As the scanner makes one full rotation, it captures a "slice" of the body from multiple angles. These images are then reconstructed by a computer into a three-dimensional model of the patient's body or the section being examined (Jung 2021).

Medical CT scanners – the most common kind – are specifically designed to scan the human body. They are calibrated to provide the best balance between image detail and minimal radiation exposure to the patient. They are routinely used for medical examination.

Industrial CT scanners are used for scanning objects for quality control or inspection of the internal structure without destroying the item. These scanners can often accommodate larger objects and use a higher power and resolution to detect finer details. They are not constructed in a manner allowing a human being to get inside.

The most important parameters of CT scanners are as following:

- *slice thickness*: determines the resolution of the scan; thinner slices provide more detail; for proper 3D printing in pre-surgery and mid-surgery use, 0.6 mm is assumed as a reasonable optimum (of the author's experience and cooperation with medical practitioners);
- *current and voltage* – affects the quality of the image and the radiation dose; higher current and voltage provide clearer images but increase radiation exposure;
- *scanning speed* – faster scanners reduce motion artifacts and the time a patient needs to remain still;
- *efficiency* – more efficient detectors can provide better images and reduce the radiation dose.

CT is long established and extensively used for diagnosing diseases and conditions such as cancers, cardiovascular disease, infectious disease, trauma, and musculoskeletal disorders. It is particularly useful in detecting tumors and evaluating their size and extent (Leksowski et al. 2011). CT imaging is also frequently used to guide procedures such as biopsies and surgeries, providing real-time images that help in accurately targeting the area of interest (Hau et al. 2002).

Another typical use of CT is to monitor the effectiveness of treatments like chemotherapy or radiation therapy in shrinking tumors (Petralia et al. 2009). It can be also used for injury assessment – in complex and severe trauma cases, CT can quickly provide images of injuries to internal organs, bones, and other internal injuries, which are critical for immediate care (Fang et al. 2006).

From the viewpoint of this book, it is however important to mention CT applications in design of customized devices, as CT scanning plays a pivotal role in the design and manufacturing of custom devices. CT data can be used to create highly accurate and individualized prosthetics and orthotic devices by capturing the precise dimensions of the required body part (Liacouras et al. 2017, Żukowska et al. 2023). As previously mentioned and earlier work of the author together with a team, surgeons use CT scans extensively to plan complex surgical interventions, often with the aid of 3D-printed models of the patient's anatomy (Banaszewski et al. 2018).

CT can be also used to design devices used directly by patients – custom implants such as dental, cranial, or joint replacements could be designed using 3D models derived from CT data, ensuring a perfect fit and reduced surgery time (Wodziński et al. 2022, Gubin et al. 2016).

CT technology, however useful and absolutely invaluable in today's medicine, has its set of limitations and problems. One of the most significant limitations of CT is the exposure to ionizing radiation. While doses are controlled and minimized, there is a cumulative risk, especially in patients who require multiple scans (Storrs 2013). That is why in the case of pediatric patients, it is only used as a last resort, if there is no other option – e.g., in treatment of life-threatening cancers. It is a strong distinction here between the CT and the optical scanning technologies, which are harmless, not costly and can be repeated multiple times until a proper effect is achieved. In the case of CT, such a trial-and-error approach is not possible, especially for pediatric and other sensitive patients, so each scan must be carefully considered and realized properly to avoid the necessity of repetition.

There is more to the aspect of patient safety – there could be a problem in use of contrast agents (iodine, usually, with others also possible, such as barium sulfate). They alter the X-ray absorption, improving contrast between different tissues and sometimes using them is essential (Lusic and Grinstaff 2013), however they must be administered – orally, intravenously, or rectally – and they can lead to allergic reactions in some patients, potentially also causing kidney problems in those with pre-existing kidney conditions (Kobayashi et al. 2013).

In the terms of the imaging precision itself, there might be some artifacts in the obtained images, making it difficult to interpret the results. Metal implants, dental fillings, or movement can create artifacts on CT images, which can obscure details and potentially lead to misdiagnosis (Lambert et al. 2015). Also, as previously mentioned, while CT is excellent for visualizing bones and detecting abnormalities in denser tissues, it sometimes has limited contrast in soft tissues when compared to other imaging modalities like MRI (Alsleem and Davidson 2013). CT scanners are also quite expensive to purchase and maintain, which can limit accessibility in resource-poor settings.

There is also the problem of overdiagnosis, as the high resolution of CT scans can sometimes detect incidental findings that may lead to overdiagnosis and overtreatment of conditions that may never have caused any problems (Gao et al. 2022).

To sum it up, Computed Tomography scans are invaluable in the medical field for diagnostics. They allow for the visualization of the internal structures of the body in great detail. It is noteworthy that standard, single CT scans can be used to obtain 3D models of hard tissues, mainly bones. To obtain proper imaging of softer tissues, such as internal organs, it is necessary to use the contrast solution appropriately before the study. As such, a 3-phase study can be conducted, of which images of arteries, veins and other soft tissues might be separated.

This book is not focused on radiological technology and methodology of realizing such measurements – this should be always decided by the doctors and radiologists, keeping in mind a specific application (pure diagnostic purposes, diagnosis and surgery planning or design of customized devices). The engineers' job is usually to state the requirements and properly process the obtained images to obtain good quality data for design.

3.2.2 Magnetic Resonance Imaging

Magnetic Resonance Imaging (MRI) is a non-invasive diagnostic tool that utilizes strong magnetic fields and radio waves to generate detailed images of the structures and tissues within the body. MRI does not use ionizing radiation, as opposed to X-ray based techniques such as CT. MRI operates on the principle of nuclear magnetic resonance (Bolas 2010). As the human (or any animal, MRI can be used also on non-human subjects (Gilbert et al. 2010)) body is composed largely of water molecules, which in turn consist of hydrogen atoms, when placed in a magnetic field, the protons in these hydrogen atoms align with the direction of the field. A radiofrequency current is then pulsed through the patient, knocking the protons out of alignment. When the current is turned off, the sensors built inside the device detect the energy released as the protons realign with the magnetic field. This information is converted into images by a computer (Sprawls 2000). Unlike the Computed Tomography (and generally X-Ray, for this matter), MRI cannot be successfully used for inanimate objects, so there are no industrial grade scanners – only medical (or veterinarian) devices.

The measurement process starts with a patient lying on a movable bed that slides into the MRI scanner main part (the gantry – similarly as in the CT apparatus). The scanner generates a strong magnetic field that aligns the hydrogen protons in the body. Pulses of radio waves are sent from an antenna in the scanner to the body, temporarily knocking the protons out of alignment and as they realign, they send out radio signals, which are received by the antenna and relayed to a computer. The computer processes these signals as coordinates of points in space and constructs a series of images of the scanned area of the patients' body.

There are several types of MRI scanner (Cleveland Clinic 2022):

- *closed MRI* – traditional scanners with a tube-like structure where the patient is inserted into a tunnel, somewhat similar in the terms of form factor to the medical CT scanners;
- *open MRI* – designed to reduce the claustrophobic experience of a closed MRI, these have open sides while still utilizing strong magnetic fields;
- *standing or seated MRI* – these scanners allow imaging in different postures to see how the body reacts to the sitting or standing positions;
- *extremity MRI* – compact devices, used specifically for scanning limbs (Wald et al. 2020).

The most important parameter of the MRI apparatus is its magnetic field strength. Measured in Tesla (T), it typically stays within range from 0.5T to 3T in clinical settings (Cleveland Clinic 2022, Runge and Heverhagen 2022). Higher strength provides better image quality but can be more expensive. Other important parameters are gradient strength, Signal-to-Noise-Ratio (SNR) and resolution, all influencing the outcomes such as imaging speed, clarity and level of detail that can be seen in obtained images.

Magnetic Resonance Imaging (MRI) is a powerful diagnostic tool with a wide array of clinical applications due to its exceptional ability to differentiate between various soft tissues of the body. One of its primary uses is diagnosis and assessment of neurological conditions, such as brain tumors, stroke, and demyelinating diseases like multiple sclerosis. It offers detailed images of brain structures and spinal cord abnormalities without the use of ionizing radiation, making it a safer option for repeated imaging, although the safety concerns of MRI are under constant consideration by medical specialists (Jabehdar et al. 2020).

MRI also is used frequently for evaluation of musculoskeletal disorders, providing clear visualization of joints, ligaments, and tendons and pathologies in these tissues. It is often the imaging modality of choice for diagnosing conditions such as torn ligaments, cartilage damage, and soft tissue tumors (Bae et al. 2020). Orthopedic surgeons rely on the detailed images from MRI scans to plan surgical interventions, including joint replacements and reconstructions (Gholamrezanezhad et al. 2018).

In cardiology, MRI contributes to the detailed assessment of cardiac structures, myocardial scarring, and function, aiding in the diagnosis and management of heart diseases, congenital heart disease, and assessment of myocardial viability after ischemic injuries (Souto et al. 2017). It is also pivotal in the evaluation of vascular diseases, including aneurysms and vessel occlusions (Kamada et al. 2022), due to its ability to produce precise angiographic images without the need for contrast agents that may pose a certain risk to patients, as mentioned in the section about CT.

Similar to the CT, MRI is applied also in monitoring the progress of treatments, such as the response to pharmaceutical interventions in inflammatory and infectious diseases. It is sensitive to changes at the tissue level, allowing for early detection of treatment efficacy or the need for therapeutic adjustments (Rausch et al. 2003).

In the realm of personalized medicine, MRI data is utilized in the design of customized implants and devices. With its three-dimensional imaging capabilities, MRI can aid in constructing accurate models of anatomical structures, which are essential for creating bespoke implants like joint replacements or dental devices that offer improved compatibility and functionality (Memon et al. 2020).

In the terms of advantages of MRI comparing to other methods (especially CT), besides the aforementioned safety (no ionizing radiation), it also provides superior contrast between different soft tissues, making it a basic choice for imaging brain, muscles, and heart. It can produce images in any plane, providing greater diagnostic flexibility, allowing easily to obtain three-dimensional images, that can of course later be used for design of personalized devices.

Considering the disadvantages, MRI is generally more expensive than other imaging methods. Similar to CT it can be claustrophobic for some patients due to the enclosed nature of the scanner. On the same note, its operation is noisy, which can be uncomfortable or alarming, especially for patients with limited awareness or other neurological conditions. The image in itself also frequently takes longer to produce compared to CT scans, which can be problematic for patients who have difficulty staying still (e.g., children, or patients with neurological conditions).

For the human body, MRI is generally harmless, but the situation changes if there are any ferromagnetic implants inside patient's body, due to the strong magnetic field that is created – any ferromagnetic part could be literally torn out of the body, with high risk of severe, even life-threatening injuries (Jabehdar et al. 2020, Cleveland Clinic 2022). That is why patients are thoroughly screened before imaging, for previous cases of implantation and in the case that any ferromagnetic parts are present, MRI is rejected as a potential imaging method.

MRI is a useful method of imaging of soft tissues, especially in the head and neck area (example in Fig. 3.X) and it is difficult to imagine realizing certain surgeries or producing anatomical models without this technology, as demonstrated in numerous works by the author with a team (Żukowska et al. 2024b, Żukowska et al. 2022). It can be also used to create images of extremities, in parallel with the 3D scanning techniques, so it can be treated as a universal method. However, it is also noteworthy that the MRI images are remarkably difficult in processing for personnel unskilled in reading of radiological data – that is why usually it is quite hard for an engineer to work out the proper shapes without aid of an experienced doctor or radiologist – which is more rarely the case considering the CT scans.

3.2.3 Other methods

The Computed Tomography and Magnetic Resonance Imaging are the most popular and useful methods of medical imaging that can generate 3D data of human internal tissues for use in design process of anatomically adjusted devices. Aside from them, however, there are several other significant medical imaging methods. Some of the noteworthy techniques include:

- ultrasound imaging,
- Positron Emission Tomography (PET),
- Single Photon Emission Computed Tomography (SPECT),
- fluoroscopy,
- Optical Coherence Tomography (OCT),
- thermography.

Ultrasound imaging, also known as sonography, uses high-frequency sound waves to produce images of structures within the body. The process involves a transducer that emits sound waves; as these waves travel through the body, they are reflected back to the transducer by the internal tissues and organs (Hollywood Diagnostics Center 2024). These echoes are then converted into live images. Ultrasound is widely used for examining the fetus during pregnancy, diagnosing conditions in organs like the heart, liver, kidney, and assessing blood flow through various vessels (Bruce et al. 2020). It is also a good method for quick screening tests, as it can be realized quickly (in less than 10 minutes), without extensive preparations and the initial diagnosis can be realized in real time. It is harmless for the human body, not excessively expensive and not overly difficult to operate.

While ultrasound is predominantly known for its diagnostic capabilities in soft tissue visualization, it also has applications in the design of custom implants and anatomically adjusted devices. In general, it is possible to make measurement on stills taken of ultrasound measurement – these measurements can form some basic guidelines for design. There is also the 3D variant of ultrasound measurement, able to generate spatial images (Zhan et al. 2020) – albeit not very precise, but still potentially useful in spite of the lack of the possibility of using other techniques. Ultrasound can provide real-time images of joints in motion, which is beneficial for designing dynamic orthopedic devices that need to accommodate a range of movements (Bimbraw et al. 2022). Additionally, for vascular implants, ultrasound can be instrumental in measuring blood flow (the so-called Doppler ultrasound) to ensure the proper fit and placement of vascular grafts or stents (Furdella et al. 2021).

Another technology is Positron Emission Tomography (PET). PET is a nuclear medicine imaging technique. Some processes in the body can be imaged this way in a 3D manner (Animated Physics 2011). A PET scan uses a small amount of radioactive materials, or radiotracers, which are typically injected into the bloodstream. The tracer travels through the body and is absorbed by organs and tissues, which are then highlighted by the PET scanner. PET is particularly useful in detecting cancer, evaluating brain abnormalities, and monitoring heart disease.

Akin to PET, Single Photon Emission Computed Tomography (SPECT) is a nuclear imaging test that uses a radioactive substance and a special camera to create 3D pictures (Ozsahin et al. 2020). While PET can measure metabolic processes, SPECT assesses the blood flow and activity level within tissues and organs, especially the brain and heart (Pelletier-Galarneu and Ruddy 2021). It is commonly used to identify clots, strokes, and areas of reduced blood flow.

PET and SPECT are more often used for their ability to visualize metabolic and physiological processes rather than direct design of implants or other customized devices. However, they can aid in the customization of implants by helping to identify the viable tissue for optimal implant integration (da Silva and Moreira 2021), especially in oncology patients, by distinguishing between healthy and diseased tissues.

Another noteworthy technique is fluoroscopy. It is an imaging technique that uses X-rays to obtain real-time moving images of the interior of the body. It is often used during diagnostic procedures and treatments to provide a continuous internal view of the body (Yamashita et al. 2016). Common uses include guiding catheters through blood vessels, directing the placement of stents, and observing the motion of joints (Bakker et al. 2001). Fluoroscopy's real-time imaging can be valuable during the surgical procedures themselves rather than the design phase. It allows for the visualization of the anatomical device (e.g., implant) as it is positioned and fitted, ensuring proper alignment and function.

Optical Coherence Tomography (OCT) is yet another non-invasive imaging test. It measures the retina tissue, by utilization of light waves to take cross-section pictures. It is beneficial in diagnosing and monitoring diseases like glaucoma, macular degeneration, and retinal detachment (Lumbroso and Rispoli 2014). OCT's high-resolution images of the retina are mainly used for diagnostic purposes in ophthalmology. However, the precision of OCT can be used for designing implants and prosthetic devices that need to interact delicately with the eye's structures, such as bionic eyes or advanced contact lenses (Opie et al. 2014).

Finally, it is worth mentioning another technology that has both medical and industrial applications. Thermography is an imaging technique that uses infrared cameras to detect heat patterns. Then they are visualized in different colors, creating characteristic images with various hues of red and blue, depending on the scale (Lagüela et al. 2016). In medicine, this can be used to detect and measure many important processes and phenomena happening in the human body. Thermography is often touted as a noninvasive diagnostic tool for cancer screening, although its effectiveness and role in comparison to other imaging techniques are subjects of ongoing research and debate (Gonzalez 2016). Thermography is not typically used for the design of implants and other anatomical devices, but may have potential applications in monitoring the inflammation and healing process around an implantation site, thus indirectly informing the design and material choice for implants that minimize inflammatory response (Christensen et al. 2012).

Examples of images obtained by 3D ultrasound, SPECT, OCT, and thermography are presented in Fig. 3.4.

Each of these imaging techniques offers a unique set of advantages and can be selected based on the specific diagnostic needs, providing a comprehensive toolkit for modern medicine to visualize and understand a vast array of medical conditions. However, it is very important to emphasize, that none of the technologies mentioned in this subchapter would be the first choice in imaging for purposes of

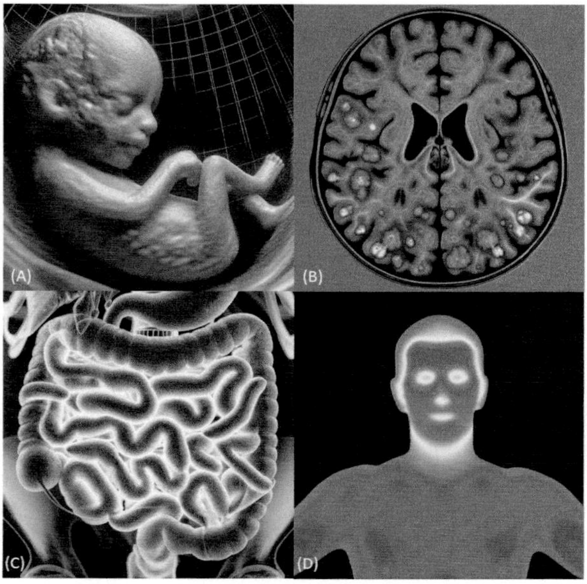

Figure 3.4 Conceptual images presenting capabilities of other medical imaging techniques: (A) 3D ultrasound, (B) SPECT; (C) fluoroscopy; (D) thermography.

design processes of anatomically adjusted devices, as their purposes are different. For visualization and creation of 3D models of human internal tissues – for various purposes – the first choice would always be CT or MRI, depending on the area, type of tissue and type of model that we want to create. Only if the constraints of the patient's safety, cost, and/or time of measurement, as well as other limitations would exclude one or both these techniques, we would need to search and potentially use one of the alternatives, to create a 3D model – although, probably, with not as clear and precise shapes as in the case of CT or MRI.

3.3 External anatomy: 3D scanning

3.3.1 Optical 3D scanning

Optical 3D scanning is a measurement process that belongs to the group of non-contact techniques. It is integrated with computer systems – the measurement is converted and recorded in software integrated with the main component of the system, i.e., the 3D scanner.

The principle of measurement with structured light is based on the deformation of light falling on an object and on the registration of images of curved so-called fringes – light has a specific structure, most often black and white stripes with a density that changes step by step (Fig. 3.5A). The scanner therefore consists of a projector that displays the image of the stripes and a camera that records their arrangement on the object and converts the flat image into three-dimensional coordinates (Fig. 3.5B).

A flat image recorded from a camera has a specific number of points — pixels. Each pixel in a flat image is assigned a set of coordinates in a spatial system, as a result of a single measurement. The collected data is therefore in the form of the point cloud, on which an automatic algorithm usually spreads a three-dimensional model, the so-called triangle mesh. Performing a single measurement (scan), which takes several seconds, does not usually allow for obtaining a representation of the entire object. Therefore, several or several dozen measurements are made by moving the scanner or the object (less often both at the same time), then all measurements are combined into one point cloud.

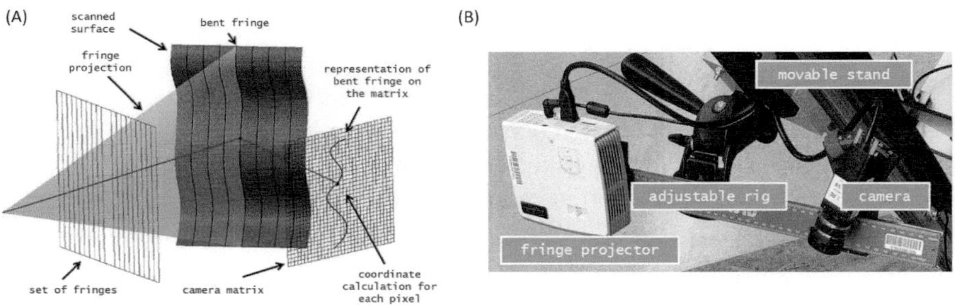

Figure 3.5 Optical 3D scanning – principle of operation (A) (based on Kowalski et al. 2011) and 3D scanner elements (B).

Currently, thanks to the rapid development of modern non-contact measurement techniques, techniques such as optical 3D scanning are increasingly used to perform anthropometric measurements (Wichniarek et al. 2020). These methods replace the previously used traditional measurement methods, such as measurement using the Martin set (anthropometer, bow compass and slider) (Goto et al. 2019) or measurements carried out on the basis of individual photos showing anthropometric points marked on the body of the measured individual. Non-contact measurement techniques such as optical scanning enable obtaining a large amount of data in a short time, and depending on the method used, it is possible to measure the entire body or a selected part (Wichniarek et al. 2020).

The small size of 3D scanners with structured light projection, very high measurement speed enabling obtaining several million points in space in a few seconds, high measurement accuracy and the universality of this type of devices have resulted in their applications in many branches of industry and medicine. In recent years, the expiration of patents for 3D scanners has resulted in the development of a market for initially hobbyist scanners, followed by low-cost scanners.

As such, currently the following classes of 3D scanners can be distinguished, based on their price and range of applications:

1. Hobby – budget scanners

   Simple hobby scanners based on devices equipped with a depth sensor (recording infrared light) and cameras. There are ready-made systems available on the market and their cost is lower than US$2000th therefore affordable for small companies, research teams and even for home uses. Devices such as standalone IR cameras with depth perception (e.g., Intel RealSense or Microsoft Kinect) can be also counted into this category, as it is possible to effectively turn them into small 3D scanners. These can work both as stationary scanners (doing series of scans from fixed positions) or as continuous measuring devices (handheld and moved around a measured object).

   A common feature of all scanners belonging to this category is obtaining data with very low accuracy (±1 mm or less), low detail in reproducing the shapes of the measured object and low resolution of the obtained point clouds. What is characteristic of these devices is that they do not need to be calibrated. Data obtained out of these devices do not necessarily need to be useless for the design of customized devices – in the author's opinion and experience, it is possible to use them for prosthetic design (example shown in Fig. 3.6), however accuracy and precision is too low to go for devices that need to have a properly defined shape, such as corrective orthoses for hand therapy.

2. Non-industrial scanners

   This group contains professional 3D scanners that are not accurate and precise enough to be used in machine production industry – however they have a very wide range of other applications, including medical ones. The construction of scanners in this group is very similar. A multimedia projector is responsible for displaying a set of bands, and one or two cameras are responsible for recording

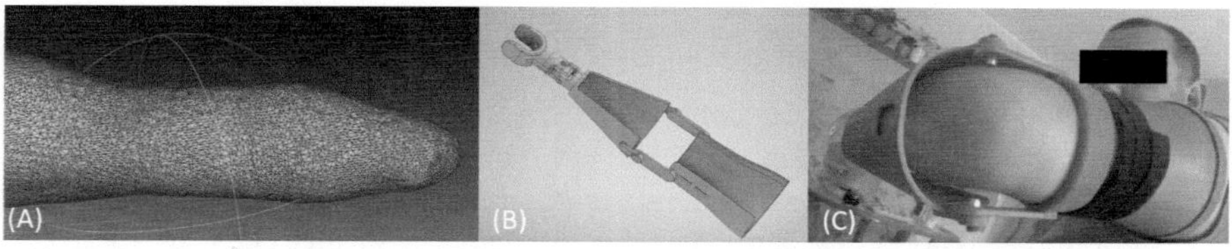

Figure 3.6 Use of low-cost 3D scanner (Structure Sensor) for design of customized prosthesis, (A) raw scan (data obtained from Dr Krzysztof Grandys); (B) automatically designed prosthesis; (C) prosthesis used by the patient.

the image. The scanners belonging to this group usually cost below or in the range of US$5000.

Thanks to much better parameters than in the case of scanners from the first group, especially due to calibration of the optical system, higher measurement accuracy and point cloud resolution, these scanners are successfully used in all reverse engineering applications in which the measurement accuracy does not have to be greater than ±0.5 mm. This is an accuracy range that is quite appropriate for measuring human body parts, such as extremities or facial features, along with gathering color data (texture). The devices belonging to this category usually are sold with dedicated software, facilitating setting up the hardware, going through the calibration and managing combining of individual scans into a single point cloud or triangle mesh.

In the AutoMedPrint system developed by the author, two scanners belonging to that category are used – David SLS-3 (Fig. 3.7B) and manual scanner Einscan Pro. Anthropometric data gathered by these devices has been used in numerous cases of practically used orthotic and prosthetic devices (Górski et al. 2020b).

3. Industrial semi-professional scanners

The third category scanners are semi-professional scanners used in industry where the measurement accuracy does not have to be greater than ±0.1 mm. These scanners have a wide range of industrial applications, but they rarely obtain certification that enables them to serve as professional quality-control devices in manufacturing processes of precision parts. As such, they are mostly used in all-purpose reverse engineering, including the medical (anthropometric) applications. They usually fall within a price below US$25000. Examples of devices belonging to this class, available as of 2024, are Artec EVA or Smarttech Skan3D Med scanners. Here, advanced algorithms are often used for reduction time of scanning, which can be realized in a couple of seconds or below. Also, wireless manual scanners are available in this category, enabling easier and more comfortable measurement from all sides.

These scanners are more than appropriate for all the kinds of anthropometric measurements realized for design of customized devices and their advanced features may help with making the process easier and faster, thus more comfortable for the patient, while obtaining a similar or better level of accuracy and precision as in the scanners belonging to the previous group.

4. Professional or specialized scanners

   This group includes professional scanners used mainly in the aviation and automotive industries due to their very high measurement accuracy and certification in accordance with the standards, realized in accredited measurement laboratories. The largest manufacturers of scanners in this class, as for 2024, are GOM, Hexagon and Breuckmann, and the prices of the measurement systems they offer range usually between US$25000 and US$100000. These professional industry scanners are used in quality control or reverse engineering of precise machine parts, requiring very high accuracy (below or in the range of 0.02 mm). They are more than capable of realizing anthropometric measurements and have been successfully used in the past for building customized medical devices – however that is not their primary purpose and, despite many advanced algorithms contained in software of these scanners, they are not adapted for measuring the human body.

In addition to industrial scanners, this group also includes special scanners for anthropometric measurements. Here, a range of specialized systems, such as Vitus from Human Solutions, can be counted. These are systems dedicated for measuring of the full human body, having a closed chamber and a number of scanners connected into a single system, that is able to make a measurement in time below one second (Jaeschke et al. 2015). These can be used for design of customized devices, however their cost makes the availability questionable, while their bulk size makes them impractical for small spaces.

Examples of scanners belonging to different groups and their use is presented in Fig. 3.7.

Figure 3.7 Anthropometric measurements realized by the scanners of different classes and types: (A) hobby scanner, manual; (B) non-industrial scanner, stationary – AutoMedPrint system; (C) semi-professional manual scanner; (D) industrial scanner, stationary.

The course of optical 3D scanning process can be divided into several distinct stages. They are detailed in the description presented below. Some of the descriptions are present in the previous works by the author, most notably (Konieczny et al. 2012) (in Polish).

1. Measurement field selection and scanner calibration

   Most commercially available optical scanners are versatile devices designed to capture and digitize objects varying widely in size. The selection of the measurement field—the dimensions of the cuboid area encompassed by a single scan—is contingent on the object's size and the required density of measurement points. The quantity of points captured in a single measurement is constant, irrespective of the chosen measurement field; this is determined by the cameras' resolution, resulting in varying point densities across different measurement fields. Modifying the measurement field necessitates adjustments to the scanner's optical system, including the replacement of projector lenses and cameras, as well as altering the cameras' and projector's relative positioning. Often, it is also essential to adjust camera settings, such as shutter speed and aperture.

   Each optical system modification mandates a calibration procedure. For most optical scanners, calibration is a meticulous process that establishes the critical geometric relationships within the projector-camera-object setup, utilizing a calibration device of known dimensions. This process entails conducting a series of measurements under various settings of the calibration device, with the precise calibration protocol depending on the specific scanner model. An example of calibration is presented in Fig. 3.8. In the example, the calibration

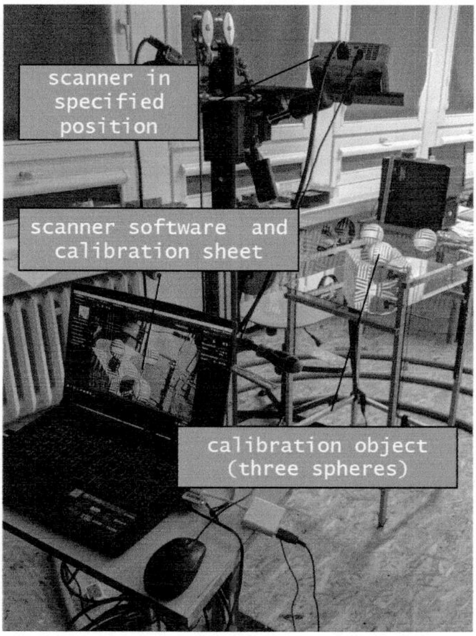

Figure 3.8 Scanner calibration in the AutoMedPrint system.

object (three spheres, connected rigidly) is scanned at a specific scanner position, one of 6 different positions used for limb scanning; the scanned coordinates in each scanner position are used later in the process of scan fitting, to calculate coordinates for translation and rotation.

2. Preparation of the measured object

   For the measurement process employing structured light to yield accurate results, it is often imperative to adequately prepare both the object under measurement and its environment (together constituting the measured scene). Scanners utilizing white light struggle with objects possessing reflective surfaces, including a majority of metal objects and a significant portion of plastic items. Moreover, scanning transparent elements is not feasible. To accurately measure an object, its surface must be properly prepared—typically by applying a matting spray powder to create a temporary anti-reflective coating.

   In professional, industrial-grade scanners, the effective and precise integration of data from individual scans requires the presence of reference points (white and black, self-adhesive markers) within the measured scene. These markers should be affixed either to the object itself or its immediate environment. The quantity of reference points must be sufficient to ensure that each scan shares at least three common points with others, facilitating their alignment and integration. Simultaneously, the number of points should be balanced to avoid overly time-consuming preparation for scanning. Insufficient reference points may necessitate reliance on automatic data matching algorithms or additional measurements to achieve an adequate number of common points for accurate data integration.

   If no reference points are possible to use, either due to measuring conditions or because of technical limitations of the used 3D scanner, there are two possibilities of merging the scanning results:

   a) scanning only in known locations (i.e., predefined, known positions of the scanner in relation to the measured object) and later using geometrical transformations to properly move and orient the scans,
   b) performing scans in a way ensuring redundancy, so that subsequent scans can be joined and oriented using similarity of parts of their surfaces (common points), by automated or manual alignment algorithms.

When positioning the object on the stage for scanning, it is crucial to ensure it remains stationary during the fringe projection. In cases where the object is a human, measures should be taken to limit the mobility of the body part being scanned, balancing the individual's comfort with the need to minimize movements that could compromise scan accuracy. Sometimes it might be necessary to utilize help of another person – e.g., physiotherapist – holding the measured body part in a specific, non-moving position. It might be useful especially when a human subject is unable to keep such a position on their own, due to either a specific condition (for example, cerebral palsy) or general lack of cooperation (mostly due to young age, e.g., children below age of four).

3. Realizing the measurement and initial processing

   After setting up the measured scene and optimizing the lighting conditions (by dimming the room where scanning occurs), a series of measurements can be undertaken. Adjusting the measurement position can be achieved through two approaches: repositioning either the scanner or the measured object. Altering the object's position necessitates modifying the scene layout, particularly when reference point markers are positioned in its vicinity. Consequently, the prevailing strategy is to first adjust the scanner's position—aiming to collect the maximal amount of data from a single object placement, especially for objects of considerable size.

Relocating the scanner is facilitated when it is mounted on a movable stand, which allows for swift and unrestricted movement and rotation to any required angle. If reference points are solely affixed to the object or if it is positioned on a mechanism facilitating movement (such as a turntable), whereby changing its position does not alter the reference points' relative locations, the object can be moved between measurements while keeping the scanner stationary—this is particularly viable if moving the scanner proves challenging. Moving of the scanner can be also realized using a mechanized rig – an example of this approach, as implemented in the AutoMedPrint system (designed by the author and his team), is shown in Chapter 6.

Typically, a single scan encompasses a sequence of light stripes (fringes) and lasts from one to several seconds. During this period, both the object and the scanner must remain immobile to avoid compromising the measurement outcomes. For exceedingly small objects, even slight displacements caused by ground vibrations (for instance, from nearby traffic) could impact the results, necessitating consideration during measurement planning.

The above-mentioned stages are valid for the 3D scanning realized in its most common, step-by-step form – where subsequent scans are specific "events", happening one after another and each of them yields a separate dataset, which can be joined with the previously measured results automatically or manually. In the continuous manual scanning, the process is realized similarly, however the light is projected constantly, without breaks, so that there are no single "scans", but rather a whole process of continuous measurement. Only one, single measurement surface is obtained as a result, which is beneficial. However, this requires not losing contact with the scanned object and keeping its surfaces visible all the time, which might be difficult when measuring a human subject and it requires a certain set of manual skills (mostly, steady hand and concentration) from the scanner operator.

An immediate outcome of the scanning process is a point cloud. Irrespective of the measurement results' intended application, preliminary processing of this cloud is essential. This involves: excising points that represent the object's surroundings, generating a triangular mesh, and establishing the object's base coordinate system. Since scanners do not distinguish between individual objects, the resulting point cloud represents the entire measured scene, including any instruments or other items within the scanning area. To limit extraneous points, employing black-colored instrumentation can be effective (as it will not reflect the scanned white light stripes, thus remaining undetected by the software). The removal of points depicting the surroundings can also be automated by the scanner software, which often includes a feature for setting a "zero plane". Points falling below this plane are automatically eliminated during scanning, a useful feature when scanning objects placed on a flat surface, like a table, to minimize unnecessary points.

3.3.2 Photogrammetry

Photogrammetry is a measurement technique based on processing photographs of selected objects to extract dimensions, shapes and locations out of them. Photogrammetry has been known for very long (almost since the beginning of photography), but officially was defined in 1988, as a discipline gathering information about physical objects and their environment by registering and analyzing images taken without direct contact with the analyzed objects (Linder 2009, Luhmann 2010).

The basic rule of industrial photogrammetry involves registering (detecting) so-called reference points on 3D objects. A series of images taken from different angles, where reference points are visible on multiple images (at least 2), enable to extract the 3D coordinates of these points and as such, create a simplified representation of the whole measured object, without coming into any direct contact with it.

Photogrammetry, considered as a measuring technique, can be basically divided into two main groups, depending on the size of analyzed objects and areas. The aerial photogrammetry is the one where large areas and buildings are inspected, usually by sensors attached to flying vehicles (hence the name), nowadays usually drones (Lee et al. 2019). The close-range photogrammetry is the technique where sensors (cameras) are operated manually, by a human operator, and the objects that are measured are usually mobile – typically these are vehicles, machine parts, daily use objects or living beings. In engineering, especially reverse engineering and quality control, this notion is much more commonly used than the aerial photogrammetry – that is why here we will use the shortened term "photogrammetry" meaning only the close-range variant.

The process of acquiring three-dimensional data in photogrammetry is basically the same, regardless of the size of the objects or areas to be measured, since a series of photographs is always used. The dynamic development of digital photography techniques and digital image transfer and processing tools have contributed to the significant development and spread of close-range photogrammetry, which, using digital cameras, is sometimes called digital close-range photogrammetry. Recent systems, made by the companies such as GOM or AICON, are fully automated and can be operated by non-specialist personnel. Data acquisition and processing can be performed at different locations, at different times and by different people. Digital off-line systems can be regarded as fully accepted 3D measurement tools that are applied in a large variety of industrial application areas (Luhmann 2010).

In practical use with typical process, the object subjected to photogrammetric analysis should be prepared by applying markers and special coded targets, which make it possible to orient individual images in relation to each other. Example of an object subjected to photogrammetric analysis is shown in Fig. 3.9.

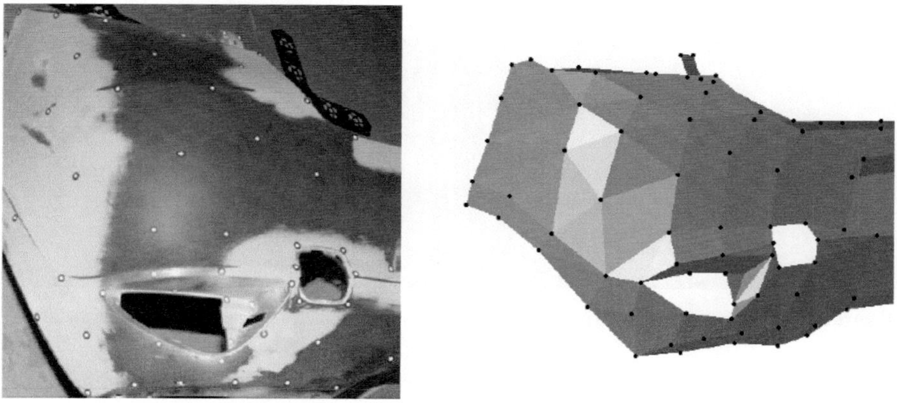

Figure 3.9 Example of measured object in photogrammetry – visible markers and their 3D point representation, along with generated mesh, based on (Konieczny et al. 2012).

In the case of objects to which markers cannot be directly applied (such as human skin), another method is used to reproduce a single point in space. This method consists in pointing out a characteristic point – it can be a corner of the object, an element of the marker or another characteristic element in terms of color, on at least two photographs taken from different angles (minimum 15 degrees). The characteristic point must be contrasting enough for the operator to locate it accurately in the said photos. This method is more time-consuming (the need for the operator to identify the points – there is no automatic determination of points here as in the case of markers). However, it is more universal – the measured object does not need to be specifically prepared for analysis, just a series of photographs must be taken from different angles and then the 3D features might be extracted. There are many semi-automated or automated algorithms available that make this sort of analysis easier (Parras et al. 2018).

Another completely non-contact method of photogrammetric measurements, which, however, requires appropriate lighting conditions (measurement in a darkened room or at night), is the projection from an image projector of contrasting points, the coordinates of which are then reconstructed automatically as in the case of non-coded targets. This method resembles methods based on structured light, but unlike 3D scanning, the projected image does not change over time. It has been developed and tested by the author and his team (Górski et al. 2010a). An example is shown in Fig. 3.10.

Photogrammetry is a well-developed and known measurement technique, frequently used for quality control and reverse engineering. There are a number of books and papers, widely covering the close-range photogrammetry details, including requirements towards sensors, analysis of different measurement and data processing algorithms and detailing instructions on how to proceed using different types of equipment (Luhmann 2010, Maas 2008, Wang et al. 2020). The general consensus is that special types of cameras must be used to obtain accurate and precise results of measurement, for industrial type of photogrammetry.

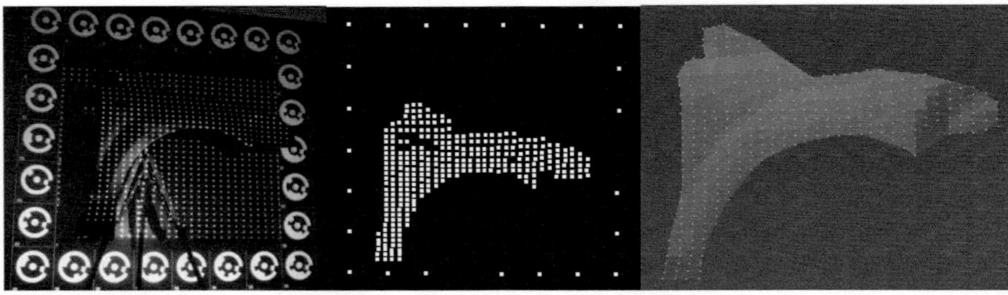

Figure 3.10 Photogrammetry with projected markers – object geometry re-creation (Górski et al. 2010a).

However, in the medical domain, for measurement of human body shapes, the industrial level of accuracy is not required. As such, it is a viable option to use typical digital cameras, available in medium or high-class smartphones, to digitize a shape of human limb or other body part, to create a dataset for design of customized devices.

The application of photogrammetry in measuring human limbs is, of course, not without its challenges. One of the primary concerns is the requirement for a controlled environment to minimize measurement errors caused by lighting variations and background complexity, which might be hard to achieve in non-laboratory conditions (such as the patient's home). Additionally, the accuracy of photogrammetric analysis can be affected by the subject's movement, necessitating stillness during the image capture process – what is also hard to achieve, especially for pediatric patients below a certain age. The texture and color of the limb's surface can also influence the precision of 3D models, with reflective or uniform surfaces posing particular difficulties.

As opposed to 3D scanning, photogrammetry offers greater flexibility and cost-effectiveness, with the ability to utilize existing camera equipment and software. It is particularly advantageous for remote projects where portability and ease of use are critical. Nonetheless, photogrammetry generally requires more post-processing and expertise to achieve comparable levels of accuracy to 3D scanning. There are numerous software solutions and studies enabling this approach to body measurement (Barreto et al. 2022) and it has become a trend that may help in remote design for patients that are unable to visit 3D scanning laboratories due to logistical reasons, as such effectively replacing 3D scanning (Stark et al. 2022). This has not yet happened, due to several reasons – measurement repeatability being one of them and lack of efficient, open source, freely available software the other.

3.3.3 Laser scanning

Laser scanning utilizes laser technology to accurately capture the spatial dimensions of a measured object. It is a technique frequently employed for anthropometric measurements. In practice, laser scanners project a beam of light onto an object, capturing the nuances of its form through the light reflected back to the scanner's sensor. The measurement of this distance is determined by the Time-Of-Flight (TOF) principle – the time taken for the light to return (Horaud et al. 2016). Sensors working with use of this technology are also known as Light Detection and Ranging (LiDAR) (Di Stefano et al. 2021). Alternatively, some systems operate on the triangulation principle, where the laser casts a dot or line onto a surface, and the spatial geometry of the laser in conjunction with the viewing angle from a separate sensor calculates each spatial point's exact location (Schlarp et al. 2019).

The advancement of Time of Flight (TOF) technology has given rise to its refined and enhanced variant, with use of multiple sensors (Bandutunga 2020). The multi sensor approach diverges from the typical TOF by deploying laser light from a generator to multiple sensors, rather than a single sensor as found in the original TOF setup.

Scanners incorporating TOF approach are especially beneficial in measuring smaller dimensions, such as mounting components, limbs, or the intricate, irregular shapes of aircraft parts, with this technology predominantly featuring in handheld scanners. This laser scanning process has become instrumental in fields such as reverse engineering, design, and medicine, distinguished by its high accuracy (Sofronov et al. 2019).

The scanning process using laser technology entails several key steps, beginning with the calibration of the laser scanner to ensure measurement accuracy. As the laser sweeps across the subject's body, it forms a detailed point cloud that delineates the body in three dimensions. This data is then processed through specialized software to construct a three-dimensional model, which, in the context of scanning of the human body, is used for comprehensive anthropometric analysis (Sacco et al. 2023).

A variety of laser scanners are utilized, as listed below:

1. *Handheld laser scanners* – portable devices that can be moved around the subject for quick and flexible scanning, an example being scanners produced by EinScan company, used in many studies by the author with the team.
2. *Stationary laser scanners* – fixed devices where the subject must be positioned within the scanner's field of view, based on either TOF or triangulation technology. They often work as rotary sensors, capturing image in 360 degrees around them.
3. *Full body scanners* – systems designed to capture the entire body in one scan, often used in the fashion and health industries, usually quite expensive, often based on triangulation principle.

A direct result of 3D scanning is a detailed point cloud, often containing color data. An example of a measurement setup – focusing on a technical object – and its raw result is presented in Fig. 3.11.

The advantages of laser scanning are evident in its high accuracy, capable of capturing the finest of details, and its rapidity, which allows for the scanning of an entire body or scene in minutes. Furthermore, as a non-contact method, it is universally applicable across subjects without the need for physical contact. Handheld devices enable to realize a continuous process, being a reasonable alternative to optical 3D scanning of patients that have trouble keeping their bodies still – continuous measurement allows obtaining meaningful data in times as short as 10 or 30 seconds,

Figure 3.11 Laser scanning, (A) measurement setup; (B) measurement result – point cloud with RGB color data.

of the author's experiences – the problem is that they need to be manually stitched together after the scanning is concluded.

There are also disadvantages to consider of laser scanning. First and foremost, laser scanning's sensitivity to movement can compromise accuracy during a scan. In the case of handheld scanners, they require a steady hand and appropriate movements of an operator, which poses a challenge – measurement results and success are related to manual skills of an operator. The technology also comes with a significant cost for high precision equipment and necessitates safety precautions to protect the eyes from laser exposure. There is a potential problem with epileptic patients (or other personnel), as laser light can alternate in quick succession, potentially triggering seizures. This problem also pertains to optical scanning, but to a lesser degree than the laser systems.

When compared with structured light scanning, laser scanning stands out in its precision for capturing even tiny details, although structured light scanners can often – considering measurement of a whole object – operate at a quicker pace due to their ability to project patterns over large areas simultaneously. Structured light systems generally provide a more user-friendly setup and lack the specific safety concerns associated with lasers, as they do not employ coherent light.

Laser scanning serves as an invaluable tool in various domains requiring accurate anthropometric data, being potentially able to replace classical anthropometry (Kuehnapfel et al. 2016). The data gained from these scanners can be used similarly as data from photogrammetry or optical scanning – from the design of custom-fit garments to ergonomic product development and healthcare solutions. The choice between laser scanning and other techniques like structured light scanning ultimately hinges on balancing considerations of cost, efficiency, ease of use, and safety.

3.4 Data processing: formats, methods, and software

3.4.1 Data formats and their conversion

The medical imaging techniques and associate measuring techniques used for digitizing human body shapes utilize various data formats to store and manage the images they produce. Each imaging technique often has preferred formats based on the nature of the data and the requirements of the imaging process.

The imaging formats can be divided into two basic categories:

1. Flat (two-dimensional) data, coming directly from typical medical imaging techniques, such as X-Ray.
2. Volumetric (three-dimensional) data, resulting in a processing of flat data or gathered directly through means of volumetric measurement, such as 3D scanning.

Additionally, both categories can be supplemented with metadata files, containing extra information, not directly containing geometrical coordinates – these are patient and case data, colors, contrast data, etc.

In terms of the first category – flat data – the most important data format is DICOM. DICOM (Digital Imaging and Communications in Medicine) is the most widely used and standardized format in medical imaging (Hosseini and Dixon 2016). It is not just a file format but a comprehensive set of rules for image data acquisition, storage, retrieval, and transfer. DICOM is used across various imaging modalities such as MRI, CT, X-rays, Ultrasound, and PET. It includes patient information alongside the image, which aids in efficient management and archival. It is crucial in radiology, cardiology imaging, radiological imaging, and in hospitals for storing and exchanging medical images.

DICOM is the standard developed by American College of Radiology (ACR) and National Electrical Manufacturers Association (NEMA) (NEMA 2024) for the communication and management of medical imaging information and related data. The DICOM Standard is designed to ensure interoperability in medical imaging. It does this by specifying network communication protocols, command syntax and semantics, media storage services, file formats, and medical directory structures for devices conforming to the standard. Additionally, it mandates the provision of certain implementation information. However, DICOM does not dictate the specific implementation details of its features, the comprehensive set of features expected in a system integrating DICOM-compliant devices, or the procedures for testing and validating conformance to the standard. This distinction between what is and is not specified underlines DICOM's role as a framework for interoperability rather than a comprehensive blueprint for device functionality or system integration (NEMA 2024).

DICOM files contain metadata that includes patient information (name, ID, birthdate), image acquisition details (date, time, device), and technical data (dimensions, modality, image format). This metadata ensures the effective organization and retrieval of medical images. DICOM images, often in 2D slices, can be converted into 3D models using specialized software. This process involves stacking and aligning 2D slices to create a volumetric representation. It is essential for diagnostics, and used for that purpose in most cases (surgery planning or decision-making about treatment (Aiello et al. 2021)), but also more and more frequently used as a basis for design of customized prosthetic devices or mid-surgery supplies (Weitz et al. 2011, Li et al. 2023).

Structured format and widespread adoption of DICOM make it a cornerstone in medical imaging, allowing for efficient and secure handling of medical imaging data across different platforms and institutions.

Apart from DICOM, there are also a number of other standard medical image file formats, partially compatible or similar to it, as presented in available literature (Larubina and Morino 2014, Sriramakrishnan et al. 2019). Among the most important ones, the following should be distinguished:

- *NIfTI (Neuroimaging Informatics Technology Initiative)* – a format specifically designed for neuroimaging, NIfTI is commonly used in MRI and fMRI (Functional MRI) data. It supports 3D and 4D (dynamic 3D) imaging and includes spatial and temporal information, which is crucial for brain imaging studies (Horien et al. 2021).
- *Analyze format* – this older format was widely used in MRI and PET imaging prior to the popularization of DICOM. It separates the image data into two files – an image file (.img) and a header file (.hdr). However, it lacks some of the metadata and standardization features of newer formats like DICOM (Sriramakrishnan et al. 2019).
- *Minc (Medical Image NetCDF)* – this format was developed by the Montreal Neurological Institute and is used primarily for MRI and PET images. It is based on the NetCDF (Network Common Data Form) and supports a wide range of metadata, making it versatile for research purposes (Larubina and Morino 2014).
- *TIFF (Tagged Image File Format) and JPEG (Joint Photographic Experts Group)* – these are widely known, general image formats, sometimes used for less complex medical images, such as dermatological photographs or scanned histology slides. They are not typically used for high-dimensional data like CT or MRI scans, however DICOM 2D slices can be, technically speaking, converted into TIFF and vice versa, using specialized software tools (Kim and Ko 2017).

Converting between these formats can be necessary for compatibility with different software tools or for specific analysis requirements. However, this process can be complex and challenging. One major issue is the potential loss of important metadata during conversion, especially when moving from a more comprehensive format like DICOM to a simpler one like JPEG. Tools like DICOM to NIfTI converters exist (Li et al. 2016), but they need to be used carefully to ensure that critical spatial and temporal information is retained. Another challenge is maintaining image quality and resolution, as some formats may compress the image data, leading to a loss of detail that can be crucial for accurate diagnosis or research. Additionally, the complexity of converting multidimensional data (like 4D fMRI scans) can lead to errors if not handled properly. It is therefore essential to use reliable software and verify the integrity of the converted data, especially in a clinical context where accuracy is paramount. As such, for reverse engineering and design of customized medical devices, as the focus of this book is concerned, the DICOM format will be assumed as standard and recommended for all uses.

In the typical reverse engineering scenarios (volumetric measurement, e.g., 3D scanning), different data formats are employed. Here, similar as in the case of medical imaging, some standardization (rather informal) is also present. The largest difference lies between the formats containing just the point cloud data (typically associated with laser scanning and photogrammetry) and the ones containing a triangle mesh, based upon the point data (used in all techniques, but basically in optical 3D scanning). The most popular format for the triangle mesh is Standard Triangulation Language (STL), formerly known as Stereolithography (Leirmo et al. 2020), as it originates from the format used to store data for the first additive manufacturing technology of the same name.

The STL format represents the surface geometry of a 3D object as a series of triangular facets, effectively approximating the shape of the object. Each facet is described by the normal and vertices (the points of the triangle). STL files can be in ASCII or binary formats, with the binary form being more compact and common due to its smaller file size, while ASCII can be easily edited and automatically created using text processing algorithms. However, STL has limitations, as it does not support color, texture, or other attributes beyond basic geometry. This simplicity can be an advantage for 3D printing applications where only the shape of the object is required, but it is less suitable for applications needing detailed color or texture information. Despite its limitations, STL remains popular due to its wide compatibility with various software and hardware used in 3D printing and modeling. Figure 3.12 presents structure and overlook of a typical STL file.

Apart from STL, other formats representing similar data are also used. The Polygon File Format (PLY) is used in both 3D optical scanning and photogrammetry and it can store both 3D geometry and color information, making it suitable for detailed models. Apart from that OBJ (Object File) is also a popular format. It is a versatile format used in various scanning techniques, OBJ supports complex geometries and, as opposed to minimalistic STL format, it includes texture maps, which is crucial for detailed and realistic models (OBJ files often have accompanying material – MTL –

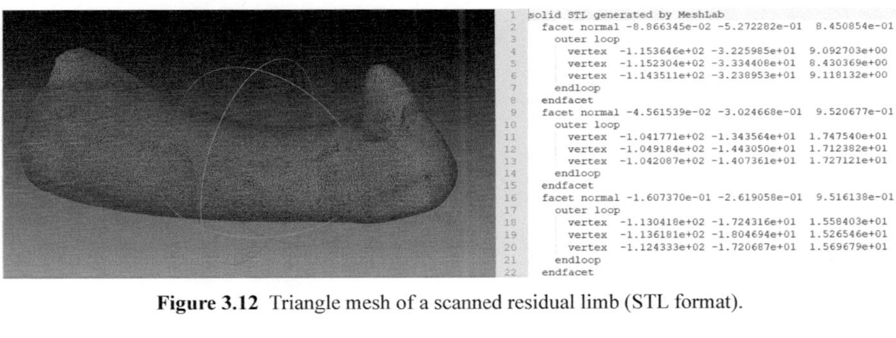

Figure 3.12 Triangle mesh of a scanned residual limb (STL format).

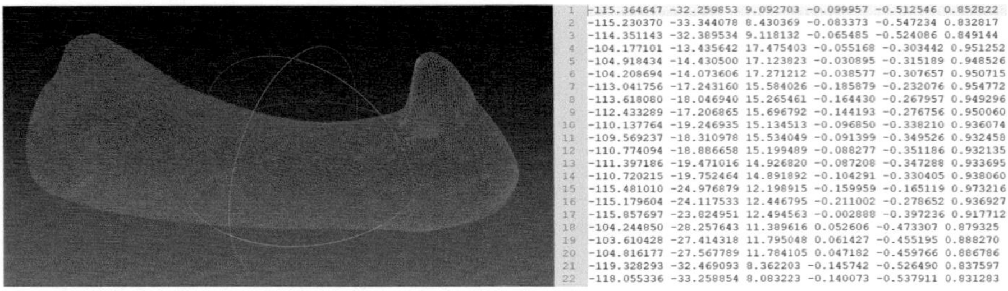

Figure 3.13 Point cloud representation of scanned residual limb (XYZ format).

files). OBJ files also allow to store multiple, yet separate geometries in a single file, which is also not possible for the STL.

The point cloud data formats, such as PTS, PCG, XYZ are similar to each other. These formats are the first output of 3D scanning processes, representing raw 3D data as a collection of points in space (Fig. 3.13). As such, these files usually contain point coordinates (one point per line), sometimes complemented with color data. They are used in laser scanning and photogrammetry.

Converting between these formats can be complex due to differences in the type of data they store. For instance, converting from a point cloud format to STL involves creating a surface from scattered points, which can lead to loss of detail or inaccuracies, and is not trivial when the point cloud contains unordered points. Another challenge is maintaining the integrity of color and texture information when moving between formats like PLY and OBJ and STL. Software compatibility and processing power are also crucial factors, as some formats require more computing resources to handle, especially when dealing with large or highly detailed models

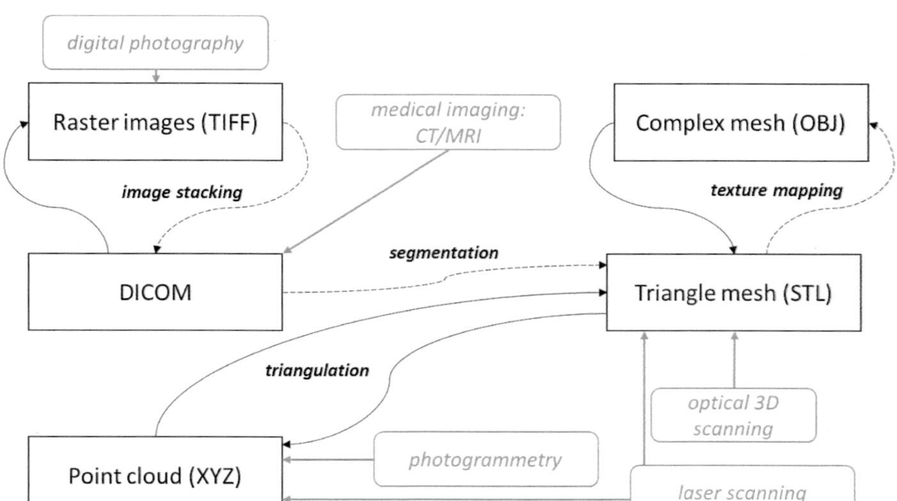

Figure 3.14 The most important data formats generated in anatomical measurement and their potential conversion processes.

– this was a big problem a decade or so prior, as with limited RAM of graphics cards and GPU capabilities, files larger than a certain size (e.g., 1 GB) could not be processed at all. Today this limitation is mostly overcome.

The most important data formats presented in this chapter, as well as their mutual conversion processes and associated measurement techniques, are presented in a schematic in Fig. 3.14.

3.4.2 DICOM segmentation methods

In this chapter, techniques of segmenting DICOM data to extract 3D meshes, crucial for applications in medical imaging and diagnosis, will be presented. Building upon basic understanding of DICOM's most important features, including its metadata richness and the challenges of converting 2D slices into 3D representations, various segmentation methods will be explored. These include thresholding, region-based, and model-based approaches, each offering unique advantages for different types of medical data, ultimately leading to the generation of precise 3D meshes. This process is pivotal for creating patient-specific models applied in modern medicine.

The strategies and algorithms presented in this chapter were first described as a part of the thesis realized by Eng. Katarzyna Turostowska (Turostowska 2016). It was made in cooperation, under instructions and supervision of the author with his team. The chapter contains their new version, updated with the author's insights after a certain number of clinical cases realized together with the team.

The process of working with medical imaging data in order to obtain a volumetric model (usually STL mesh) begins with the acquisition of DICOM files by performing an actual measurement. These files are then imported into a software program, capable of reading the DICOM format and working with it to export the 3D data as a mesh. In the program, checking for completeness of data and the presence of any artifacts is realized (automatically and/or manually, by the engineer). If the data is complete and artifact-free, the input data analysis phase is commenced, which, depending on the type of data, may require assistance of a radiology specialist, a surgeon or other medical specialist. As a result, a segmentation method is selected, and segmentation is performed. If the segmentation is successful, parameters for generating a voxel model are introduced. The voxel model is then generated. Upon successful model generation, the model is converted into an STL file format (triangle mesh). The process ends once the STL file is obtained.

A properly imported dataset should appear in the main window of a DICOM-processing software in three views, corresponding to the coronal, sagittal, and transverse planes (an example is shown in Fig. 3.15). With basic information about the area under examination, it is advisable to verify the correctness of the import by checking the completeness of the data, assessing the quality of the tomogram, and ruling out the presence of artifacts that could impede further analysis and data processing. This verification should include comparing the study area with reference data, recognizing the position and shape of anatomical structures using medical literature (anatomical atlas), and ensuring that the cross-sections fully and clearly display the objects of study for the user.

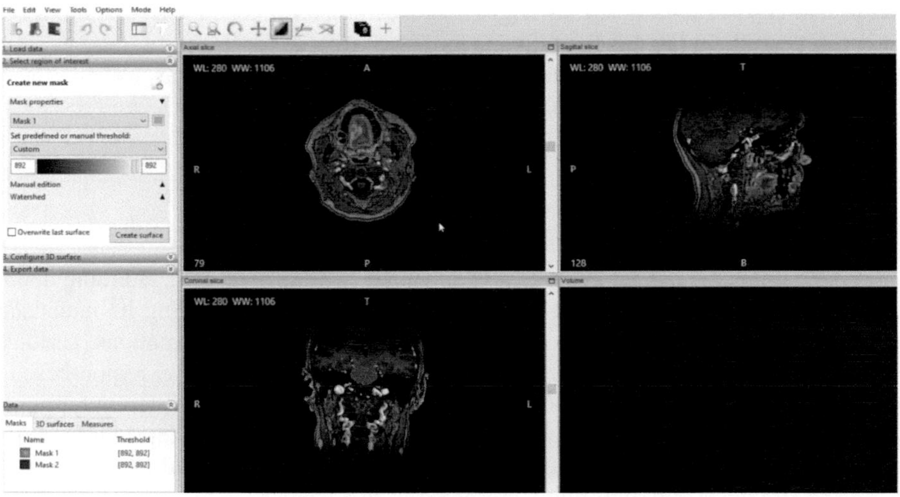

Figure 3.15 A dataset of anatomical measurements (MRI) imported into segmentation-capable software.

The basic algorithms for segmentation assume partially manual or automated selection of a Region Of Interest (ROI), usually representing a given type of tissue (e.g., bone or muscle) or organ, on multiple sections of a given set of data in DICOM format (which are just sets of flat images, taken by making a series of images in different locations, as described in one of previous subchapters). The two basic methods of segmentation can be considered:

- *Thresholding* – this function uses the fact that, depending on the source of DICOM data (CT, MRI), it is possible to assign a specific tissue to one of the shades of gray. In most software applications (3D Slicer, InVesalius), the thresholding tool works automatically (either based on entered value or controlled by a slider), and a one-time operation results in marking the area on all sections. The method is more effective the greater the contrast between the given tissues; it is mainly used when working with data from a CT scan, for hard tissues (bones).
- *Growth Algorithm* – this involves defining two input areas – the foreground (the tissue intended for extraction) and the background, using samples of any shape and color. If *n* elements of a given tissue occur within a section, similarly *n* areas must be defined. Based on the markings made, the growth algorithm automatically calculates the size and shape of the area of interest and fills it with the foreground color, while the rest of the pixels are filled with the background color. To extract the tissue of interest, the layer editing function can be used, the background can be defined as transparent. The profitability of using the growth algorithm increases with the decrease in the number of defined foreground elements and reaches maximum efficiency in the case of segmenting structures with a compact structure, such as tumors or organs.

The proper selection of segmentation methods is a key factor in the efficiency of DICOM data processing. The decision should be preceded by a process considering random conditions dependent on input data. To build a decision-making scheme, the main goal must be established: this would be, in most cases, selecting the optimal segmentation method (process, consisting of specific actions) for a given anatomical structure. This includes methods such as thresholding and growth algorithm. These solutions are the most universal – the complexity and duration of the process are not dependent on the number of sections and the type of input data (registered object).

Anatomical structures, due to their specificity, do not undergo any universal classification, based on which it would be possible to organize them in terms of size, shape, or location. Therefore, for the purpose of choosing a segmentation method, the input data can be classified based on the following criteria:

1. The degree of complexity of the object's geometric structure.
   a. An object with a compact structure: concerns structures with a simple geometry (e.g., long bones, soft tissue organs and body parts such as the liver, kidney, tongue),
   b. An object with a dispersed structure: concerns structures with varied geometry, having numerous branches, pores, constrictions (e.g., most of the skull bones, blood vessels, bile ducts, etc.).
2. The form of spatial organization of the object.
   a. An object with a continuous structure: structures that can be determined using a single segment on each section (mainly refers to clear, not diffuse tumors, individual bones, and organs with a compact structure, differing from the neighboring areas),
   b. An object with a discontinuous structure: structures that, in a significant number of sections, are impossible to determine using a single segment (e.g., complexes of several bones, muscle tissue organs such as the tongue, diffuse tumors affecting several areas, etc.). There are discontinuous objects that are uniform – possible to determine by choosing one additional segment – and multiple – possible to determine only by choosing two or more additional segments.

For each case, risk factors should also be considered, which may affect the final form of the segmented element. These are:

- *Choosing a non-selective method*: mainly occurs in the thresholding process. It concerns the case where, during segmentation, both the object constituting the study area and other elements corresponding to its physical characteristics are extracted (e.g., extraction of all skeletal elements during the segmentation of a single bone)
- *Excess or deficiency error*: occurs when using the growth algorithm. It involves assigning elements of the section that are not part of the object constituting the study area for extraction or omitting elements that should be segmented.

Performing segmentation results in the creation of a set of flat structures corresponding to the extracted area. These form the basis for building a voxel model, in which pixels in individual segments will constitute the dimensions of the voxels in the X and Y directions, and the pitch of the data set (resulting from its resolution) will determine the dimension in the Z direction.

Within the scope of generating a voxel model, the following activities can be distinguished:

- defining input and output data,
- introducing a smoothing filter (optionally),
- generating the model,
- exporting to a file representing a mesh of triangles (STL format).

A simple example illustrating how the segmentation result should look like before exporting to STL is shown in Fig. 3.16. More details on the course of the performed segmentation are presented in further chapters (case studies) of the book.

To realize the DICOM segmentation, proper software is necessary. The main tasks realized in such programs are the following:

1. Importing DICOM data from various imaging techniques.
2. Visualizing data coming from different series in uploaded DICOM imaging.
3. Applying diverse segmentation methods to isolate given tissues in selected images at once.
4. Generating a 3D shape in form of voxels and conversion into a triangular mesh.
5. Exporting data into STL, OBJ or another known format of storing 3D triangle mesh data.

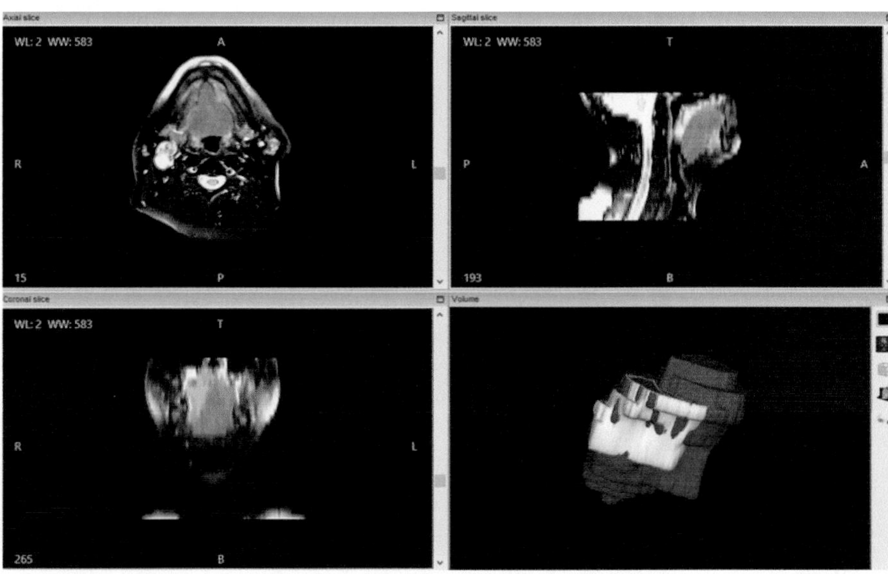

Figure 3.16 Result of segmentation on example of tongue with cancerous tumor, segmented of MRI imaging.

Some examples of the software are described below. There are free software packages, such as 3D Slicer and InVesalius (Matsiushevich et al. 2019), as well as more advanced, commercial ones, such as Materialise MIMICS.

3D Slicer is a very popular free software designed to work with DICOM data, mainly dedicated to educational and research purposes. However, the suite of features it offers allows for a complete transformation from DICOM to STL, thus enabling design activities. The evolution of 3D Slicer is an ongoing endeavor, with its expansive development leading to a rich array of supplemental applications. Segmentation tools within 3D Slicer are quite versatile, enabling realization of automated and manual segmentation processes. The level tracing effect, for instance, assists in the task of manually delineating structures, leveraging grayscale shades to detect and extract target areas. This feature is particularly useful for low-resolution datasets or when isolating coherent structures. Another useful feature is the universal threshold effect, previously mentioned technique that proves invaluable when working with CT data. Similarly, the grow cut effect is a growth algorithm that offers a dynamic approach to segmentation. By defining distinct foreground and background regions and applying intuitive color-coding, the software autonomously determines the contours and fills of the designated area.

3D Slicer's open-source ethos is focused on only on its cost-free accessibility but also on creating a collaborative environment, where users can contribute custom applications and algorithms, further enhancing its functionality, e.g., by adding AI algorithms (Zhang et al. 2024) – one of examples is the MONAI toolkit. Compatibility across major operating systems like Windows, macOS X, and Linux ensures broad accessibility. The current version is already a fifth generation of the software. Examples of 3D Slicer images are shown in Fig. 3.2A, as well as in Chapter 7 (case study #2).

Another, very popular solution in the field of free software for DICOM processing is InVesalius. InVesalius is an entirely free software tool developed for processing DICOM data, fostered by the Research and Development Center of the Brazilian Ministry of Science and Technology (MCT) (Lopes da Silva 2013). The open-source nature of InVesalius allows users to modify the code freely. Despite this, medical entities worldwide effectively utilize the software in its original form for a variety of medical applications.

The array of tools within the InVesalius environment, while not extensive, is quite comprehensive, ensuring ease of use and effectiveness. The main window features a dropdown menu with tabs on the left, and the ribbon at the top houses buttons for essential functions such as zooming and rotating the view. The main, simple yet efficient, segmentation tool of the software is based on thresholding. Users can utilize predefined options for the fully automatic extraction of different types of tissues, including bone, fat, and muscle. If necessary, it is possible to define a custom threshold range or modify an existing one. Extracting data from a single cross-section automatically applies the segmentation to the entire data set, significantly accelerating the segmentation process.

InVesalius is a versatile and cost-free program, offering flexibility in source code modification. The developers have ensured compatibility across various operating systems, including Windows, Linux, and selected versions of MacOS X. The simplicity of InVesalius's interface means that operational efficiency is not contingent on the user's level of expertise. However, by the author's experience it must be stated that the effectiveness of the tasks performed with InVesalius is inherently dependent on the quality of the input data; the tools available may render the segmentation process less effective or even impractical in certain cases. Despite this, the user-friendly design and accessibility of InVesalius make it an invaluable resource in the medical community, particularly for those seeking a cost-effective solution for DICOM data manipulation.

In contrast to the 3D Slicer and InVesalius, Materialise MIMICS is a powerful commercial software with various, versatile applications in the realm of medical engineering. Materialise Mimics is an advanced software suite crafted for medical image processing, incorporating capabilities for computer-aided design (CAD) and Finite Element Analysis (FEA). This multifaceted tool is composed of multiple modules, which, when used collectively, enable the creation of comprehensive 3D models of human organs, including implants (Comaneanu et al. 2012).

Mimics facilitates the import of medical images via intuitive wizards, supporting a wide array of formats such as DICOM, TIFF, BMP, and RAW. The most common practice involves importing DICOM images, a format integral not only to Mimics but also to other software. DICOM files encompass image data for each recorded examination, such as CT slices, and can also contain additional patient information.

After the import, Mimics offers a volumetric display of the scanned data, with body positioning options in multiple orientations. The software visualizes the patient's scan data through three projections: anterior, right, and top, with an additional capability to generate a 3D model from the volumetric data, viewable in a dedicated window. Image reorganization and manipulation tools are accessible if necessary, with a range of editing features.

The grayscale intensities in the projections are delineated by a histogram, setting the Hounsfield units range, critical for defining contrast. The histogram allows customization below the minimum as black and above the maximum as white. Users can adjust the Hounsfield range through the histogram, a specific projection window, or by entering data directly. Segmentation in Mimics is accomplished using segmentation masks, defined as pixel sets bounded by threshold values. The software facilitates the creation of multiple masks, which can be tuned for different tissue types. Lower threshold values are geared towards soft tissue selection, while higher values cater to denser bodily structures like bones. A comprehensive model may call for a lower threshold for pixel color intensity to capture the entire body's representation (Dotremont 2023).

Thresholds in Mimics are managed via the Thresholding command within the Segmentation toolbar. The process defines the corresponding masks across model projections, based on minimum and maximum values chosen either from a dropdown list of predefined values or manual input. A confirmation of the defined value range results in the creation of a green mask encompassing the selected range. Another segmentation technique involves the Profile command, which permits the definition of a value set along a profile line drawn within a projection. After drawing the profile line, adjusting a horizontal line within a diagram sets the range values. The mask generated encapsulates the defined range, ready for further manipulation.

Mimics, just like the other software packages, is able to transform segmented images into polygonal models suitable for subsequent CAD processing or further refinement within the Mimics environment itself. By employing the Calculate 3D command, Mimics renders a 3D representation of the segmented mask. The newly created object's quality is adjustable to meet varying needs and allows for transformations, volume shading, and more.

Exporting a volumetric model to an STL file suited for CAD software is quite simple and straightforward. Selecting a mask and adding it forms an STL file model, stored in a pre-defined folder. For higher quality demands, the software can be configured to generate high-resolution 3D objects during creation.

Mimics software, therefore, stands as a robust and dynamic software platform, empowering medical professionals, and researchers with tools for precise image processing and model creation, fostering the bridge between medical imaging and practical application in patient care and device fabrication. However, its costs of licensing are remarkably high and not all applications necessitate them, especially for smaller companies or healthcare facilities. That is why although it is superior to the freely available software packages in almost every aspect, for economical and practical reasons, the formerly described software is often more frequently used, especially for research, education, and medical training.

3.4.3 Triangle mesh processing methods

The point clouds and triangle meshes can be obtained through different ways of conduct, using various measurement techniques, as described in previous chapters. The triangle mesh (usually STL file) is almost always a basic, universal target format, used to unify measurement results and 3D images of various human body parts coming from different setups. If the point cloud is a basic format (as in laser scanning and a number of other methods, such as photogrammetry), a triangle mesh must be first created upon it, using the triangulation process (Bern and Eppstein 1995), with raw point clouds rarely being used as a basis for design of anatomical shapes.

The main requirements towards the triangle mesh, for it to be usable in further stages of design of anatomically customized devices, are as following:

1. *Completeness* – the measured body part has to be structurally complete, containing all the surfaces needed for design of a specific medical device. Empty (not measured) areas are acceptable, as long as they can be automatically or semi-automatically filled without significant changes in shape of the limb, that may lead to design inaccuracies.
2. *Unambiguity* – the mesh resulting of measurement has to represent a single state of the measured limb, that is supposed to be static (not moving) for the period of the whole measuring operation. A common error here is obtaining a mesh with double features, e.g., more than five fingers of a hand, or double surfaces representing a limb that moved.

3. *Clarity* – all the measured surfaces should be clearly visible, without any large artifacts or objects covering the body features. All the environment elements (such as parts of the scanning rig) and unnecessary areas (e.g., fingertips in the wrist hand orthosis) should be removed and a single solid representing the region of interest should be of main focus.
4. *Accuracy* – the measurement should be as accurate as necessary, faithfully representing the measured body part, with maximum possible resolution. To achieve that, sometimes sensor fusion is required (connecting data from two 3D scanners, or combining images from MRI and CT, etc.). Any smoothing or triangle reduction operations cannot affect the measurement accuracy.
5. *Smoothness* – the data should be as smooth as possible without losing important features, so that it is easy to get proper measures needed for the design operations. All the surface artifacts should be removed or repaired.
6. *Waterproofness* – the obtained mesh should be "watertight", so its inside parts are not visible from the outside. In other words, it should be possible to easily convert it to a solid model, without any ambiguities whether which part is "internal" and which is "external". Waterproofness is usually obtained at the end of mesh processing.
7. *Lightweightness* – the final obtained mesh should be as lightweight as possible in terms of file size – compressive formats should be used (binary, not ASCII STL format), but also triangle mesh should be properly decimated, re-organized and optimized in size and proportion of triangles, so that it is easy to open, process and re-use in any software or exchange between process participants (e.g., sent by e-mail). As such, file sizes of meshes representing body parts should be rather counted in megabytes, not gigabytes.

To fulfill all the requirements, certain operations must be undertaken, in order to process the raw dataset. The detailed activities and operations realized with the dataset will be, however, dependent on the source, i.e., measuring technique. In general, the following scenarios belong to the most frequent starting points:

a) CT or MRI scan was subjected to segmentation and STL file was exported, containing volumetric representation of segmented tissues;
b) multi-positional 3D optical scanning was realized without using markers – raw triangle meshes of single scans are available,
c) manual (continuous), marker-based or fully stationary 3D scanning was realized – a single, combined mesh of an object is available,
d) laser scanning was commenced and a colorful point cloud is available, to be converted to a triangle mesh (automatically, by triangulation, a process not covered in this book),
e) photogrammetric analysis was realized and a set of points is available, again possible to be converted to a triangle mesh as in laser scanning.

Using the most popular mesh editing software, we can organize the tools available for mesh processing in the following main groups:

1. *Scan fitting and orienting* – a set of tools that help properly orient the scans with each other, join them together and set them in a common, properly oriented coordinate system.
2. *Mesh cleaning and adjusting* – tools and algorithms (partially or completely automated) used to clean mesh of artifacts, cut it to the desired area, fill missing spaces, as well as repair and smoothen the surfaces.
3. *Data extraction* – algorithms used to extract data necessary for further design operations – measurements, sections, point coordinates, curves, selected mesh regions, etc.

The main idea is to go from the set of raw scans to a cleaned, properly oriented mesh that can be used as a basis for design of anatomically adjusted devices, such as prostheses or orthoses. The idea is presented – as an example of the upper limb – in Fig. 3.17.

In the first stage, depending on the input data, the scans have to be joined together and oriented properly in the coordinate system, setting the base of it. When using low-cost marker less systems for optical scanning (e.g., David SLS-3), each scanning position is a separate mesh, located in a different coordinate system than the other meshes. Scanner locations may be predefined – then, based on known scanner positions and rotations, proper transformations can be calculated and applied, manually or automatically. If the scanner positions are undetermined, the so-called "best fit" algorithm may be applied, with the software trying to find out proper alignments of the scans automatically, by geometrical similarity between their surfaces. The algorithm might be manually controlled, e.g., by manually indicating three identical characteristic points on each of two joined scans.

In the case of professional systems, where markers (targets) are usually applied (e.g., GOM Compact Scan 5M), alignment of subsequent scans is usually done automatically and the exact position and rotation of scanner does not need to be known, provided that part of the markers (minimum of 3) are visible in two subsequent scans. A result is a single mesh that is already joined of subsequent scans

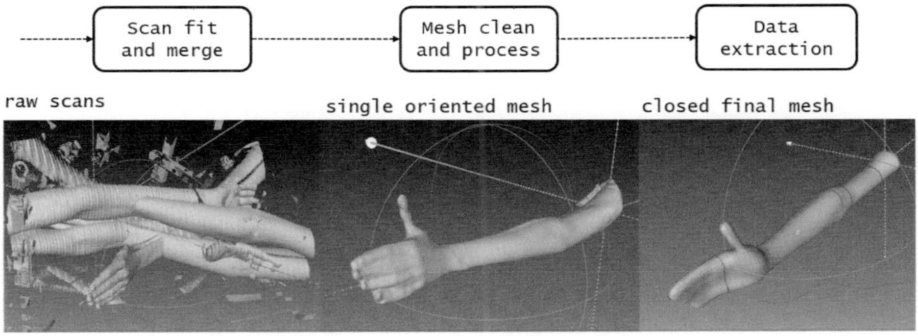

Figure 3.17 Triangle mesh processing stages.

based on measured marker positions. In the case of continuous scanning (e.g., using laser-based devices, such as EinScan Pro), a single mesh is also obtained, with no need of joining the scans.

However, regardless of the method, there are situations when both marker-based and mark less solutions require manual alignment and joining of different scans. These situations frequently involve scenarios when the measured object was moved, such as a patient involuntarily changing limb or body position during scanning. Then usually the automated algorithms will not work and two or more partial scans must be combined to achieve a single image of a scanned object. By the author's experience, this happens frequently when measuring children aged 6 or below, as they have problems with keeping their bodies still for longer periods of time, but also in a number of musculoskeletal or neurological conditions. There can also be situations, where the scanned object is moved on purpose, e.g., to scan the other side of it. In such cases, also scanning of mirror image could be used, although that is possible only in the optical systems (not laser-based), where fringes are projected on the actual limb, while camera shots are taken of its mirror image, generating a mirrored mesh in the process. This mesh must be then (manually or automatically) flipped and oriented properly towards other scans. This approach is utilized in the AutoMedPrint system, as shown in Chapter 6.

After this stage is finished, the stage of cleaning and preparation begins. The following operations may or may not be realized, usually in no particular order:

1. Removing smaller, separated chunks of the mesh (artifacts or elements of environment – not the main object).
2. Cut-off by a plane or cuboid, limiting the measured area only to the one that is interesting for further use.
3. Removing bad data (double vertices and triangles, triangles with inverted normal vectors, etc.).
4. Closing holes resulting from lack of measurement in difficult to reach places.
5. Bridging larger gaps in the case of huge openings to create the possibility of further hole closing.
6. Repairing measured surface in the case of unevenness, artifacts, badly looking areas, etc.
7. Smoothing out surfaces, thinning or reducing mesh, optimizing the triangle size to be more evenly distributed.

Finally, a single, enclosed, smooth mesh representing a given object (a body part, in the context of this book) should be achieved and exported for further use. One final operation that might be performed upon it is reconstruction – an automated generation of a new, optimized mesh, based on provided point or triangle data. There are many algorithms for automated reconstruction, one of the most useful – in the author's opinion – for work with anatomical meshes is Screened Poisson Surface Reconstruction (Kazhdan and Hoppe 2013), available in MeshLab software. The reconstruction can bring remarkable results, obtaining meshes that are very lightweight, optimized yet still accurate enough to generate proper results in terms

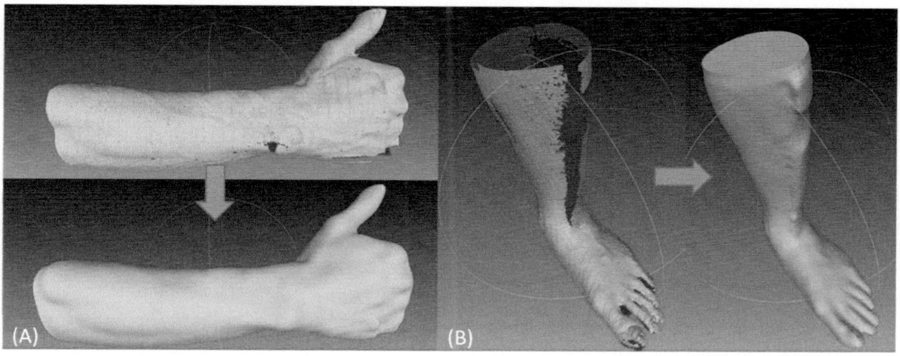

Figure 3.18 Mesh reconstruction – (A) proper data, success; (B) improper data, failure.

of fitting anatomically shaped devices. However, it requires a proper data set – if data is complete and properly processed beforehand, the results will be also useful (Fig. 3.18A), but if the input mesh has large missing areas and unequivocal (doubled) surfaces, as well as other artifacts, the reconstruction will be a failure, meaning that obtained mesh will not resemble a real object (Fig. 3.18B).

The last stage of work with the mesh is preparation for extraction of data for CAD design. This might involve various measurements (manual or automated), cutting the mesh to export just its selected regions to CAD software and other operations, depending on the context and the need. Based on a taken approach to anatomical modeling (more on that in Chapter 4), the used tools might be completely different, and in some approaches, finalized, reconstructed mesh is a proper dataset in itself, not requiring to do anything beyond its importing into appropriate design software.

There are many software tools that can be used for triangle mesh processing, coming from various producers and created for different applications. Here, notable examples, used in daily work by the author and his team, will be described.

GOM Inspect is a software designed for processing and analyzing 3D data, facilitating comprehensive analysis of meshes or point clouds derived from 3D scanners or computed tomography. This program enables the generation of geometric meshes with capabilities for smoothing, hole filling, reducing the number of triangles comprising the mesh, and optimizing their edges. It also provides mesh editing through data selection, aimed at isolating specific structures and removing the rest of the model. GOM Inspect is a free version of the GOM Professional software, which is bundled with professional 3D scanning and reverse engineering equipment from GOM GmbH company.

GOM Inspect offers tools for both rough and precise processing. Rough processing involves trimming the mesh model along the X, Y, and Z axes, while the precise one entails removing unnecessary or damaged mesh fragments by selecting individual triangles. Due to its origins and original purpose, the program contains many tools that facilitate the alignment of geometries relative to each other, assigning coordinate systems, and generating inspection reports for quality control, such as verifying the scan's conformity with the standard CAD model.

Various data formats can be imported into GOM Inspect – apart from the standard STL, also including STEP representing 3D solids. STL file transfer to GOM Inspect is done using a dedicated Import tool. Before loading a file, one must ensure that the predefined unit is in millimeters, and the target element type is a polygonal mesh (mesh).

GOM Inspect has advanced methods for selecting specific mesh triangles. These include:

- Surface select/deselect;
- Through surface select/deselect;
- Select separate, unconnected elements;
- Select boundaries;
- Select along a drawn line;
- Select single triangles;
- Select reference points;
- Increase/decrease selected area;
- Select all/deselect all, invert selection.

The mesh editing tools available in the program enable the following operations:

- Automatic or interactive (manual) hole filling (Close Holes);
- Mesh smoothing (Smooth Mesh);
- Mesh thinning (Thin Mesh);
- Creation of mesh bridges to fill missing areas, used in conjunction with Close Holes (Create Mesh Bridge);
- Mesh repair to remove artifacts and irregularities (Repair Mesh);
- Mesh refinement for further smoothing (Refine Mesh).

Beyond the basic tools for processing triangular meshes, GOM Inspect also possesses tools for conducting inspections and generating measurement reports, such as superimposing meshes from a 3D scan onto a nominal CAD model and generating color deviation maps, often used for accuracy evaluation (Górski et al. 2019a). It is also possible to base meshes on parametrically defined features (line, point, cylinder, plane, etc.), including imported solid models. The software also facilitates the diagnosis and automatic repair of damaged (error-containing) triangular meshes, which are sometimes impossible to open with other software (e.g., MeshLab). In the free version of the software, scripting and automation are not possible (available only in the Professional version delivered with 3D scanners).

The most important advantages of GOM Inspect compared to competing programs are as follows:

- Clear and user-friendly interface;
- Numerous options for precise geometry selection on the mesh;

- Quick and effective execution of basic operations: hole patching, cleaning, reconstruction, and repair;
- Ability to generate measurement reports and color deviation maps;
- Easy basing and modification of the coordinate system.

GOM Inspect represents a powerful, user-friendly solution for professionals in various fields requiring accurate 3D data analysis and manipulation. Its expansive feature set makes it an invaluable tool for quality control, product development, and reverse engineering tasks.

Another free to use software is MeshLab – different in many terms and properly complementing the GOM Inspect in its approach to data processing and work with files. MeshLab is a comprehensive open-source software geared for processing and editing of 3D triangular meshes, point clouds, and raster objects. Developed by the Visual Computing Lab of ISTI-CNR in Italy since 2005, MeshLab is adept at handling large unstructured triangle meshes and includes a suite of tools for editing, cleaning, healing, inspecting, rendering, texturing, converting meshes, and preparing models for 3D printing (Cignoni et al. 2008).

As a robust mesh viewing application, MeshLab facilitates the loading and interactive inspection of 3D objects in a variety of formats. It allows users to manipulate a loaded mesh through interactive tools or a vast array of parametric filters, which perform automated tasks on selected parts or entire meshes. A significant feature of MeshLab is its extensive range of filters, capable of executing advanced operations underpinned by complex mathematical algorithms (Callieri et al. 2012). Another unique aspect is the software's ability to automatically record filter usage, enabling the auto-creation of macros that can be saved as hypertext files and reused either manually via the script interface or through a console-based tool (formerly MeshLab Server, now replaced by PyMeshLab).

MeshLab's primary uses encompass the following tasks (Cignoni et al. 2008, Callieri et al. 2012):

1. Aligning 3D scanned or medical imaging data in the form of point clouds and triangle meshes. Tools provided allow for the alignment of different meshes into a common reference system, facilitating integration of data from various sources, including active scanners and 3D-from-image technologies.
2. Basic mesh processing including transformation, rotation, mirroring, scaling; manual or semi-automated selection and removal of vertices, triangles, and surfaces; and automated mesh error removal and cleaning.
3. Advanced cleaning operations, such as filling holes, artifact removal, partial or total mesh repair, and tasks like mesh division, simplification, refinement, or remeshing.
4. Mesh reconstruction utilizing diverse algorithmic approaches, from volumetric methods like Marching Cubes to implicit surface algorithms like Screened Poisson.

5. Features tailored for 3D printing, including offsetting, resampling/remeshing, automatic hole closure for watertight meshes, flattening the bottom area, among others.
6. Measurement and analysis tools for manual measurements on the mesh surface and automated geometric and topological computations, as well as planar sections.
7. Conversion and interchange support for numerous 3D and 2D formats.
8. Raster image integration for the utilization of 2D imagery or data layers.

Operated via a GUI, MeshLab includes various windows and toolbars in its design. The main component is the 3D viewer, which allows for inspection and manipulation of meshes in 3D space, offering features such as camera perspective toggling and visualization options. Meshes are organized into named layers within projects that can be saved, duplicated, hidden, or reordered.

MeshLab's wide range of filters for mesh processing and advanced algorithms for measurement, analysis, and comparison set it apart from competitors like MeshMixer and GOM Inspect. Its capacity for task automation at various levels and the optimization of its engine for rapid operation, even on less powerful computers, are also notable benefits.

MeshLab was selected for the "Automation of design and rapid manufacturing of personalized orthopedic and prosthetic supplies based on anthropometric measurement data" project at Poznan University of Technology for its programmability and automation potential (Górski et al. 2020a, b). Operations on meshes can be recorded as control macros, which can be saved, reused, and run remotely, streamlining the entire process. MeshLab's role as the only open-source and automated mesh software makes it a pivotal tool in efficient automated systems like AutoMedPrint, as demonstrated in several publications (Górski et al. 2020b, Górski et al. 2022a), as well as in case studies in the Chapter 6 of this book.

Other noteworthy tools that can be used to handle triangle meshes are:

- *MeshMixer* – a free software from Autodesk company, with a set of features similar to what can be found in GOM Inspect or MeshLab,
- *Blender* – a free 3D modeling software, used mostly for creating visualizations for rendering purposes, but also capable in working with mesh geometries,
- *CATIA v5 Digitized Shape Editor* – module of known professional CAD software – CATIA – enabling streamlined process of design based on processed triangular mesh,
- *Rhinoceros 3D* – all-purpose 3D modelling software, capable of processing spatial meshes.

To summarize, triangle mesh processing is a well-known process pursued both in industry and medicine. There are a number of known methodologies, tools and ways of conduct, depending on the original data source and type of object that it depicts. Freeware and open-source software is widely available and capable of realizing even the most demanding tasks, although sometimes requiring tedious, manual work. The prepared data can be then used for further design activities in CAD systems – which is a main focus of the next chapter.

CHAPTER 4

CAD Design of Anatomical Models

4.1 Types of CAD models

4.1.1 Overview and basic concepts

Modeling in CAD systems is one of the basic tasks realized by design engineers. During this phase of lifecycle of a new product, its shape is created, in a multi-stage process, sometimes involving various types of design (as mentioned in Chapter 2 – routine, creative or innovative design). The modeling phase is aimed – as its name suggests – at creating a digital, geometrical model of a designed product. Usually, the model should be as detailed and realistically representing the geometrical features as possible.

In typical engineering scenarios, many simplifications are allowed (e.g., representing threads by merely denoting them in a flat drawing, not by actual 3D helix cut along a cylinder or hole). However, when designing anatomically shaped products, especially those which will be later 3D printed, faithfulness is the key and no simplifications are allowed – the model will come out of a 3D printer just as it was created. Moreover, geometries that will be dealt with while designing anatomically shaped products are usually very complex and not defined in simple mathematical terms (i.e., non-defined curves rather than lines, arcs, and simple splines). As such, focus should be put on proper selection of modeling methods, first not to lose any features when going from anatomical data through the design stage, second – to be able to perform the modeling, without excessive workload and overloading of central or graphic processing unit of a computer that is used during the design process. The basic methods of CAD modeling – in general – are shown in Fig. 4.1.

The scheme presents the most popular basic types of 3D models used for professional product development. They have been divided into three main categories:

a) *models based on mathematical operations* – these are solids, surfaces, or wireframes, created on the basis of geometrical shapes – lines, circles, curves, forming 2D or 3D shapes, of which the model is composed;

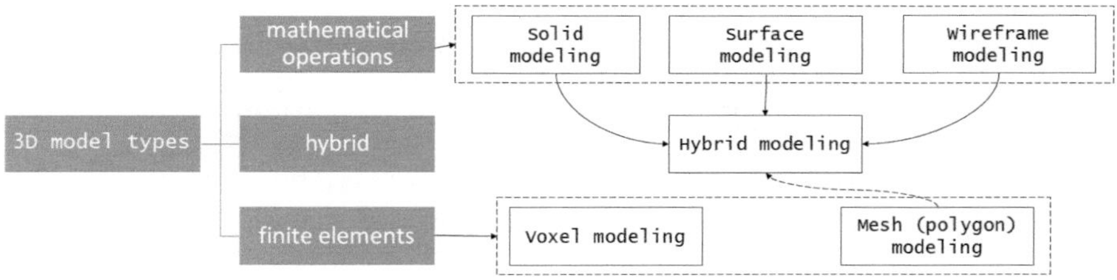

Figure 4.1 Basic division of CAD modeling methods.

b) *models based on finite elements* – a whole geometry is divided into many small pieces, based on similar, simple 2D or 3D shapes, e.g., triangles, quads, cuboids, tetrahedrons – they are connected to each other and usually can be edited together to manipulate the geometry;

c) *hybrid models* – models, in which there is a simultaneous use of many modeling techniques, from both groups, e.g., meshes connected with parametrized surfaces, solids with wireframes, etc.

A typical CAD design process for mechanical engineering usually involves a lot of solid modeling (Cucos et al. 2018), unless more complex geometries are needed, either of aesthetic or functional reasons. Then, especially in automotive or aeronautical branch, a lot of surface modeling is involved (Harries et al. 2019), naturally leading also to hybrid modeling (Salehi and Wang 2019). However, professional CAD users creating new products in their companies rarely go outside the mathematical models – they don't use any types of meshes unless it is necessary due to specific requirements.

These requirements can be a result of an inverse, or reverse process – when geometry is first acquired from an existing, physical object, what we have already defined in this book as reverse engineering. As such, 3D scanned or otherwise recorded geometry is, most of the time, a triangle mesh. As such, if there is a need to create a new part based on existing, 3D scanned geometry, hybrid mesh-solid or mesh-surface-solid modeling may be utilized. This is true also for medical models, through the same methodology – first acquiring geometrical data of a patient, then creating a surface (or wireframe) model around it, then converting it to a solid, which can be further processed in order to be manufactured (e.g., using the previously mentioned CAM software). That way of conduct is known to be efficient, as demonstrated in some previous work by the authors' team (Górski et al. 2022b) and in literature as well (Zhang 2018, Ciobanu et al. 2013).

Typically, solid and surface modeling in CAD systems was associated with the so-called parametric approach (Camba et al. 2016). Parametric modeling is an approach that allows full control over the geometrical features of a created model – basically, it is created out of a set of parameters, that are used in specific operations, performed in a defined order. Changing these parameters, or operations, or their order will result in a change of the model – sometimes even of its very basic features.

In general, parameterized modeling is based on operations, performed in a specific order, thus creating a history of the model. The operations are of following types:

1. *Sketch-based features* – these operations involve creating first a 2D or 3D vector line drawing (also parametrized), a.k.a. sketch, which is then used in combination with an additional parameter, set of parameters or another sketch to create a 3D feature (solid or surface). The sub-types of this features are:
   a) *single-sketch features* – these involve extrusion and rotation, where a closed (in solid modeling) or open (in surface modeling) profile, composed of a set of lines and/or curves, is extruded along a straight line in a specified direction, at a specified distance, thus creating a 3D geometry; examples involve creating a cuboid by extruding a rectangular scan or a shaft by rotating a closed broken line profile; these operations can also be positive ("adding material") or negative ("removing material");
   b) *multiple-sketch features* – the most typical features here are known popularly as "rib" – extrusion of a closed profile along an open line (or open profile along a line in surface modeling) – and "loft", extrusion between two or more different closed profiles (or open, in solid modeling);
2. *No sketch features* – the geometry modifications that do not require any drawings, e.g., chamfers, fillets, holes – creating these features involves only delivering a set of parameters (e.g., chamfer size and angle, fillet radius, hole size, location, and type, etc.).
3. *Transformation and multiplication features* – these involve transforming existing geometries to obtain a specific effect, such as:
   a) *rectangular pattern* – cloning objects/features along a plane of rectangle (rows and columns),
   b) *circular pattern* – cloning objects around a specified point, defining angle, radius, and amount,
   c) *irregular pattern* – cloning objects along a curve or a set of points,
   d) *symmetry/mirror* – creating an object mirrored along a plane, deleting or leaving original object intact,
   e) *transformation* – translation, rotation or scaling applied usually to the whole object,
   f) *Boolean features* – adding, intersecting, or subtracting features from one another.

Usually, creating a parameterized 3D model (a "pure" one, either solid or surface) involves the following steps:

1. Selection of a starting plane.
2. Drawing a sketch (or sketches).
3. Creating a first 3D sketch-based feature (must be "positive", with adding of the material).

4. Drawing more sketches and creating more 3D features (after the initial geometry is created with a first feature, the subsequent ones can be positive or negative).
5. Applying no-sketch features.
6. Applying transformation and multiplication features.
7. Repeat steps 4–6 until the model is finished.

The strategies here involve working with a "subtractive" approach – where one large piece of "material" is created (like a cube) and then the "material" is "subtracted" in subsequent operations – or an "additive" approach, where features are added step-by-step, increasing the volume of the created model along its history.

The key point of parametrized modeling is that any parameter – since the beginning of model's history – can be altered at any time, having influence on the rest of the model. The history can also be "reversed", so the creator or editor of the model can always go any number of steps back, effectively reversing the operations done in the most recent history of the model (however, if they are not deleted, they stay in the tree but remain inactive). An example of a history model is presented in Fig. 4.2.

Parametric modeling is widely recognized as a trademark of "engineering" approach to design of 3D objects. It enables full control of the model, as well as global modifications to its core features, streamlining the process of generating new parts based on the same root geometry, or modifying the root geometry to create a new product family – in this regard, this method is unmatched with non-parametric methods (historically regarded as "artistic" or "free" shaping methods). The problem with parametrized modeling is that it becomes cloggy and unclear with the very

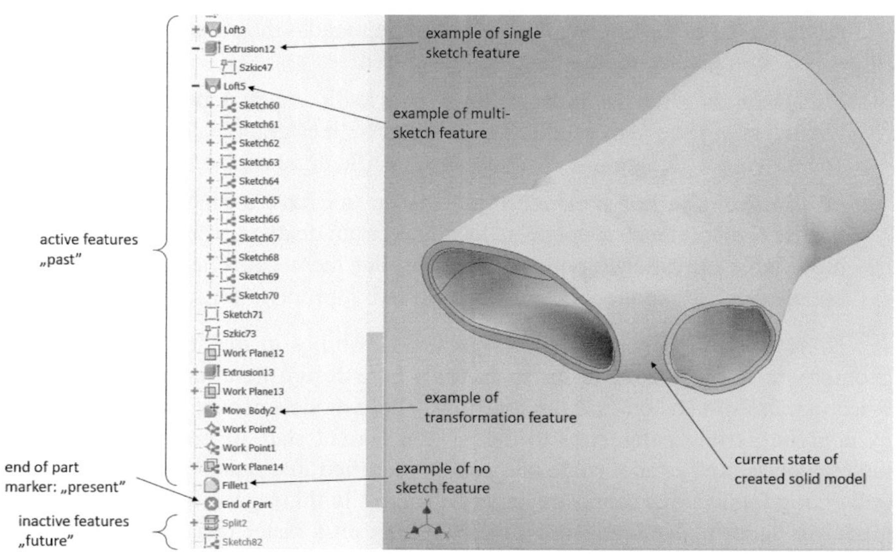

Figure 4.2 Parametric modeling with history – examples of features on the basis of anatomically customized wrist hand orthosis model.

complex geometries, especially when these geometries have an "organic" background. The number of parameters describing an anatomical geometry quickly skyrockets when building such a model, the more faithful – the more parameters. Then, the models can become prone to the "house of cards" effect – change of one parameter or operation results in multiple errors, as certain geometries cannot be generated due to topological restraints. Examples of these errors are: too a small radius of a surface to which a fillet is applied, too thin a wall between two complex surfaces or entwisting of an extrusion between two heavily modified splines making the feature impossible to properly compute. Sometimes, inexplicably, as of the authors' practice, these errors are unable to be reversed simply through getting parameters back to their original state – that is because certain CAD systems (especially free or low-budget ones) are not prepared for a certain level of complexity and start generating bugs – which is not uncommon.

As such, non-parametric modeling methods have recently reemerged as an alternative approach to creating of geometries in professional design engineering – however, these methods have been always known in the non-engineering world. Non-parametric modeling refers to a modeling approach that does not rely on predefined parameters or constraints to define the geometry. Instead, it often involves techniques like direct modeling or mesh-based modeling (Hakala 2015). The point of origin usually involves creating a basic solid – like a block of material – and then shaping it by series of operations, much akin to a process performed by a sculptor (that is why non-parametric modeling is frequently referred to as "digital sculpting" (Alcaide-Marzal et al. 2013)). The key points of this approach are:

a) *no predefined parameters* – unlike parametric modeling, in non-parametric models a list of parameters is nowhere to be found – the parameters are applied *ad hoc* to a given operation and are not stored;
b) *direct modeling* – this approach allows users to directly manipulate the geometry without being bound by predefined parameters, like "grabbing" a point or surface and moving it along a line, without any specific dimension or angle applied;
c) *no history* – the non-parametric models do not have a global history like in the case of parametric models; locally, certain features can have a stack of modifiable operations applied into them, but these can often be subjected to collapsing, leaving the whole geometry "as is", prone to further modifications.

Non-parametric modeling offers greater flexibility in making changes to the geometry, making it suitable for scenarios where design iterations are frequent or where precise dimensions are not essential. These models cannot become "broken" by applying wrong parameters to them – the current state of geometry, visible to the creator, is always accessible and can be modified in a number of ways, but the underlying foundations cannot be largely changed. In this regard, manipulating these models is more realistic and resembles working on a raster image, or a sculpture, thus being more approachable, especially to the creators without an engineering background and technical mindset. One does not need to think about the "house of cards" of the stack of parameters, but rather introduce more and more gradual modifications, to obtain a final result.

However, the non-parametric modeling process is difficult to reproduce and precisely describe, just like real sculpting. As such, it is also quite difficult to automate or present in an algorithmic fashion, making it longer to realize subsequent variants of the same type of product. However efficient it is in a single case of a product with a complex, organic geometry, creating a streamlined, repetitive design process out of it, possible to be implemented in a modern, digital factory is also a problem.

That is why for efficient and productive industry-scale creation of anatomically shaped devices, such as orthoses, prostheses and implants, a hybrid approach is required. Depending on the desired shape and source of data, the design process should be composed of different modeling techniques and their use should be properly justified. However, an engineer dealing with design of anatomically shaped models must have significant expertise in all modeling types. That is why in this chapter, all the notable techniques are shortly presented. In the further chapters, case studies based on various techniques and their combinations are shown for full comprehension.

4.1.2 Solid modeling

Solid modeling is the basic process used in CAD systems, for creating industrial grade models of products which are later manufactured and distributed to clients. The modeling is an initial phase of a product lifecycle. It is also a basic technique taught at technical university courses, right after the basics of technical drawing, and an essential skill that any engineer dealing with production processes and industry should possess.

In the course of the solid modeling process, a solid is created, a single or a composite one (a part created of many separate or connected solids). It means that the 3D representation of an object is volumetric by nature. This has certain implications – an obvious one is that if the model is cut in half, two new solids will be obtained. Also, the model can attain a material. "Material" in this case is a set of properties, affecting not only visual qualities of the model (such as color, texture or light reflection and refraction properties), but also the physical ones, such as density and other accompanying characteristics that may later be used for various calculations – relatively simple ones, like calculating the volume, mass, center of gravity and moment of inertia – and the more complex ones, realized through advanced CAE simulations.

The solid models can be either parametric or non-parametric. The parametric ones are created according to the guidelines mentioned in the previous chapter. Historically, solid modeling was achieved usually by performing various Boolean operations on primitive shapes, such as cuboids, spheres, or cylinders. First, a basic shape was defined (e.g., a cuboid of specified size – three lengths), then, another one, with a specified Boolean operation (e.g., a cylinder of a specified length and diameter, subtracted from an original cuboid to make a cylindrical hole). This approach is still perfectly viable and capable to be utilized in known CAD systems, however, it is ineffective, requiring sometimes hundreds of operations to obtain the desired effect. It has evolved into the process that is well known today, based on 2D parametric features – sketches – which are used to create 3D features, as mentioned in the previous chapter.

Non-parametric solids can be acquired either through direct modeling (already mentioned) – starting from a primitive shape and then adding more and more features to it, sometimes even without precisely specified dimensions or directions – or by conversion from a non-parametric model of another type, or a parametric model coming from a different source (e.g., non-native software format). If the solid model is achieved by means of conversion from a mesh model, for instance by careful processing of the mesh to make it watertight (as explained in Chapter 3), it is often called the "dumb" model (Walloch 2023), as it has no available history and its surfaces and features are meaningless (even if they resemble flat surfaces or cylindrical shapes, geometrically they are not and will not be recognized as such). Examples of both types of solid models with anatomical geometry are presented in Fig. 4.3.

The "dumb" models, when converted between CAD systems, often take a form of a STEP model (files with STP extension), holding information about solids (parts), their connections, often materials and other data useful in design. Thanks to STEP models it is possible to interchangeably design one large product (e.g., a car) in different CAD systems and them assemble them together, while still working on solids (enabling some common operations on various parts, such as drilling a hole along several assembled parts, or deriving a shape of one part from another one).

Solid modeling by itself is a basic skill desired from a design engineer, working in any branch of industry. As such, in this book we will not delve deep into technicalities of the process and possibilities of various CAD systems – there are numerous textbooks – older (Farin et al. 2002) and newer (Land 2023), as well as online courses and video tutorials, both free and commercial, that explain different problems in greater detail, step by step. Here, in later chapters, certain cases of building parametrized, anatomically shaped 3D models will be explored.

Solid modeling is often associated with the more "mechanical" design, being extremely efficient with straight geometrical features, such as flat or circular shapes and operations made on them, while the more sophisticated surface geometries, based

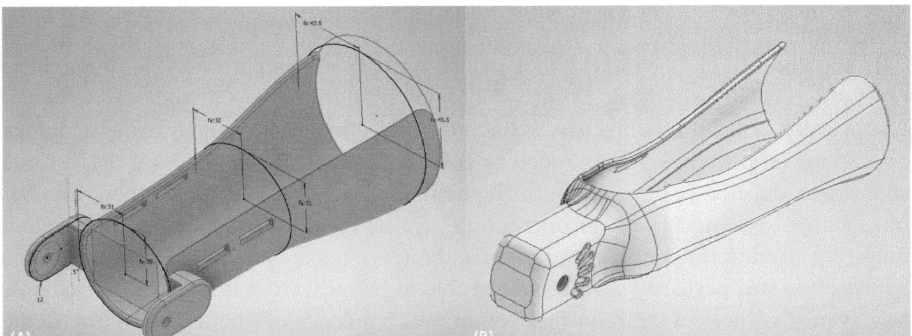

Figure 4.3 Solid model on example of an open prosthetic sockets, (A) a parametric model, some sketches visible; (B) a "dumb" solid model resulting of file conversion.

on various types of curves, are usually easier to design and process using the surface modeling techniques. This does not mean that solid modeling is not suitable for more advanced, organic shapes – however, it may not always be the best technique to start when working with certain classes of anatomical geometries. Of the author's experience, it might be stated that solid modeling is suitable for most (not all) cases of externally used medical devices, such as prostheses or orthoses – while it is not as good when considering more intricate internal shapes of the human body, such as customized implants or models representing softer tissues. In these cases, usually mesh models or surface models are used.

4.1.3 Surface and wireframe modeling

Surface modeling is a process which is usually noted as secondary in terms of popularity and the range of usage in the engineering design, when considering "manufacturable" products, right after solid modeling. Surface modeling is a broad, umbrella term covering a wide range of activities, not all of them strictly related to work of engineers. That is why, for the sake of this chapter, it will be assumed that whenever the term "surface modeling" is used, we would mean the subtype of modeling, usually parametrized, similar in approach to solid modeling and realized in professional CAD systems.

A result of surface modeling is always a set of surfaces – geometrical features that are defined in three dimensions, but do not have thickness. A real-life analogy of surface models and their creation would be creating a model out of a sheet of paper, cut, and bent into various shapes. Of course, in real life this is limited due to physical possibilities of the paper, the limits that are not present in the digital world. Surface modeling gives much more freedom to the designer than solid modeling – it is possible to build very complex, advanced shapes based on fully parametrized operations. However, solid modeling is also much more difficult to properly perform, as the paradigm is different and less "realistic", so building such models from the scratch requires the designer to have a considerable amount of imagination and spatial sense.

In terms of creating such models in modern CAD systems, the methodology is quite universal and it was shown already in the initial subchapter. The parametrized modeling requires sketches, just as solid modeling, but unlike solid modeling, most of the used sketches are open profiles – lines, curves, or combination thereof. The basic operations involve creating a flat surface on the basis of a closed profile, extruding one open profile along another (closed or open) or along a specific direction, or creating a surface which is a blend between two or more open profiles, or two or more surfaces. There can be also more sophisticated operations, such as capping an open profile with a curved surface, or various forms of blending. There are also operations like chamfering, filleting, stitching, or cutting of the created surfaces.

A surface model is not a solid – even if it resembles one, it will be empty after cutting. When creating a surface model, there are no constraints that would impose the need of waterproofness – the surfaces can cross each other or be open (freely visible from both sides). As such, these models do not have properties such as density, volume or gravity center and any material applied to them will only affect the appearance of the created shape. Conversion from a surface model to a solid one requires obtaining the watertight shape, without any holes, extruding surfaces or other ambiguities – which is possible over the course of two main operations: mutually cutting the surfaces by themselves and closing the gaps between surfaces. A watertight model can be then converted to a solid one. In the other direction, it is always possible to convert the solid to a surface model. Often it is not done for a complete solid, but rather some specific surfaces are "extracted" from it to build some advanced features. Such models are considered as hybrid ones and are described in one of the further subchapters.

Non-parametric surface models (other than polygonal meshes, that are described in the next section) are created similarly as in the case of the non-parametric solids – either by direct modeling or by conversion from another type of model. One of such subtypes are mesh models, widely described in other places of this book.

The surface modeling is usually associated with more sophisticated forms of design where aesthetics or streamlined shapes are important. Surface models are therefore used especially in architecture, automotive or aeronautical cases, but also in the medical design – everywhere that solid models become difficult to efficiently implement due to the complexity of used shapes, the surface modeling techniques can step in and replace or complement the solid modeling. They are also invaluable in reverse engineering processes, when modeling on the basis of 3D scan data (Sofronov et al. 2019, Kumar et al. 2013).

Similar to the case of solid modeling, there are endless sources of technical knowledge regarding detailed processes of creation and editing of such models (Land 2023, Lee 1999), that is why here no more details will be given in this book, apart from some use cases in the latter chapters. A surface modeling skill is not essential for all engineers and competences in this area are usually built on the foundations of existing skills in solid modeling. However, it is also important to notice that surface modeling exists and functions quite well outside the engineering context – mostly for creating artistic models used in various Computer Generated Imagery (CGI) for movies, games and other forms of entertainment – in such a case, usually entirely different software tools are used. It is noteworthy that biomedical engineering, especially when considering the area of building anatomically customized models, extensively uses both approaches – the engineering one (based on CATIA, SolidWorks, Rhinoceros 3D and others) and the artistic one (based on Blender, 3DS Max, Maya, and others). It is not uncommon to find medical engineering cases when two different software classes are used, mostly due to flexibility and effectiveness issues (especially when a given software is both free to use and has easy entry threshold).

The third basic type of modeling used in the CAD systems is wireframe modeling. These models are one of the most basic and essential types of representations. Essentially, a wireframe model is a visual representation of a three-dimensional object using solely lines and vertices (no 2D or 3D features). These models do not display surfaces or solid forms. Instead, they depict an object through the skeleton (the eponymous "wireframe") of its edges and corners, resembling a three-dimensional mesh made of wires.

The construction of wireframe models involves defining the key points in the 3D space (usually by direct coordinates or in relation to existing elements) and

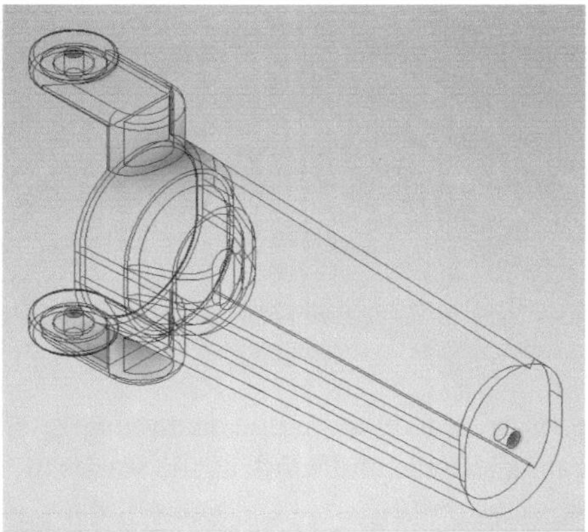

Figure 4.4 Wireframe representation of a prosthetic forearm model.

connecting these points with lines to outline the shape of an object. This process can be relatively simple for basic geometric shapes but becomes increasingly complex as the intricacy grows. Wireframe models created in CAD systems (Fig. 4.4) are parametrized – designers can input dimensions and relationships between various elements of the model, automating the generation of the wireframe.

Wireframe models serve several important purposes in the realm of design and engineering. Firstly, they provide a straightforward method for conceptualizing and visualizing complex structures, enabling designers to draft the layout and spatial relationships of various components quickly. This is particularly useful in the early stages of design, where the focus is on the arrangement and interaction of different elements rather than their physical properties.

What is even more important in some cases, wireframe models require significantly less computational power to generate and manipulate compared to more detailed representations. This efficiency makes them ideal for preliminary design iterations, allowing for rapid modifications and adjustments without the overhead of rendering full textures or surfaces. Also, complex solid or surface models can be converted to wireframes for simplification and used for visualization or exchange, especially when larger structures are of concern (e.g., in civil engineering). Wireframe models are also essential in certain types of engineering simulations, e.g., when computing deformations and stresses in complex lattice structures (such as scaffolds, sometimes used in bioprinting or tissue engineering).

In relation to solid and surface models, wireframe models represent the most basic form of 3D modeling. Solid models provide a complete representation, including volume, mass properties, and material specifications, making them suitable for detailed analysis and manufacturing processes. Surface models describe the exterior surfaces of an object, offering a balance between detail and computational efficiency, useful for aesthetic evaluations and complex surface interactions. Wireframe models, on the other hand, lacking the detail of solid and surface models, are not used for manufacturing or detailed analysis but are invaluable for initial design conceptualization, layout planning, and quick visual communication.

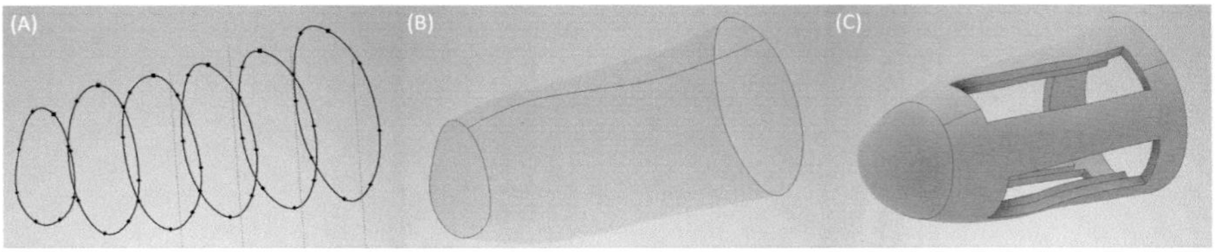

Figure 4.5 Models on various stage of part development on example of upper limb prosthetic socket: (A) feature layout, wireframe; (B) basic shapes, surface; (C) detailed structural features, solid.

Figure 4.5 presents comparison of a single, anatomically shaped object in three different forms: wireframe, surface and solid, as the stages of modeling progressed.

4.1.4 Mesh and voxel modeling

Mesh or voxel modeling belong to the group of modeling techniques which are non-parametric by nature. These modeling methods, by principle, involve representing the desired 3D (or flat) shape in the form of a usually large number of small, primitively shaped elements, connected together to form surfaces (in the case of meshes) or whole volumes (in the case of voxels).

The mesh modeling is a separate class of both 3D data storage and exchange and creative processes. In the realm of digital modeling, polygonal mesh models are akin to the scaffolding of virtual structures, forming the backbone of 3D graphics in video games (Merlo et al. 2013), simulations, and CAD, especially intertwined with reverse engineering processes.

A polygonal mesh is a collection of vertices, edges, and faces that define the shape of a 3D object in digital space. These basic elements are polygons, typically triangles or quadrilaterals, which when connected, create a mesh-like network that represents the surface of an object. The intricate web of these polygons allows for the detailed rendering of complex shapes and surfaces with finite, flat faces (Botsch et al. 2010).

The structure of a polygonal mesh is usually systematic. Vertices are points in 3D space with x, y, and z coordinates. Edges are the lines that connect pairs of vertices, and faces are the flat surfaces bounded by edges. The more polygons a mesh contains, the higher its resolution, and the more detailed and smoother the model appears. Polygonal meshes are stored in various file formats, each with its unique features. Popular formats include:

- *STL (Stereolithography, also Standard Triangulation Language)* – an already mentioned format originating from 3D printing and widely used there, STL files contain only the surface geometry of a 3D object without any representation of texture, or other common CAD model attributes (Conejo et al. 2021);
- *OBJ (Object File)* – a simple data format that represents 3D geometry: the position of each vertex, the UV position of each texture coordinate vertex, vertex normals, and the faces that make each polygon; they can be accompanied with MTL files containing material information (different materials associated to various objects);

- *FBX (Filmbox)* – a proprietary format that can store complex information such as animation, 3D models, 2D imagery, and audio, making it a versatile format for multimedia production.

Polygonal meshes are very popular in different applications, for several reasons. Their geometric simplicity makes them versatile and computationally efficient, which is crucial for real-time rendering in video games and simulations. They are also hardware-friendly, with Graphic Processing Units (GPUs) optimized for processing triangles, the most basic polygonal shapes.

From the viewpoint of this book, the polygonal mesh models are crucial in:

- *engineering design* – for visualizing designs and collaborative work, as well as for the simulations (universal exchange format); also, as a basic format of contactless measurement results (i.e., 3D scanning),
- *manufacturing* – in 3D printing and CNC machining to prototype and fabricate parts (STLs are universally used),
- *medicine* – for creating 3D visualizations of anatomy, based usually on DICOM segmentation (as described in previous chapter – STLs are used for exporting the 3D medical imaging data),
- *games and interactive experiences* – meshes are universally used in game engines (such as Unity or Unreal Engine) for depiction of all the 3D content (FBX and OBJ formats are extensively used, as they carry information about materials),

An example of a polygonal mesh of a complex anatomical object is shown in Fig. 4.6. The object is a hepatoblastoma (liver tumor), segmented out of MRI DICOM along with veins, arteries, liver tissue and spleen. It has been used for visualization in Mixed Reality and 3D printing, for the purpose of pre-surgery preparation. In this case, no parametric modeling was used and only free software for segmentation and mesh processing was used (namely, 3D Slicer and MeshLab). The case is described with more detail in Chapter 7.

The advantages of polygonal mesh models are notable. They require less computational power to render than the volumetric, parametric models, which is critical for interactive applications like Mixed or Virtual Reality. Most 3D software supports polygonal meshes in one of the available formats, making them a universal

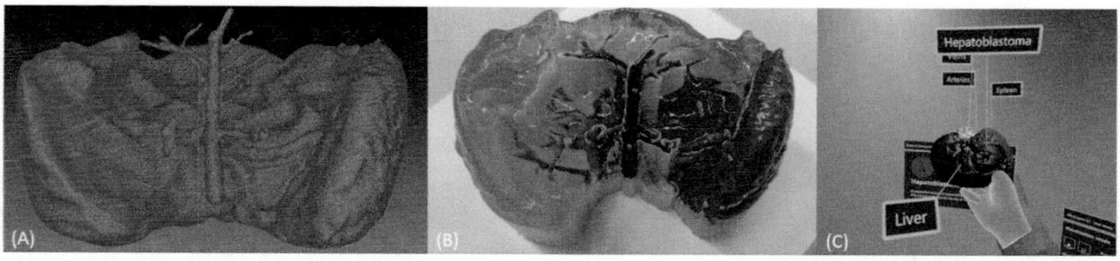

Figure 4.6 Anatomic mesh model (A) and its usage: (B) multi-material 3D printing (realized using Stratasys MediJet J5 device); (C) interactive Mixed Reality (realized in Unity game engine).

tool for 3D modeling. They can be easily processed in a freeware or open-source software (e.g., MeshLab, Blender, Meshmixer and many others). Meshes can be easily manipulated, allowing for a vast range of shapes and designs.

However, there are also disadvantages. The quality of a mesh is dependent on the number of its polygons, which can limit the detail in models without making them computationally heavy – there is always a compromise between mesh size, detail quality and computational power. In the above-mentioned example of a liver tumor, the original, segmented mesh had approximately 3.5 million triangles, which was perfectly acceptable for 3D printing (Fig. 4.2B) – unfortunately, Mixed Reality application was not able to run smoothly with that amounts of triangles on a provided device (Meta Quest Pro), so for that the purpose of the model had to be decimated to 10% of its original density. This introduced visible simplifications in more delicate structures, such as veins. Furthermore, in mesh models, curved surfaces can only be approximated, not accurately represented, leading to a less realistic appearance without a high polygon count. Also, making changes to a high-polygon, complex model can be challenging and time-consuming and it is very difficult to introduce significant shape changes quickly (unlike in the parametrized models).

Still, it must be emphasized that the mesh models are here to stay – their range of use and easiness in exchanging them between various software is uncanny and some of their limitations will be gradually fading away as computing power increases and more and more smart algorithms (including AI) will be available to help process them with more ease. On the other hand, viable alternatives are searched upon for more efficient rendering techniques, such as gaussian splatting (Wu et al. 2024).

The second class of "finite similar element" models are the voxel models. These models are to 3D spaces what pixels are to 2D images—a fundamental unit of graphical representation, offering a different set of advantages and challenges compared to their polygonal mesh counterparts.

A voxel, or volumetric pixel, is the smallest distinguishable part of a three-dimensional space, just as a pixel is the smallest part of a two-dimensional space (Joshi and Jain 2016). Voxel models are constructed from these volumetric elements, much like building blocks, to form a complete object or scene. Each voxel contains volume information and data about its location in space, as well as some other attributes – like color and material properties, depending on their usage and environment in which they are built. Usually voxels are considered to be box-shaped elements, however they can take various forms, such as tetrahedrons or parallelepipeds.

Voxel data can be stored in various file formats, some of which are VOX (MagicaVoxel) – a format associated with the MagicaVoxel software, it is designed to store data for rendering voxel models – and VDB (OpenVDB), used primarily for visual effects and simulation, highly efficient in storing data for voxels that occupy a lot of empty or homogenous space (Lacewell 2021).

Voxel models are beloved for their intuitive representation of 3D space, making them popular in medical imaging, where CT scans and MRI data are naturally represented in three-dimensional grids. Their popularity also extends to gaming—especially sandbox games like Minecraft, where the environment is manipulated at the voxel level (Lacewell 2021).

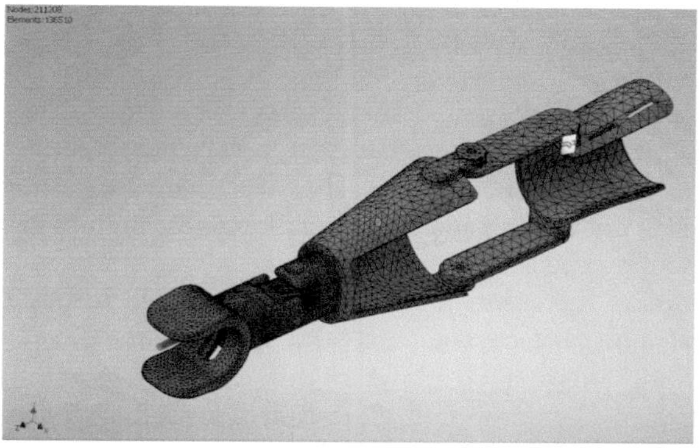

Figure 4.7 Use of tetrahedral mesh for finite element analysis of an upper limb prosthesis model.

Anyhow, from the engineering standpoint, the most frequent use of voxel models is for carrying out various simulations, especially static strength calculations with the use of finite element method (FEM, mentioned in Chapter 2). Geometrical elements used in FEM calculations to build models are never explicitly called voxels in the popular CAE software (the notion used there is usually "mesh"), however they are just that – small, regular, three-dimensional elements with basic spatial and material properties. An example of such use is presented in Fig. 4.7.

The voxel approach to 3D modeling comes with its set of advantages. Modifying voxel models is similar as in the case of polygonal mesh – akin to sculpting with digital clay, offering a straightforward way to build and carve models. However, unlike polygonal meshes, voxel models do not suffer from non-manifold edges or holes, which can create issues in rendering and simulation – they are always volumetric and, as such, easy to, e.g., cut in half (which is difficult in mesh models, as they are empty inside). Also, voxel meshes are the only sensible way to realize certain computations, making them a must-use format of data in non-real-time simulations performed in CAE systems.

Voxel models also have some disadvantages. The biggest one is related to their size. Higher resolution in voxel models leads to exponential growth in data, which can quickly become unwieldy and demand significant computational resources. Traditional rendering techniques are also less efficient for voxel models, often requiring specialized or hybrid rendering methods. Voxel data can be voluminous, necessitating large amounts of storage space, especially for detailed models and requiring large amounts of Random Access Memory (RAM), otherwise work with them will not be smooth.

In summary, voxel models offer an alternative to polygonal meshes that is particularly well-suited to certain applications. They are not as computationally efficient as their polygon-based relatives, their intuitive structure and the granular control they offer over 3D space make them an indispensable asset. For a typical application in medical engineering, FEM analyses are the first things that come to mind, while other applications are more rare (such as, e.g., simulation of bone cutting process in virtual reality training (Arbabtafti et al. 2008)).

4.1.5 Hybrid modeling

The modeling techniques presented in the previous chapters are quite versatile, giving plenty of possibilities to design even the very complex models resembling organic structures and anatomical shapes. However, there are design tasks for which a single approach is not enough and a combination of techniques is required. That is when hybrid modeling needs to be implemented.

Hybrid modeling represents a sophisticated approach in CAD integrating the strengths of wireframe, surface, solid and other types of modeling techniques within a unified framework. This method allows for different sections of a single model to be developed using the most appropriate of these three modeling strategies, tailoring the construction process to the specific needs of various parts of the model. For instance, a model might employ wireframe modeling for its preliminary layout and conceptual design, surface modeling for complex exteriors or interfaces, and solid modeling for core components requiring detailed analysis and manufacturing information.

The versatility of hybrid modeling becomes particularly advantageous in scenarios like reverse engineering. In these scenarios, the design starts with an empirical data collection, obtained through 3D scanning or other measurement techniques. These measured points and surfaces form the basis for a digital model, in the form of a mesh. Hybrid modeling facilitates this process by allowing these imported datasets to serve as the foundation for a comprehensive 3D model. Designers can use wireframe or surface modeling techniques to recreate the object's external contours before transitioning to solid modeling to add volume and mass properties, thereby constructing an accurate and fully-realized model. The mesh can be used "as is", or converted into a "dumb" solid or surface model.

Building customized medical devices is a special case where hybrid modeling shows an exceptional utility. In the creation of custom prostheses or orthoses, the exact topography of a patient's limb can be captured and used to generate a surface model. This model serves as a template, upon which a solid model of a prosthetic or an orthotic device is designed. Hybrid mesh-solid models can be also utilized for digital fitting, as shown in Fig. 4.8.

In various CAD systems there are various possibilities of including hybrid models – but the trend is to integrate the surface and solid features whenever possible.

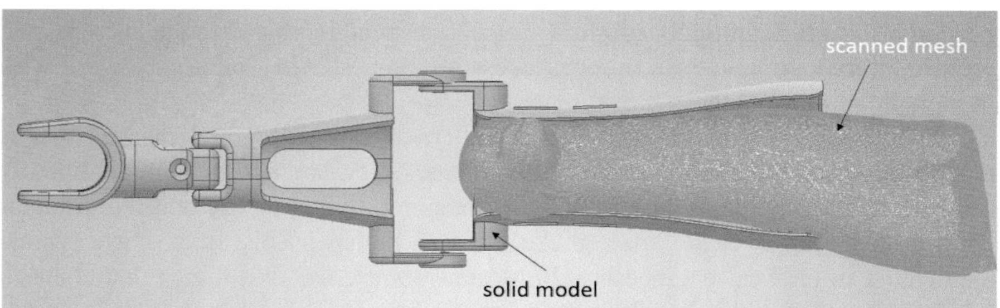

Figure 4.8 Hybrid modeling on example of prosthetic fit (visible locations of stump and socket misfit).

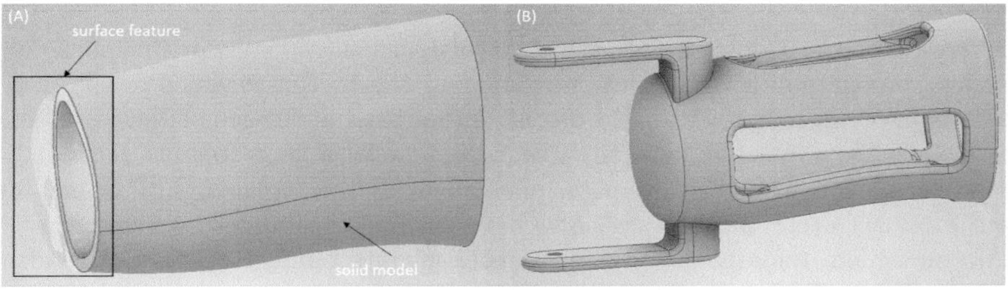

Figure 4.9 Hybrid modeling on example of prosthetic socket – (A) surface feature based on solid's edges; (B) seamless integration of a surface into the solid.

An example of such an integration is presented in Fig. 4.9 – a surface feature is created at the end of a prosthetic socket (Fig. 4.9A) and then seamlessly converted into a solid feature. The building of the model is continued until a desired shape is obtained (Fig. 4.9B).

Hybrid modeling's ability to seamlessly combine different modeling techniques within a single model streamlines the design process, enhances flexibility, and facilitates a high degree of precision. It enables designers and engineers to exploit the specific advantages of different modeling types in a coordinated manner, optimizing the design and manufacturing process across a wide range of applications. This integrated approach significantly expands the possibilities for innovation and customization in product development and design and is practically unavoidable in design of anatomically customized orthoses, prostheses or implants that contain complex mechanisms or advanced structural features.

4.2 Review of CAD software

4.2.1 Overview and software types

As already established, CAD involves using computer software to create, modify, analyze, or optimize a design. This technology significantly enhances the designer's productivity, improving the quality of design, communications through documentation, and creating a database for manufacturing. In this chapter, current state of the art – specific software, available for the users in the mid-2020s – will be presented. CAD and 3D modeling software can be classified based on their specific functions and industry applications.

The very basic division involves separation between 2D and 3D tools. The 2D tools are primarily used for drafting and documentation. They are simpler than their 3D counterparts and are typically used where depth and perspective are not primary concerns. In this book, we are not really interested in them, as anatomical design must involve working with intricate, organic 3D shapes. That is why we need 3D tools: they allow for the creation of three-dimensional models, often very complex, using techniques discussed in the previous subchapter.

The 3D modeling software is a wide concept, involving both CAD and non-CAD solutions. The CAD solutions are primarily created for manufacturing purposes – the basic intended use is to create a documentation of a product, that will be later brought to life by various production processes. That is why the "physical" side is very important – we need precise information about geometrical features, as well as about material, generally treating the modeled geometry as a potentially physical object, with density, mass, moment of inertia, volume, surface and so on. Most 3D CAD software tools also have drafting capabilities, to help creating documentation, traditionally used in typical manufacturing technologies (drafts, schemes, lists of parts, etc., rarely needed in 3D printing).

Apart from the universal, all-purpose CAD systems, there are also programs dedicated for medical use, in specific branches of medical industry. Most examples can be found for dental implant design (Simplant, Planmeca Romexis), but there are also dedicated solutions for designing orthoses or limb prostheses (MediACE3D CAD, Canfit). There are also systems that have DICOM processing capabilities, but go far beyond that and enable also realizing certain anatomical design procedures – example of such software is Materialise Mimics (described in Chapter 3). Here, we assume that the result of work of this software will be ultimately converted into a physical product, by means of additive manufacturing or other technologies.

The non-CAD solutions involve software, in which there is no such assumption and the 3D models are created for various purposes. One of them might be manufacturing, for instance by 3D printing, but usually models in this type of software are created for visual purposes – such as Computer Generated Imagery (CGI) in movies, video games, virtual and mixed reality, etc. That is why there is less precision involved there and physical approach is not as stressed as in industrial CADs. Solid modeling is also not prevalent and drafting capabilities are minimal or non-existent. This software might be very useful as part of medical design, though, as some features might not be possible to obtain in simple CADs, especially when considering cosmetic personalization, or potential parallel use of physical and virtual products (as in the case of surgical supplies in form of holograms, with an exemplary case shown in Chapter 7).

In the further considerations of this chapter, solely the CAD software will be described, as they are more suitable for obtaining physical products. However, in the case study chapters, multiple types of software are demonstrated to be used in practice, especially when building Automated Design Systems and their visual interfaces.

4.2.2 Autodesk Inventor and Fusion 360

Autodesk Inventor is a robust and sophisticated 3D mechanical solid modeling design software used extensively by professionals for product development. It is a suite tailored mostly for mechanical engineering and product design, offering precise and detailed control over 3D mechanical designs, simulation, tooling creation, and design communication. It is one of the basic products by Autodesk company, which holds many various engineering and artistic software solutions aimed at work with 3D data (Waguespack 2013).

Autodesk Inventor is architected to provide an in-depth set of tools for 3D mechanical design, documentation, and product simulation. Unlike cloud-based platforms, Inventor is primarily a desktop application, enabling intensive processing and graphic tasks to be handled locally – it is also capable of working offline, what may prove itself important in certain situations (e.g., use in literal or metaphorical "battlefield", zones of catastrophes, etc.). The software facilitates various stages of product development, from initial concept through to the detailed engineering, analysis, and manufacturing stages.

The user interface of Autodesk Inventor is divided into task-based modules that specialize in different areas of product development. These modules are accessible through a comprehensive interface that allows users to navigate between different stages of design and development efficiently. Inventor's environment includes sketching, part modeling, assembly design, and drawing environments, among others. The approach to UI is that everything is integrated – there are no separate "modules" (as in CATIA), but rather different parts of interface pertaining to solid, surface and simulation features are accessible using the MS Office ribbon-style groups of buttons (as presented in Fig. 4.10).

Inventor's assembly modeling employs a Constrain feature, which allows for precise control over parts alignment and movement within an assembly. The software offers a wide array of standard joints and constraints to simulate real-world mechanical movements and relationships between components. The engineers can select whether they prefer to work with the constrain system (simpler, geometrical attachments) or the joint system (more complex, feature-based connections).

Inventor supports traditional design methodologies, including bottom-up and top-down design approaches. Inventor enhances these methodologies with parametric

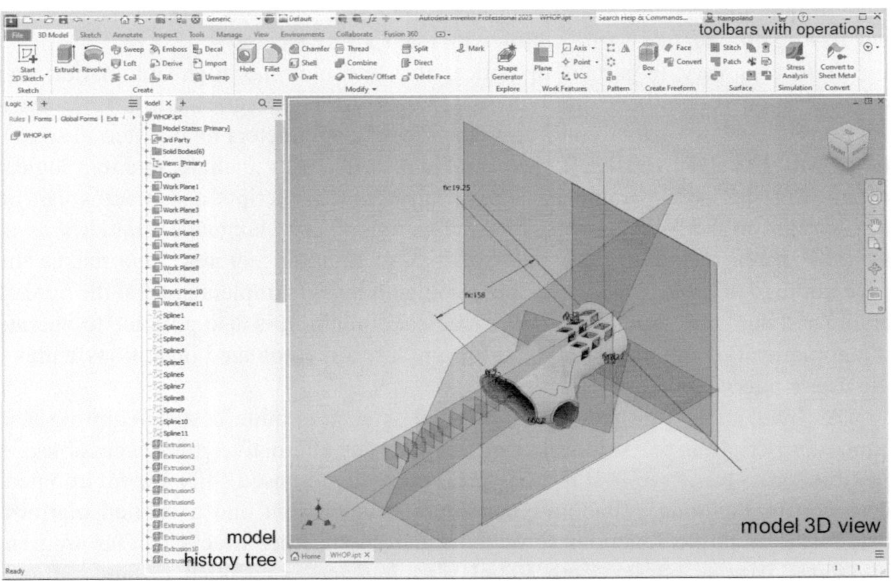

Figure 4.10 Autodesk Inventor – basic interface (solid modeling).

and free-form modeling tools, enabling designers to create complex shapes and surfaces. As in any other parametric CAD, the parametric modeling core allows for design changes to be propagated automatically throughout the associated parts and assemblies. It is important to notice, that surface and solid modeling features are not separated – in a single operation (e.g., extrusion) the user can select if the result should be a solid or a surface, thus facilitating the hybrid approach (no need to switch between toolsets).

Autodesk Inventor has a high degree of interoperability, with native support for various CAD file formats and data exchange standards. This ensures seamless integration with other Autodesk products and third-party CAD (and non-CAD) systems, for flexible design methodologies. Inventor also integrates with Autodesk Vault, a data management tool that helps organize design data and track revision history.

Inventor extends its capabilities to the manufacturing process with tools designed for the creation of tooling for custom parts. The software includes powerful sheet metal and frame design tools that facilitate the creation of complex mechanical components ready for fabrication. These can be used in a non-orthodox fashion – the sheet metal tools can be used for any kind of thin-walled products (including medical devices).

A significant aspect of Inventor is its built-in simulation and Finite Element Analysis (FEA) capabilities. These tools allow engineers to validate the performance of their designs under real-world conditions, such as stress, strain, and heat transfer analysis. This simulation suite is directly integrated into the modeling environment, streamlining the design validation process and albeit simple, can be of added value due to – again – seamless integration with the design process.

It is also very important to notice, that Inventor has high automation and customization capabilities, through its iLogic system (Waguespack 2013). On a very basic level, it enables connecting the model's parameters to an external source (e.g., MS Excel spreadsheet). More advanced features include making logical connections between parameters and certain features. Scripts and macros can be also written in the Visual Basic for Applications (VBA) language, similarly as in the MS Excel, enabling more sophisticated automation features. This makes the Inventor a good basic tool for creating intelligent models implemented at the core of automated design systems, also in the medical domain. It is also possible to operate it automatically from the outside, performing certain automated operations without the user's intervention.

As Inventor is a professional CAD and as such is quite costly, requiring also annual subscription, the Autodesk company offers an alternative, more accessible and flexible tool – Fusion 360. It is a comprehensive cloud-based software environment designed to facilitate collaborative product development and a unified platform that serves multiple functions throughout the engineering lifecycle. The program integrates various phases from initial planning and conceptual design, through prototyping and testing, to final production.

The architecture of Fusion 360 allows for the combination of different stages of product development into a single, cohesive workflow. The software negates the requirement for multiple disparate systems, thereby streamlining the design process. It supports various aspects of product development such as parametric CAD modeling, CAM programming, simulation, and analysis via finite element methods (FEM), generative design, visualization, rendering, and animation, as well as 3D printing (Verma 2018).

Fusion 360 presents a user interface that is compartmentalized into several modules, each corresponding to a different aspect of product development. These modules are accessible via a selection panel typically located in the upper left corner of the interface. The primary CAD module, termed "DESIGN", is structured to be approachable for individuals new to CAD modeling. It encompasses a comprehensive set of tools that align with industry standards and allows for the importation of 3D models from other CAD systems.

The assembly features within Fusion 360 operate using a system of joints. This approach simplifies the process of component assembly by allowing two parts to be linked using a single reference point. Different types of joints cater to various articulations required in an assembly, and these can be managed within the browser window of the interface.

The Fusion 360 interface is composed of the Data Panel, Toolbar, Browser, View Cube, Timeline, Navigation Bar, and Display Bar, including Profile & Help sections. The workspace and the corresponding set of tools adapt to the module currently in use. Navigation within Fusion 360 is designed to be straightforward. Users can navigate the workspace using the navigation bar, the View Cube, or a combination of the keyboard and mouse inputs. These methods allow for efficient model manipulation, including orbiting, panning, zooming, and fitting to the screen. In general, the user interface is very similar in its look and feel to the Autodesk Inventor, making switching between two software solutions seamless for more advanced users. In general, Fusion 360 users can easily get into work with Inventor and vice versa.

Fusion 360 supports various methodologies of CAD design including top-down, bottom-up, and middle-out approaches. Its cloud-based nature ensures that users have access to models and components without being limited by their local hardware resources. Changes to models are synchronized in real-time, which is especially beneficial when working in collaborative environments. All components of an assembly are contained within a single file, with the option to link multiple files for larger assemblies.

Post-design, the CAM module of Fusion 360 can be utilized for programming cutting operations and generating toolpaths based on part geometries. This module enables the specification of cutting parameters and the export of toolpaths as G-code, which can be interpreted by machine controllers.

Fusion 360 is accessible free of charge for educational purposes and for startups generating less than a specified annual revenue. The software's compatibility with other CAD formats facilitates a smooth transition for users migrating from different systems. Fusion 360 represents Autodesk's approach to an all-encompassing product development environment. It provides a platform that encompasses multiple functions of the engineering process, enabling a unified workflow from conception to production. It is also a remarkably good solution for non-industrial uses and can be successfully implemented outside production companies, e.g., in smaller or larger healthcare facilities.

To sum it up, Fusion and Inventor are software applications that have a similar feeling and way of work. Fusion is a bit easier to start working with, both from a technical perspective and the economical perspective, as its license is cheaper and free plans are also available – so it could be a viable option for small-scale endeavors. The Inventor is a bit bigger software, with more capabilities and modules, increased automation capabilities, but also not without flaws and with some capabilities – especially regarding surface modeling or advanced simulation – much worse than in the case of more advanced CAD systems such as CATIA or NX.

4.2.3 SolidWorks

SolidWorks is an advanced solid modeling computer-aided design (CAD) and computer-aided engineering (CAE) software package and a software which is, on many levels, very similar to Autodesk Inventor and as such it has been its basic competitor for many years. Developed by Dassault Systèmes (a company also owning CATIA and 3D Experience solutions), it operates primarily on Microsoft Windows. SolidWorks is widely recognized for its solid modeling capabilities, with a primary focus on engineering design and development.

SolidWorks software is structured to provide engineers and designers with a robust suite of tools for mechanical design, simulation, and production. It is designed to offer a comprehensive and integrated solution for the entire product development workflow, encompassing the initial concept, detailed design, validation, communication, and manufacturing. In this aspect, it is very similar to Inventor, although many things are realized quite differently.

In terms of the interface, the software is organized into separate modules, each dedicated to different aspects of the design process. These include Part, Assembly, and Drawing environments, along with additional tools for simulation, motion, and design validation.

The SolidWorks interface is comprised of the Command Manager, Feature Manager design tree, Property Manager, Configuration Manager, and Task Panel, among other elements. These components are integral to the software's user interface, providing quick access to tools, features, and configuration options. In general, SolidWorks user interface is viewed by many as very intuitive, as compared to extensive capabilities of this software. Navigational ease is a hallmark of SolidWorks, with a user-friendly graphical interface that allows for intuitive model manipulation. Users can perform actions such as rotate, zoom, and pan using mouse gestures, keyboard shortcuts, or the standard view manipulation tools provided within the software.

SolidWorks supports diverse design methodologies, including both top-down and bottom-up design approaches. The software's parametric and feature-based approach allows for the creation of detailed 3D models and 2D drawings. SolidWorks facilitates a seamless transition from the design stage to manufacturing, with capabilities that extend into generating computer-aided manufacturing (CAM) data.

In the context of assembly modeling, SolidWorks employs a "Mate" system, which is pivotal for defining the relative motion and position between individual components. This system is sophisticated, providing a wide array of mate types to simulate real-world mechanical interactions accurately. Users can manage and review these relationships through the feature manager design tree, which is a central component of the SolidWorks user interface.

Integrated within SolidWorks is a suite of simulation tools (Fig. 4.11), allowing designers to test and analyze their designs under real-world conditions. These tools enable the prediction of the behavior of components and assemblies, informing design decisions and improving product reliability. Compared with the other popular simulation tools available in other CAD systems, SolidWorks package is the most diverse and reliable (Magomedov and Sebaeva 2020).

SolidWorks offers broad compatibility with various file formats, ensuring interoperability with different systems and facilitating collaboration across diverse platforms. It is also part of the Dassault Systèmes 3D EXPERIENCE platform, which provides additional collaborative and data management capabilities.

SolidWorks is recognized for its strong performance in 3D design and engineering. With its comprehensive toolset and simulation capabilities, it continues to be a preferred solution for professionals seeking an efficient and reliable CAD/CAE platform (Kurowski 2016). Its use in design and simulation of medical, anatomically customized devices is also possible and in many aspects – easier than using Inventor, especially concerning advanced shapes and assembly tools. It also contains a basic toolset for creating intelligent models, so it can be used similarly as the Inventor – in the core of an automated design system.

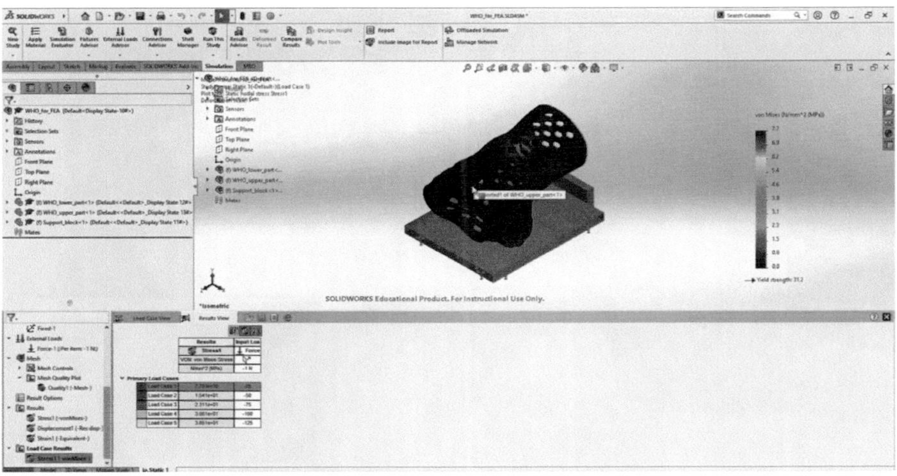

Figure 4.11 SolidWorks – simulation module.

4.2.4 CATIA

CATIA V5 is an integrated system that incorporates modules for both 2D and 3D modeling, as well as functionalities for machining process planning, product lifecycle management, numerical analysis, and the generation of digital product mock-ups. This system is designed to support a wide array of engineering tasks, facilitating the design of complex products through its comprehensive toolset (Tickoo 2018).

For surface modeling, CATIA V5 includes several modules, such as Generative Shape Design and Free Style, which provide tools for the creation of a variety of surface models. These tools are capable of producing surfaces that fulfill high-quality and aesthetic criteria, including A-class surfaces. This functionality demonstrates CATIA V5's role as a tool in engineering design, offering capabilities for a wide range of design and modeling requirements without automatic error detection, emphasizing its utility in creating detailed and complex designs. It must be underlined here, that CATIA's algorithms can handle complex shapes and geometries (especially those based on curves) much better than the Inventor or SolidWorks, by the author's experience – so for a skilled engineer it is much easier to proceed with some sophisticated design tasks using CATIA than other software, which could potentially generate errors or problems, while CATIA would not.

The software possesses the capability to import meshes (STL files) of considerable size, with the success of this operation dependent on the computing power available, particularly the amount of RAM memory. Polygon meshes can be converted into solids ("dumb" models) using the DMU Optimizer module, resulting in .CAT part files, recognized natively by the system and possible to utilize in further design stages. The Digitized Shape Editor module (Fig. 4.12) provides tools for

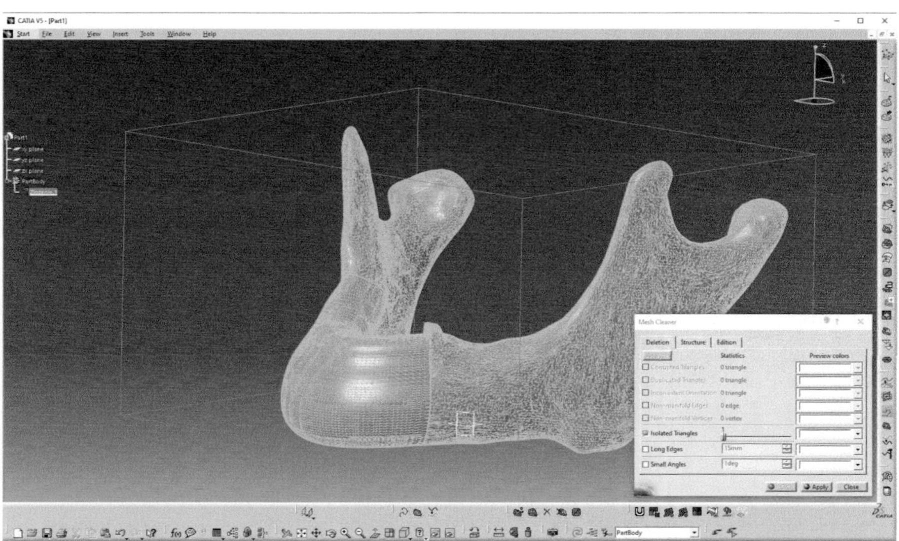

Figure 4.12 CATIA v5 – triangle mesh editor.

editing at the polygon mesh level, although the software lacks an automatic model diagnosis feature (except for basic Mesh Cleaner tool), requiring manual checks for model integrity.

CATIA V5 stands as a complex, advanced system, offering an extensive suite of tools. It goes far beyond just a CAD system – it is a full CAx solution, containing CAM, CAE, CAPP, and many other capabilities. CATIA is also customizable and programmable, containing the possibility of writing customized scripts and interfaces. It has a knowledge management module, making it possible to create fully customized and very complex intelligent models that can be then used for building advanced automated design systems, as proven in earlier work by the author's team (Górski et al. 2011).

Apart from its lack of detailed documentation and problematic approach to newer versions (including an enforced transition into a cloud solution – 3D Experience – by the Dassault Systemes company, CATIA v6 is available only as a cloud solution), as well as many smaller errors and incompatibilities, CATIA has the widest range of features, making it a great, universal tool for all sorts of anatomical design. It is unfortunately not possible to implement it in all scenarios, as there are no free versions and the license cost is very high, making the system not affordable by small or even medium-sized enterprises, let alone medical practitioners or freelancers.

4.2.5 Free CAD systems

Unlike in the world of tools for mesh processing (described in Chapter 3), most computer aided design tools for parametrized solid and surface modeling are commercial solutions, with considerably high license costs, impossible to sustain by single persons or micro enterprises. The only exception is the Fusion 360 software, which can be used for free for recreational (hobby) purposes and the license cost is not high when smaller enterprises are concerned.

Open source or freeware CAD solutions are known, but most of them are not extensively popular in professional applications. They have certain limitations and professional engineering companies rarely use them in the integrated chains of product lifecycles. However, medical use can be a different case – that is why it is important to notify their existence and consider their use if no option of using a professional CAD is available.

Some examples of freely available CAD and CAD-like solutions are described below.

1. Tinkercad

   An easy-to-use, browser-based 3D design and modeling tool aimed at beginners and educators. It allows users to create simple designs from a library of shapes and can be used for 3D printing. It is best for beginners, educational purposes, and hobbyists.

2. FreeCAD

   A parametric 3D CAD modeler suitable for designing real-life objects of any size. Its parametric modeling capability allows for easy adjustment of designs. It is an open-source solution, which offers a modular architecture, enabling users to install plugins for additional features. It is the best for intermediate users (already knowing CAD basics), mechanical engineering, and product design.

3. SketchUp Free

   A web-based version of SketchUp, a 3D modeling tool known for its ease of use. It offers an intuitive interface and is suitable for architectural design, interior design, and woodworking. It is frequently used by beginners and professionals in architecture and design, looking for a simple tool.

4. Blender

   Blender is primarily a 3D modeling and rendering software, but it also includes features for CAD and can be used universally instead of CAD for some design tasks (examples shown in Fig. 4.13). It is well-suited for creating detailed models and visual effects. Although it has a steep learning curve, it has powerful capabilities overall. It is mostly used by artists, designers, and 3D printing, game development, and animation enthusiasts.

Free CAD software often provides basic functionality suitable for hobbyists, beginners, or for educational purposes. It may lack the advanced features, precision, and complexity available in professional software. Professional CAD software offers a wide range of advanced features and tools designed for specific industries, such as mechanical engineering, architecture, and automotive. These include complex simulation, analysis capabilities, and comprehensive support for standards and documentation, which the free systems lack.

Typically, free CAD software solutions have community-based support with limited access to professional training or customer service. Development might be slower, relying on community contributions. The professional CAD software

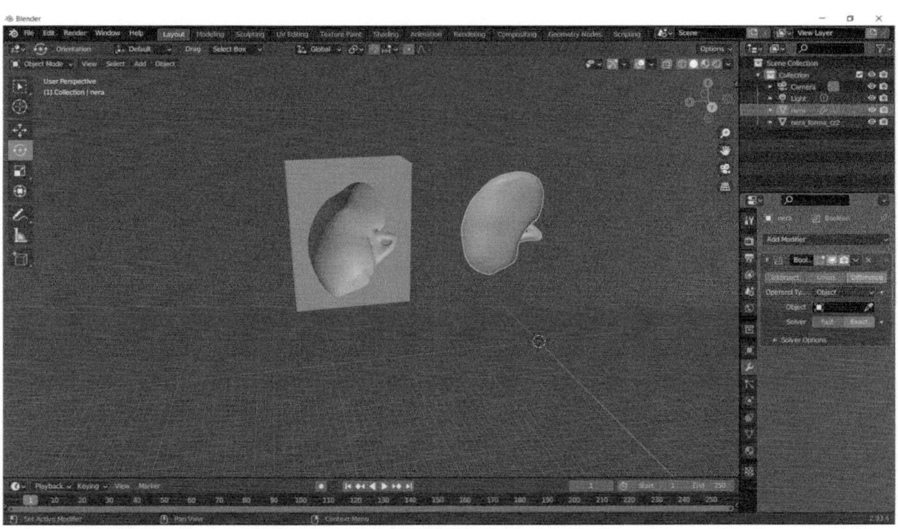

Figure 4.13 Blender – processing of a medical model (mold for kidney casting).

comes with professional support, including customer service, training resources, and regular updates to improve functionality and security – that is why the license and subscription costs are so high. In production companies, any hiccups can generate heavy losses – that is why software companies behind professional CAD software often aim at fixing errors in time of hours.

Also, it is important to notify that free CAD software is based on open-source visualization engines. That means that it may not perform as well with large or complex designs compared to professional software. Scalability can be limited, affecting collaboration on large projects. Integration with other tools and workflows can be also limited. File compatibility with standard industry formats is often supported but might not be as seamless as in professional CADs. However, certain free solutions can be a good place to start and a viable option for generating medical designs, especially for smaller companies and freelancers. Free CAD systems are viewed as key technology pushing forward the expansion of 3D printing beyond typical factory floor uses (Junk and Kuen 2016).

4.2.6 Other CAD systems and software comparison

There are numerous other professional CAD systems available in the market, that can be used for building various types of models, also for medicine, based on the human anatomy. Some selected software packages, with known uses in medicine, are listed and shortly described below.

1. AutoCAD

   Developed by Autodesk, this highly versatile and powerful software offers a wide range of features for 2D drafting (that is what it is mostly known for), detailed 3D modeling, and architectural drawing. AutoCAD is renowned for its precision and advanced tools, making it ideal for professionals – mostly in architecture, civil engineering, and construction, rarely in production engineering or manufacturing. It has an ability to handle complex projects and is compatible with various file formats. Rarely used for medical purposes mostly due to other Autodesk software doing it better (Inventor, Fusion), but there are no contraindications against it either.

2. NX CAD

   Developed by Siemens and formerly known as Unigraphics, NX CAD is a comprehensive solution for product design, engineering, and manufacturing. It is recognized for its advanced capabilities in parametric and direct modeling, simulation, and manufacturing solutions. NX CAD supports the entire product development process from concept design to engineering and manufacturing, making it a useful tool for large companies. It has integration with digital twin technology and additive manufacturing processes. Its uses in medical engineering are known (Buonamici et al. 2020), but in general these are rather implemented by large industry players or researchers than for smaller enterprises.

3. Solid Edge

   A versatile parametric 3D CAD modeling software also developed by Siemens and a direct competitor to Inventor and SolidWorks. Solid Edge is designed to cater to the needs of smaller and medium-sized manufacturing companies, especially in areas such as assembly design, part modeling, drafting, and simulation. With its synchronous technology, Solid Edge combines the speed and simplicity of direct modeling with the flexibility and control of parametric design (McGill University 2024).

4. PTC Creo

   A comprehensive, yet less known CAD solution that provides a wide array of capabilities for product design and development. Developed by PTC, Creo includes parametric and direct modeling, 3D printing capabilities, and advanced simulation and analysis tools. Creo is scalable and has an extensive suite of features. It is also not targeted to a particular industry, with known implementations ranging from manufacturing and automotive to electronics and healthcare (Torok et al. 2024). It has capabilities of integration with Augmented Reality (AR) and the Internet of Things (IoT), compatible with the Industry 4.0 roadmap.

5. Rhinoceros 3D (Rhino)

   A highly flexible and powerful 3D modeling software known for its mathematical precision in creating and editing curves, surfaces, and solids. Rhino is favored for its versatility, allowing users to model complex shapes with accuracy and ease. It is widely used in architecture, industrial design, jewelry design, automotive design, and multimedia. Rhino's open SDK and support for a vast array of plugins make it highly customizable, catering to the specific needs of various design disciplines, including medical design (as demonstrated in some publications (Cazon et al. 2014)). It is best suited for designers seeking an intuitive interface and the freedom to explore complex geometries without the constraints of traditional CAD software, as the modeling methodology is quite difficult than the one used in typical parametrized CAD systems.

Popular CAD systems have similar, yet frequently different features. That is why it was decided to make a comparison between the most important features, advantages and disadvantages of particular CAD software packages and present it here (Table 4.1). Only the software known and practically used by the author and his team in design of medical products was taken into this comparison, to avoid imprecise evaluation due to lack of practical knowledge. As such, the comparison involves Autodesk Inventor, Fusion 360, SolidWorks and CATIA – however, the methodology can be expanded to other above mentioned professional CAD systems, such as Siemens NX, SolidEdge or Rhino.

Comparing with the available, popular solutions, several conclusions can be drawn. First of all, Inventor and SolidWorks are very comparable and it is only a matter of preference and details to select between these two (or three, if counting SolidEdge as the third equivalent medium-enterprise CAD). CATIA (on par with

Table 4.1 Comparison of popular CAD systems.

Feature/software	Inventor	SolidWorks	CATIA	Fusion 360
Type	Parametric, feature-based CAD/CAE		Advanced Multi-platform CAD/CAM/CAE	Cloud-based CAD/CAM/CAE
Industry focus	Product design, engineering, and manufacturing across various industries		Aerospace, automotive, industrial design, and complex systems	Product design, mechanical engineering, manufacturing
Primary use cases	Mechanical design, simulation, visualization, and documentation		Complex and advanced product development, including systems engineering and shape design	Integrated product development: CAD, CAM, and CAE
User interface	User-friendly, customizable interface, moderately high entry threshold	User-friendly, with context-sensitive toolbars and customizable interface	Complex with a higher learning curve, highly customizable	Intuitive and streamlined interface, accessible for beginners
File compatibility	Good within Autodesk, moderate with others	Overall versatile compatibility, with some limitations	Superior, extensive range of formats	Very high, versatile conversion tools
Cloud services	Limited	Possible through 3D EXPERIENCE platform		Fully integrated with cloud
CAE capabilities	Available and simple to use but limited FEA and other tools	Comprehensive simulation suite with a variety of analysis tools	Advanced simulation capabilities, including multi-physics and systems engineering	Integrated FEA, CFD, and motion studies in one package, limited range of tools
Strengths	Detailed documentation, strong part modeling, and assembly tools	Intuitive design, strong community support, and extensive third-party tools	Advanced modeling, large project management, systems engineering	Flexibility, cloud collaboration, and broad toolset for product development
Weaknesses	Less suited for industrial design and surface modeling	Less optimal for managing very large assemblies	Steep learning curve and high cost	Limited simulation tools and parametric modeling
Price	High initial cost, with maintenance subscription; educational licenses available		Premium pricing, aimed at enterprise-level solutions	Subscription-based, affordable with free options for startups/education

NX CAD, not taken into comparison) is a large, professional solution, with a very advanced suite of tools – it is safe to assume that if a certain function is not available in CATIA, it won't be available in most other CAD systems. However, the cost of purchase of license of this system effectively renders it unavailable for smaller or medium-sized companies, as well as healthcare facilities. In other words, CATIA is very useful and invaluable, yet unaffordable outside production companies where daily use justifies its purpose. On the other end of the spectrum, there is Fusion 360 – a low-cost CAD from a well-known and established company (Autodesk), which should be the first consideration for small-sized enterprises, freelancers and semi-professionals seeking something better than the free solutions.

Anyhow, depending on the size of enterprise and planned scope of design activities, a certain solution can be selected out of the plethora of the existing ones. If the scope is just a single type of medical devices (dental implants, leg orthoses, etc.) it would be probably possible to find a dedicated solution. If not, a universal CAD should be selected, in connection with a mesh processing software and (optionally) DICOM segmentation software, to form a full suite for realizing anatomically customized designs.

4.3 Design automation and intelligent CAD models

4.3.1 Basic definitions and concepts

Automation is applying technology in order to perform tasks and activities in a way that does not require any engagement from the side of a human operator (Nof 2009). The automation is usually applied to reduce labor consumption (both physical and intellectual), as well as significantly increase productivity and efficiency of certain processes. It also involves realizing activities that would be otherwise impossible to perform by a human, due to certain physical requirements (such as strength or precision), as well as intellectual ones (complexity of a problem and large amounts of data to process).

The automation in itself is a very general, umbrella term, that can be applied virtually everywhere in human professional activity, and it can have different meanings to different people (Nof 2009). It should not be confused with "automatics", also known as control engineering, which is use of sensors and actuators to control certain processes realized by machines. It also should not be confused with "autonomization", which is aimed at making certain systems or machines fully autonomous in their operation, also including the decision-making processes, without not only engagement, but also supervision of a human operator. However, it is a part of a larger trend of digital transformation, where certain forms of work or services are converted into digital ones, and human labor is replaced with work of automated, computer-based systems realizing algorithms previously followed by humans.

The automation in industry can be applied throughout the whole manufacturing process, mainly including the following stages (Fig. 4.14):

1. *Manufacturing automation* – collaboration and production robots, automated centers: assembly, connections, regular and hybrid machining, etc.
2. *Production management automation* – the use of automated IT systems to plan, coordinate and monitor production processes.
3. *Quality control automation* – automated algorithms for measuring product quality indicators at every stage of production (vision systems, automatic measurement systems).
4. *Safety automation* – automatic monitoring and control of safety in the production plant (e.g., air composition, temperature, health status).
5. *Transport and storage automation* – the use of robots and transport systems to move and store products.

In recent years, automation started to rely heavily on various artificial intelligence (AI) techniques. AI is a concept of building machines and systems that think, behave and act in a way that is similar to humans – the AI systems are able to realize tasks that normally would require human-level intelligence. AI is a separate scientific discipline and has been described in numerous books and papers (Garg 2021, Aphirakmethawong et al. 2022, Sarmah 2019). Basic AI techniques used in industry involve neural networks, machine learning including deep learning, fuzzy logic (Garg 2021) and the most recently emerged natural language models, hailed by some authors as beginning of a new era (Waisberg 2023). It all boils down to teach machines how to recognize certain patterns and make conclusions on the basis of the large sets of teaching data. Depending on the need, the AI systems can be applied to almost any type of problem-solving activities in manufacturing, including those related to design of new products – which is the main focus of this chapter. The AI systems, in general, are often classified into three main groups (Sarmah 2019):

a) *Artificial Narrow Intelligence (ANI)* – the basic and well-known AI techniques, available for decades now, where certain techniques (such as machine learning) are used for solving certain, usually repeatable problems in a narrow area;

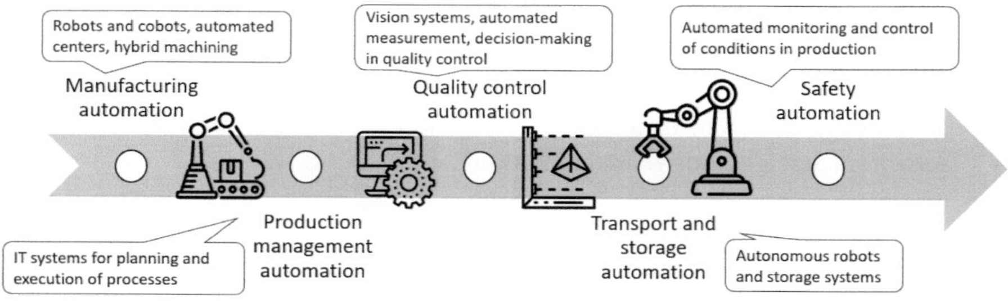

Figure 4.14 Types of industrial automation in manufacturing processes.

b) *Artificial General Intelligence (AGI)* – the relatively new invention – AI system that can be used to solve general problems, appearing as smart as a human being and possible to communicate in a similar way;

c) *Artificial Super Intelligence (ASI)* – not yet existing systems (but a possible prospect) that will be far superior to humans in terms of intelligence and problem-solving capabilities, across all fields.

AI is widely used both in industry – for optimizing and increasing effectiveness of many processes in product lifecycle, including the design – as well as in medicine, mostly for diagnostics, design of new drugs, decision-making and in medical robotics (Aphirakmethawong et al. 2022, Mintz and Brodie 2019).

In the context of industrial automation systems, aimed for intellectual work (especially in design and production planning processes), a notion of expert system is used quite often (Yusupbekov et al. 2019). An expert system is a computer software that realizes complex tasks of high intellectual requirements, to replace or aid a human expert in a specific discipline, especially during decision-making (Liao 2005, Yusupbekov et al. 2019). In the field of engineering, expert systems offer significant potential to enhance decision-making processes, optimize designs, and improve the efficiency of various operations.

Engineering expert systems are applied to tackle problems that require significant human expertise, ranging from diagnostic tasks to design optimization. The fields of civil, mechanical, electrical, and chemical engineering have seen the integration of expert systems in various capacities, such as:

1. *Design and analysis* – expert systems assist engineers in designing complex systems, components, and processes by providing recommendations and identifying optimal design parameters.
2. *Fault diagnosis and maintenance* – in manufacturing and process engineering, expert systems are employed to diagnose equipment failures and recommend maintenance procedures. By analyzing symptoms and operational data, these systems can pinpoint potential issues and suggest corrective actions, minimizing downtime and maintenance costs. This approach is known as predictive maintenance (Alvares and Gudwin 2019).
3. *Quality control* – expert systems in quality control processes analyze product characteristics, detect deviations from standards, and recommend adjustments to manufacturing processes to ensure product quality. This application is particularly prevalent in industries with high precision requirements, such as aerospace and automotive manufacturing.

Expert systems rely heavily on Knowledge Engineering (KE) – it is a field of computer science, dealing with the design, creation and application of information systems that use the collected and processed knowledge of experts to solve practical problems or make decisions (Tripathi 2011, Yusupbekov et al. 2019, Górski et al. 2016). KE utilizes various AI techniques and methods, as well as knowledge bases, ontologies, and formal languages, to construct systems that contain expert knowledge and can be operated by workers that do not possess this knowledge, effectively replacing or complementing the work of a human expert in a specific field (Zawadzki 2018, Górski et al. 2016).

Despite their benefits, the development and implementation of expert systems in engineering face challenges, such as the difficulty of capturing tacit knowledge, maintaining the currency of the knowledge base, and integrating these systems with existing workflows. It is especially evident in medical engineering and replacing traditional manufacturing techniques with digital ones, where there are no knowledge bases to draw from, to build these systems. Current research directions include the incorporation of various AI techniques to enhance the adaptability of expert systems, improving user interfaces for better interaction, and developing standards for knowledge representation in specific engineering fields. This book does not focus on knowledge engineering and methodologies of building knowledge bases – they can be found in available literature (Patalas-Maliszewska and Kłos 2018) and applied to medical cases presented here.

In this context, design automation can be defined as building expert systems, that allow elimination of routine, repetitive activities performed by humans when designing variant products. Design automation encompasses the application of computational techniques to streamline the design process, enhancing efficiency, accuracy, and innovation in engineering tasks. This type of automation unifies and replaces the sequence of these activities with an automated computer program, giving this program the ability to make decisions about the shape and functions of the product based on the parameters entered by the user. The design automation is mostly based on CAD systems and modeling operations, thus heavily relying on these systems' capabilities for introducing automated algorithms, macros, scripting, and more advanced AI-based operations (Górski et al. 2016).

The design automation, when understood as a process, leads to the creation of the so-called automated design systems. These systems use intelligent CAD models as a basis to create a complete IT solution, integrating input from a potential client or recipient of a product, with output in the form of a CAD model along with a technical documentation needed to manufacture the product (e.g., a CAM-based manufacturing program for a CNC machine or a 3D printer) (Zawadzki 2018).

Automation of design and its efficiency and potential is related to repeatability and difficulty of the design processes realized for a specific product. In general, there are three types of design, which are differently approached when introducing automation. These have been introduced in Chapter 2, so here we just briefly remind them:

a) *routine design* – a known solution is re-designed to match a new set of requirements, but fitting in an already known scope of a problem (e.g., redesigning simple stiffening hand orthosis for a new patient),
b) *innovative design* – a new solution should be prepared on the basis of existing ones, due to the set of new requirements outreaching the known scope (e.g., creating a posture correction orthosis on the basis of a simple, trauma orthosis, requiring new geometrical features in a known design),

c) *creative design* – an entirely new design is created, aimed at solving an entirely new problem, only loosely based on a given, existing set of previous solutions (e.g., designing an entirely new type of orthosis, for helping with a given condition, not based on previous designs).

These three types of design problems, applied to a customized medical product are illustrated in the Fig. 4.15. Successful application of design automation to these design tasks will depend on the level of creativity required of the automated design system, as well as size of its training data and possible scope of modifications to be introduced within a defined class of a product. As such, routine design is mostly feasible for automating (not in all cases), not requiring AI (rather simple decision tree algorithms with conditional statements and parametric formulas). Innovative design is much more difficult, requiring AI algorithms to propose new classes of solutions (Khaleel et al. 2023). Creative design is by far the most difficult to automate and the only chance of doing it is relying on AGI models – recent advancements are quite spectacular as regards text, numbers, images, and video (Hubert et al. 2024). However, generative AI solutions aimed at creating new 3D models are much less advanced at this point, although they are subjected to intensive development (Kreuzahler 2023).

Creating design automation systems is not easy and requires significant time and cost, as well as expertise, often interdisciplinary (when considering medical products, for example). That is why it is only feasible when the design process has a certain proportion to the total time and cost of the product lifecycle. It is not advisable to automate design of parts that take a reasonably short time to design or re-design manually and then are produced in massive quantities for a long time – in

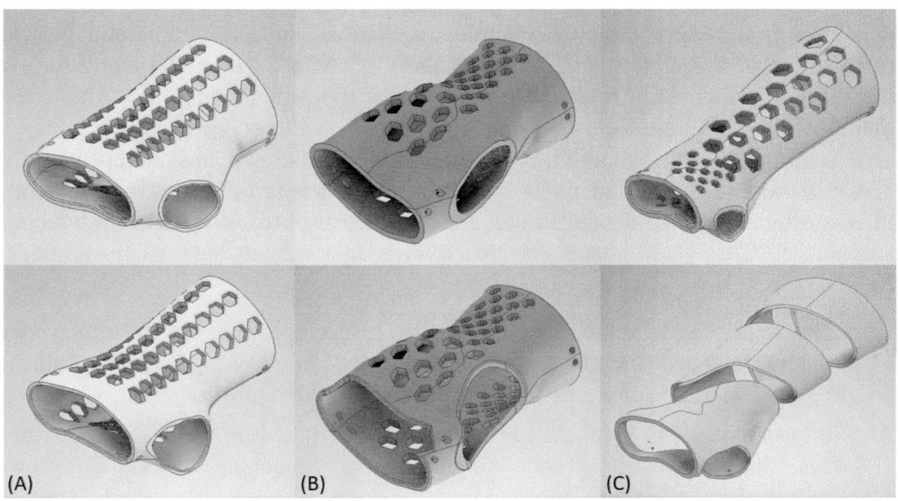

Figure 4.15 Types of design tasks, on example of hand orthosis: (A) routine: the same product, different patient; (B) innovative: the same patient, different product function and features; (C) creative: the same patient, different product.

such a scenario, process improvements and automation techniques should be rather applied to the manufacturing phase, which has lots of potential to reduce. But in the case of highly individualized products – subjected to anatomical individualization, or high customization, or both – the design of a product or its part may take a significant portion of the whole production process. In such a case, assuming certain repeatability of the design activities (a large portion of routine design activities), it would be reasonable to invest time and money into creating an automated design solution, to eliminate redundancies and routine tasks in the design phase and giving the design engineers more creative space to work on novel solutions.

The design automation mostly applies to CAD systems, with the focus put on obtaining capability of time spent on routine design activities. The automation is then mostly focused on creating intelligent (autogenerating, or "generative" (Skarka et al. 2023)) 3D models, which are described in greater detail in the next subchapter. The intelligent models can pertain to:

a) *single part models* – for products with non-complex hierarchy (such as hand orthosis visible in Fig. 4.15 or other products with just a few parts), sometimes a simple part model is enough; then, the focus is mostly put on proper parametrization of the part and division it into appropriate solids (sub-parts); the "generative" part model can be also incorporated into a "non-generative" assembly, meaning that the assembly itself is not automated, just the part shape is subjected to automated changes;
b) *assemblies* – in a case when assembly structure, as well as component selection and joining are dependent on entered parameters, coming from customization processes (e.g., out of visual product configurator), the automation must be also included in the assembly, usually requiring complex logic to control activity of certain configurations and connections between parts, based on design tables containing Boolean or integer values indicating selection of a specific module (an example of such automation can be a modular mechanical prosthesis, as shown in Fig. 4.16),
c) *flat drawings* – it is needed when the 3D model is not used directly, but rather is a base for technical documentation used in further stages of the process – then the "generative" drawings must be also created; for single parts this is a relatively non-complex task, but for the assemblies, the number of drawings and information contained in them will be varying and should be subjected to logical constraints.

In terms of model types, the most basic form of a generative 3D model is a solid model with parametrization. Of course, automation can be applied the same way to surface models, as well as wireframe models. This means that automation of hybrid modeling is supported and is relatively straightforward if the modeling techniques used in a given model are all parametric – an example is presented in Fig. 4.17, where a leg orthosis is modeled in a three-stage process, starting from a wireframe, through surface model and then finally a solid.

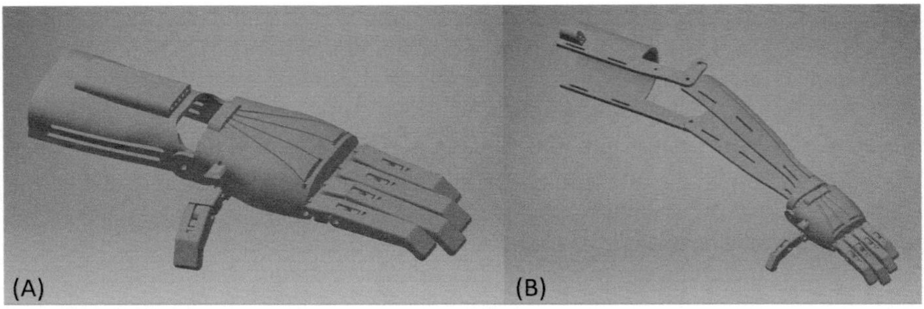

Figure 4.16 Automated assembly, on example of mechanical hand prosthesis based on UnLimbited Arm and RoboHand concepts: (A) RoboHand mode; (B) UnLimbited Arm mode, various patients' data used.

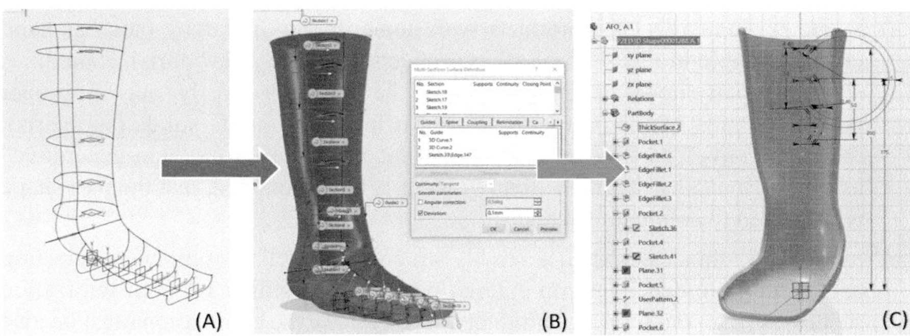

Figure 4.17 Automated hybrid modeling on example of leg orthosis: (A) wireframe model; (B) surface model; (C) solid model.

The situation gets far more complicated when hybrid mesh-solid or mesh-surface models need to be created. Automated usage of triangle mesh and getting parameters of such a mesh to create a solid or surface model is quite difficult and rarely applied in practice in generative models. It is better to first extract all the necessary parameters (point coordinates, diameters, lengths, angles) of the mesh, store them in a knowledge base or design table and then build the solid (or hybrid) model using the extracted data for parametrization. An alternative solution is to convert a mesh into the "dumb" solid or surface model and then use it in combination with parametrized modeling operations to create a complete model – but it also requires a lot of effort to build a model that would not be prone to errors when loading various size and type of meshes.

The design automation can be partial or complete. In the partial design automation, only some activities performed by the designer in the CAx system are automated, while the others are not and manual work is required to finish the model. Such models require supervision and/or modification of automatically generated designs by engineers. An example might be generating a highly personalized orthosis, where the main shell will be first automatically generated (saving lots of work to the engineer), while additional features are added manually later by the engineer, depending on the severity of a current patient's case. That type of automation is easier to reach and usually is a first stage, before complete automation is achieved. In such a way, the work of design engineers is still required to complete the design and get it into production, but the required knowledge, skills and time are much lower than before the automation.

On the other hand, there is complete design automation. In such a process we assume that there is no constructor in the process – generation is realized using only an external, friendly interface and can be operated by anybody – customers, patients, medical personnel, or salesmen (depending on various scenarios of implementation of automated design systems). Here, supervision and/or modification of the model is only achievable with the participation of recipients (customers).

The complete design automation is much harder to reach. First of all, it requires the intelligent model to be almost flawless, always generating proper configuration, regardless of entered data. As such, the models and algorithms must be foolproof and thoroughly tested – that is fully achievable in the case of machine parts (where geometries are simple and input data are usually within certain limits) than in the case of anatomical models (where geometries can vary drastically, are very complex and usually there are no hard limits or relations between various dimensions). Secondly, complete automation cannot be implemented just having the intelligent CAD model – the whole environment must be constructed around it, starting from a friendly UI (a configurator), through client database, server, and many others. At the same time, partial automation is possible to implement just on the CAx level and as such is much faster to realize.

On the other hand, complete automation, if successful, has a phenomenal potential. In theory, it is possible to completely expel the routine (or partially even innovative) design processes outside design offices and, in connection with 3D printing potential, enabling product design and production outside the manufacturing companies, totally changing the paradigm. This is important especially in medicine – joining complete design automation with cheap 3D printers and scanners, automated design and production systems could be placed anywhere – in bigger or smaller healthcare facilities, but also in schools, hotels, etc. – anywhere there might be a need for quick production of a personalized medical device (such as trauma orthosis).

In technical terms, the automation is usually applied to CAx systems. The three main automation types corresponding to main types of these systems are as following:

a) *CAD automation* – the most basic and typical type of CAx design automation, where data about the product (coming from customization and/or anatomical individualization) are used to generate a 3D model;
b) *CAM automation* – the second type of CAx design automation, where the product model and data about the process (technology, tools, requirements, etc.) are used to generate 3D models, documentation, and programs for production (e.g., customized tooling and NC programs for manufacturing);
c) *CAE automation* – the third type of CAx design automation, where the product model and operational requirements (e.g., standard strength test data, or other use conditions, e.g., temperature) are used to automatically construct simulations,

testing generated variant's capability to sustain certain loads or survive applied conditions; this type is the rarest and almost nowhere to be found in existing scientific literature.

In the manufacturing industry, especially in automotive and aeronautical branches, the design automation systems have been known for a long time and in literature there are plenty of examples of such solutions enabling almost unimaginable 95% and more reductions of design times – from weeks to hours, from hours to minutes or from weeks to minutes (Stokes 2001, Chapman and Pinfold 2001, Górski et al. 2016, Zawadzki 2018). CAM automation systems are also known, allowing for high reduction of time spent designing customized tooling or preparing programs for NC machines (Kowalski et al. 2021).

In medicine and medical engineering, use of design automation techniques is far less common. One of examples is the AutoMedPrint system prepared by the author of this book alongside a team (Górski et al. 2024c), yielding very good results, enabling reduction of time of orthotic or prosthetic design to mere minutes after loading patients' data into system. In further chapters, some technical details of AutoMedPrint system structure and operation will be presented, so that the reader can get acquainted with way of conduct with the individualized automated design in medicine.

4.3.2 Intelligent models – basic concepts

Intelligent models, also known as autogenerating models (Skarka et al. 2023, Zawadzki 2018) are CAx models (mainly CAD) enriched with knowledge about features, relationships, and rules for their selection. The autogenerating model is basically a template, that adapts under the influence of the data fed to it (one can also use a metaphor of a "shape-shifting" model). These models, in their basic state, usually do not have a "default" or "standard" shape. Instead, they fluently change their features when new sets of parameters are read from the data source. Autogenerating models represent the entire product families, not just single products – using simply a properly parametrized CAD model hooked up with a design table (stored in a spreadsheet, data base or even a simple text file), series of unique designs can be created, suited for the needs of individual recipients.

The three most important foundations of intelligent CAD models are:

a) *features* – quantitative and qualitative functions and parameters describing the structure of product elements,
b) *relations* – numerical (mathematical) and logical relations between features, determining their geometry and state,
c) *rules* – mostly conditional instructions or AI algorithms, used for selecting individual features, activating interdependencies between them and altering values of parameters.

Intelligent models lie at the heart of automated design systems and are their core components. Their most important quality is that they usually do not need to be "operated" using CAD (or CAx) methodologies – the user of such models needs solely to prepare the external file with appropriate dataset for a given variant and then the model prepares itself. If a proper level of automation is to be achieved and scripting is involved, as well as external programs for management of work of a given intelligent model, its user may even not interact with a CAD system at all, or even know that one is in use – if everything happens in the server-based mode, or at a console level, the only tangible effect for the user is generation of a new output model, ready for further processing (for example, STL model of a part, ready for preparation of 3D printing processes – and that is only if the 3D printing preparation itself is not automated).

The key difference between regular CAD models and the autogenerating, intelligent ones is that the latter contain embedded specialized knowledge, both tacit and implicit. Such knowledge is extracted from available "hard" resources, such as previous projects and their files, usage reports and scientific studies, as well as from "soft" sources, such as experience of design engineers and other specialists, extracted through methodologically performed interviews and surveys. Usually, such extraction of knowledge leads to obtaining a proper scheme, or decision tree, which is followed through by human engineers when creating a new instance of a model belonging to a selected product family, down to a single modeling operation and its parameters.

That scheme, decision tree and parameters (along with their relations and rules for their selection) are then used to create a new parametrized CAD model, where features and the whole history are not random, but meaningful – each feature and parameter has its description and purpose and could be potentially connected to each other. On a superficial level, these models are very similar – but deep down, instead of just entering numbers into fields that require them (when constructing sketches or various 3D features) – we enter links to specific parameters or formulas to calculate them on the basis of other parameters. Moreover, the whole model can also have a programming (scripting) layer, where an external code can run and alter values of parameters and the existence of certain features, depending on the same embedded knowledge. The knowledge can be stored internally, inside the model (e.g., as a sequence of 3D operations and connected parameters), or externally, inside a knowledge base, or a design table.

The autogenerating models can be built on various levels of "intelligence". In the current state of the art, five levels can be distinguished, corresponding with layers of an intelligent model. These are:

1. *Parametrization* – here, the basic "variant switching" functionality is available within the models, through easy change of interconnected parameters.
2. *Design tables* – on this level, all the important parameters are available externally – usually in a spreadsheet, often containing its own logical relationships between parameters.
3. *Decisions and sequences* – on this level, additional scripting layer is present – various conditional statements, as well as action sequences are launched from the underlaying scripts whenever dataset is changed.

4. *AI variability* – here, narrow AI techniques (such as genetic algorithms, or simple neural networks) are utilized to select proper variation, i.e., optimal values of all parameters for a given set of general requirements towards the given variant. This level is a current accepted state of the art of autogenerating models.
5. *AI generation* – this is a future possibility, already pursued by some experimental solutions, where the models are built in a quasi-creative way – on the basis of text prompts, which are used to generate the whole underlaying structure of 3D features and parameters. So far, no known software exists for creation of such models.

The first level of building an intelligent CAD model is parametrization and internal CAD automation based on forms and type series. Here, a number of parameters are named and properly linked to certain features of a part or product. Then, two things can happen. In the "type series" approach, a table can be created, with predefined amounts of variants of a given part (leaving the user a possibility to add new ones) – a predefined set of these parameters. This approach is good for parts with known variability – simple examples are springs, screws, seals, and other simple parts. By one click, the user of such a model can switch between different "recommended" configurations of the designed part (or assembly).

The "forms" approach is a bit different and more complex for the user – here, a UI is created, where various values can be entered by the user (manually, by writing in numbers, or using UI elements such as sliders or drop menus) and after entering them, the model is regenerated. An example of this approach applied to a parametrized model of a medical device is shown in Fig. 4.18.

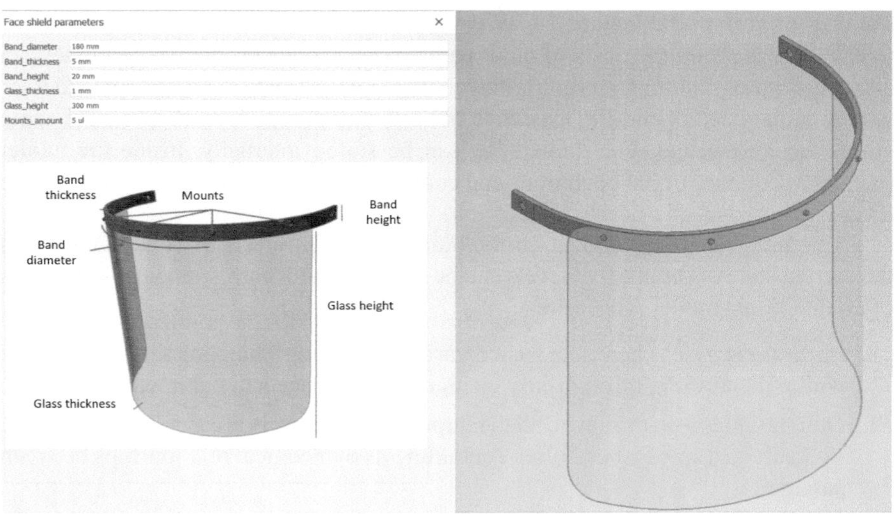

Figure 4.18 First level of intelligent models – automation by forms, on example of customizable face shield model.

The first level models can be used only for partial automation, as usually the algorithms work only inside the CAD system, requiring the operation by a user with CAD experience and access to the software. The second level of intelligence of autogenerating models is different – it is the first step towards complete automation. Here, the intelligence of the model is distinguished by use of design tables and formulas. Basically, design tables are equivalent of forms, but as the forms are used internally, the design tables are external files, able to be edited outside CAD system, usually in a non-specialized software, requiring much less experience and no need to access professional engineering systems. If this level of automation is done properly, a design table contains all information about a single "member" of a product family, effectively being its full representation. Among other, already mentioned advantages, (capability of full automation), it also allows to easily exchange data about product variants – instead of sending bulky 3D models (in native or universal formats), elegant, small-sized files can be exchanged, or stored in a cloud manner, while assuming that every engaged party has the same version of an original intelligent model linked to a given design table.

The most typical manner of storing a design table is in a spreadsheet, that can be opened in MS Excel or other software (e.g., Google Docs). The spreadsheet is often thought as a "native" format, but the design tables can also be created in other formats. The basic available formats and their pros and cons are discussed in Table 4.2.

The most popular and recommended default format for building most of autogenerating models is MS Excel. It is relatively accessible and can be easily

Table 4.2 Formats of design tables used in intelligent CAD models.

Format	Advantages	Disadvantages
plain text TXT ASC	• accessible everywhere • simple, possibly automated creation • simple editing	• no relations and formulas • lack of clarity with large amount of parameters • prone to errors
spreadsheet XLS	• good accessibility • simple editing • clarity even with big amounts of data • relations and formulas • full programmability (scripting)	• requires external software (MS Excel or similar) • automated creation is not trivial • slightly prone to errors
XML	• accessible everywhere • readability and understandability • compatible with programming languages and spreadsheets	• difficult to program • not trivial to automatically generate • no automated logical relations and formulas
database (e.g., MDB, SQL)	• clarity and comprehensibility • advanced relations between objects • compatible with various software, depending on database platform use • full scalability and programmability • can be linked directly with knowledge base	• viewing and editing requires external, licensed software • automated generation and editing are quite difficult • lots of effort needs to be put to create

edited also in free-to-use software (Google Docs). Spreadsheets are great for numbers and their relations – such a file is not only a bunch of parameters put together and linked to a CAD model (although it can also be like that). It can be stated that use of spreadsheets allows considering all the three characteristics of an intelligent model – features (reflected in numerical and non-numerically defined parameters), relations (reflected in formulas) and rules (reflected in conditional statements and, possibly, macros). The MS Excel files can be very advanced and scripting enables converting the spreadsheet into an interactive application, with buttons, forms, etc. However, most CAD software (such as Inventor or CATIA) cannot make use of scripted spreadsheets (namely, they do not read .xlsm files), so if the design table is meant to contain any scripts, in the final step of its preparation, a clean spreadsheet must be generated anew and saved.

An example of a design table, pertaining to an autogenerating model of the wrist hand orthosis (one of the basic products in the AutoMedPrint system (Górski et al. 2020b, Górski et al. 2024)), is shown in Fig. 4.19. This example is further described in greater detail in the Chapter 6.

Building design tables, as mentioned many times already, is a quite simple thing that has powerful repercussions if done properly. When the model is finished and no more innovative design tasks are realized within it, its users may not ever need to interact with the CAD system again. That is when the paradigm shift occurs, as the scheme of use of such model, and the whole design process, is changed. In medical models, customized and individualized anatomically, it has even a deeper meaning. Design table, if equipped with a set of tools facilitating its creation on the basis of a scan (more in the next subchapter), allows a different approach, requiring much less competence and experience for most cases. Instead of manual work with the scans (or medical imagery) and then tedious 3D modeling using various techniques of solid, surface or wireframe modeling, we just generate the design table on the basis of the given patient's data and go on with the project, just checking if the generation was realized properly. This difference between the traditional CAD approach and

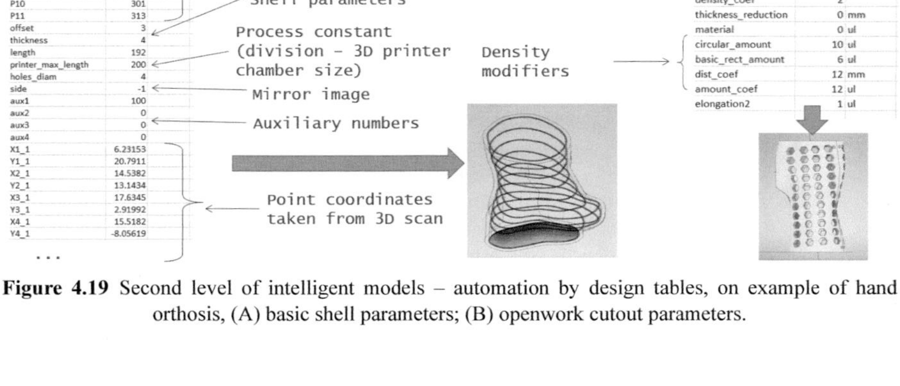

Figure 4.19 Second level of intelligent models – automation by design tables, on example of hand orthosis, (A) basic shell parameters; (B) openwork cutout parameters.

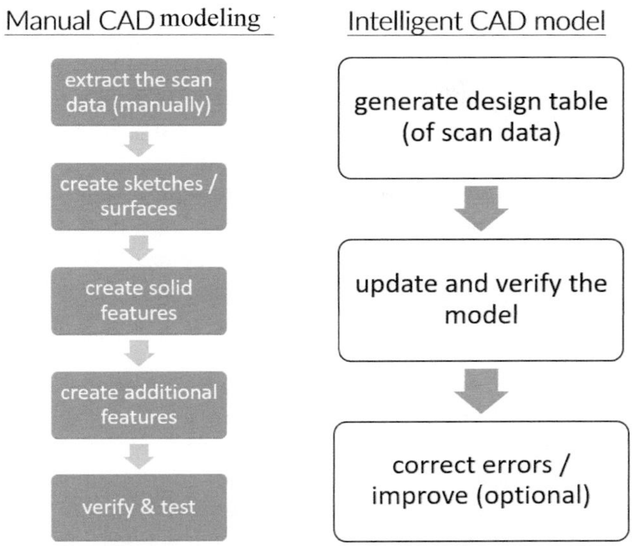

Figure 4.20 Paradigm shift – difference between manual CAD modeling and use of level 2 intelligent CAD models.

the intelligent model approach (starting from intelligence level 2 – design tables) is presented in Fig. 4.20.

However, level 2 models, with just plain design tables do not allow to make full use of capabilities of intelligent models. They are easy to create (no programming required) and very suitable for products of low structural complexity (without assembly hierarchies), such as trauma orthoses, or simply shaped implants. In the case of modular, configurable devices with multitudes of variants and wide decision trees, logic stored just in Excel tables may not be enough. In particular, it is very difficult to create programmable assemblies using design tables alone.

Level 3 of intelligent models introduces deeper complexity in relations, as well as sequences and logical implications deciding about the number and type of design operations included in the model tree, although it requires a higher skillset to implement in a proper manner. In level 2 solutions, all the magic is introduced in the design table, which enables a very basic change of paradigm, as discussed above. However, level 3 models take this approach far beyond – of course, also relying on design tables, but more sophisticated rules and restraints of model construction are embedded into the CAD models themselves, usually in form of a code. Typically, event-driven programming is applied, when certain events happening in the model (on the most basic level – opening the model or activating its parts) are used to construct functions, that allow sequential and conditional operations – most notably, the *if... then* statements that enable controlling activity (visibility) of certain components or features, based on parameters or states of other components and features. The code can be used to directly operate on the model parameters, adding, modifying, removing, and adjusting different parametrized features. All the good stuff available in modern programming languages is available – for example, Autodesk Inventor uses Visual Basic for Applications for these scripts (in this CAD, the whole ecosystem is known as iLogic), so there could be loops, switch instructions, file reading and saving, interacting with Excel sheets, complex mathematics, and many other things.

The one thing that also is available when introducing programming to the intelligent models is sequential model building – it allows adding new components and parts to an assembly, starting from a clean sheet of a model. On this level of model intelligence, another classification can be introduced – between configurable and additive models. The configurable models always have "some" variant applied into them – they can be reconfigured by the means of design tables and logic, but they cannot be "clean". The additive models can start out empty and then, based on parameters extracted from user's choices (customization) or measurements (individualization) can be filled with various elements.

A typical example of the difference between configurable and additive intelligent models is illustrated in Fig. 4.21. A prosthesis can be reconfigured totally – each of its main parts can be switched into something different, but it will always be a prosthetic limb consisting of functional parts with certain roles – socket, end effector and the middle part (usually forearm). On the other hand, additive models are usually applied when there is a need to customize a space – like a surgery room. An example of such customization is also shown in Fig. 4.21. The work starts with a blank space, which is gradually filled up with objects and configured. Such an approach is rarely useful in anatomical models, but can still be incorporated if the need arises.

Level 3 models also enable relatively easy integration between visual product configurators (e.g., in virtual reality) and CAD models. The selections made by the clients (patients) may be translated into numbers stored in design tables, which in turn might be fed to conditional instructions, automatically transforming the whole assemblies basing on what was selected in a product configurator. As such, only on level 3 both customization, individualization, as well as personalization may be achieved, as more sophisticated algorithms can be used to introduce features requested by the potential clients.

An example of the level 3 model is upper limb hand prosthesis, built as a part of AutoMedPrint system under supervision of the author and described in previous

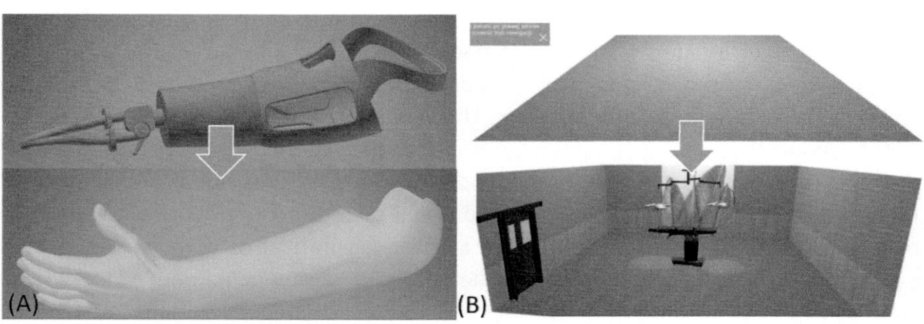

Figure 4.21 Configurable (A) and additive (B) intelligent models, on example of a prosthetic limb and an operating room.

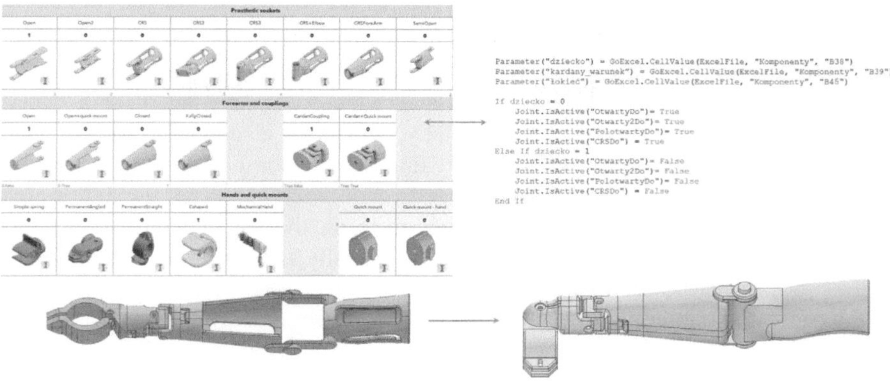

Figure 4.22 Intelligent model of level 3, on example of modular prosthetic limb – relations between design table and scripts, as well as their implications on final product shape.

publications (Górski et al. 2023a). The basic idea behind the prosthesis model is shown in Fig. 4.22. The configuration values stored in Excel table are extracted and used to form logical conditions, deciding of activity of certain components in an assembly and their joints. The design table is created by the means of a visual configurator, along with an automated algorithm of 3D scan processing. This model is described in greater detail also in Chapter 6.

The further levels of intelligent models are more speculative ones, as research in that area has only recently advanced beyond purely theoretical approach. Level 4, according to the presented classification, incorporates models that have all the features of level 3 and below and additionally they make use of artificial intelligence algorithms, such as neural networks, machine learning or cognitive computing. The range of applications of AI techniques in scope of development of a new variant of an anatomically individualized models for medical purposes is very wide.

First of all, AI can be used for diagnostics and data processing. The medical imaging can be processed by specially trained neural networks that are able to recognize the pathological tissue (e.g., a cancerous tumor) and suggest an appropriate solution (Najjar 2023). The AI-based diagnosis is an already known process, still under development by many researchers worldwide (Hafizović et al. 2021). AI can be also used for image processing, once the diagnostic part is over – the algorithms can answer the questions such as what part of the body is analyzed, what is the type of tissue and degree of damage it sustained (e.g., as a result of trauma), allowing to select the features of a designed medical device (e.g., orthosis). Properly trained AI can also extract the most important parameters and dimensions out of the analyzed data set containing anthropometric measurements (Najjar 2023, Koohi-Moghadam and Bae 2023).

Then, AI can be used in various forms to aid and automate the design process itself. Genetic algorithms can be used to compare different variants between themselves and select the one most fit to the set of requirements (Xue 2019). This iterative selection of the variant can take into account the criteria of strength, weight, fit, etc. The neural networks can also be used in connection with a proper knowledge base to propose a proper shape of a given product, e.g., an implant or a dental prosthesis (Revilla-León et al. 2023). Finally, AI can be also used to improve the manufacturing processes, e.g., to iteratively find the best combination of 3D printing process parameters based on a set of general requirements towards the 3D printed product (Rojek et al. 2021). What is common to all the autogenerating solutions belonging to the level 4 is that AI is used only partially, narrowly, at selected stages of the process and the whole process is not strictly governed or realized by any form of the general AI model.

The highest form of intelligent, autogenerating 3D models is still yet to come in form of level 5 solutions – the AGI solutions, based on language models (LLMs) generating whole 3D models on the basis of text prompts from a user. The generative AI models are developing really fast and, as of 2024, are already widespread and will continue to develop and find their uses in more and more places, including medicine and medical design (Thirunavukarasu et al. 2023). So far, attempts at creating generative artificial intelligence engines for creation of 3D models have only been launched without largely spectacular results, but there are certain platforms that allow such activities, especially targeted at creating unique models for creative videos, gaming, etc. (State 2023).

In the most advanced form, it would be expected from such generative AI tools to work in a following way – similar to interaction with various modern generative AIs such as ChatGPT or MidJourney:

1. *Taking a prompt* – the prompt (inquiry) should be constructed in a way to unequivocally define the product that should be an effect of work of a given AI model, its features, structure, and target. The prompt should be supplemented with 3D data (3D scans or medical imaging). It would most likely be a command from the user formulated in natural language, e.g., "lightweight two-piece brace for the treatment of wrist fracture for a 10-year-old child".
2. *Automated generation* – here the user would wait for one or several proposed shapes of a product based on the prompt. Probably, due to high requirements and need to process 3D data, the initial proposals would be of low resolution and quality, and it would be up to user to decide which proposition should be further processed to obtain its hi-quality, fully detailed version.
3. *Validation* – here user checks the obtained model, decides to continue generating variants or changes the query.

What is quite important to ponder here is if the generated models would be parametrized solutions or simply the mesh or voxel models, composed of smaller structural elements, just like the text or image AI models generate their content by puzzling together very small elements. Probably the first versions of generative intelligent models will export their results in form of STL or OBJ files (or similar), while more advanced, further solutions will be able to construct the whole, elaborate, sophisticated CAD parametric models – it will be all dependent on the size of data sets that will be used to train the neural networks behind these AI engines.

Anyhow, for now, level 5 solutions are unavailable for medical design and hopefully they will be introduced to work in the scope of the next decade. Solutions from level 4 are more and more popular and partial solutions, like AI-based iterative variant selection or medical image processing are already accessible and can be incorporated in professional work of medical engineers. The levels 1–3 remain the industry standard, state-of-the-art regarding techniques that can be relatively easily and quickly incorporated into CAD models to make them intelligent and enable automated generation. However, their popularity in machine industry is not yet followed by usage in medical engineering and recent scientific work of the author is mostly directed at changing it, as described in this and following chapters.

4.3.3 Methodology of creating of intelligent models

In the development of autogenerating models for designing anatomically individualized products through 3D printing techniques, one fundamental premise of the methodology proposed by the author is the incorporation of Knowledge Engineering (KE) framework tools. Knowledge Engineering is an interdisciplinary field that addresses the complexities involved in the acquisition, representation, and application of knowledge within computer systems. It encompasses the development of knowledge bases and employs semantic technologies to enable computer systems to process knowledge efficiently (Schreiber 2007).

The appropriation and documentation of knowledge in a structured knowledge base are critical for the design, production, and successful operation of technical products. This necessitates a meticulous process of knowledge acquisition and organization. The principal reservoirs of knowledge within engineering endeavors to encompass a wide range of sources, including technical documentation (such as drawings and diagrams), industry standards and catalogs, insights derived from literature reviews, outcomes of prior projects, extensive libraries of documents, books, and articles, personal notes, and the specialized, often tacit knowledge of experts, as well as reports from patent offices and research institutes. Additionally, feedback from customers and suppliers plays a crucial role in enriching the knowledge base.

The construction of Knowledge-Based Systems (KBS) faces a big challenge in the accurate recording and organization of knowledge. The quest for effective methodologies to support the creation and maintenance of KBS has sparked significant interest by various researchers, working closely with the industry, resulting in a variety of methodological approaches over the past two decades. Notably, the CommonKADS methodology (Schreiber et al. in 1999), offers a broad framework for knowledge acquisition and documentation structuring. Building on the foundation laid by CommonKADS, the MOKA (Methodology and software tools Oriented to Knowledge-based engineering Applications) framework was developed, as detailed by Stokes (2001).

MOKA emerged from the collaborative efforts of academic researchers and industry professionals from the aviation and automotive sectors. Its formulation aimed to address several key objectives, as identified by Oldham and colleagues (Oldham et al. 1998). These objectives include the reduction of risks, time, and costs associated with the development of knowledge-based engineering (KBE) solutions by approximately 20–25%, the provision of support for the ongoing development and maintenance of KBE applications, the preparation of tools to facilitate the method's application, and the promotion of international standardization in KBE practices.

Through the synthesis of these methodologies, the methodology under discussion seeks to leverage the collective intelligence encapsulated within diverse knowledge sources to streamline the design and manufacturing of personalized products using advanced 3D printing technology. This approach underscores the importance of a systematic and structured knowledge management framework in fostering innovation and efficiency in the engineering domain.

The MOKA, itself, can be used as a framework for building many other methodologies of building various knowledge-based systems. Previous research by the author included using MOKA assumptions to create a methodology of building professional, industrial Virtual Reality (VR) applications with open access to knowledge embedded into them, enabling easy changes of contents of a given application without need of programming skills (Górski 2017). Most notably, it resulted in building a couple of advanced industrial VR systems (Maik et al. 2022, Górski et al. 2016b) – so it has been successfully proven that the basics of MOKA can be used for a wide range of purposes.

Here, a more basic usage of MOKA is proposed. The methodology of building the intelligent CAD models of anatomically shaped products for medicine is based on the four basic stages of work:

1. Learning the requirements and needs and answering if it is feasible and possible to build a KBE solution within a defined set of constraints.
2. Acquiring and processing knowledge necessary to create an intelligent model and define the assumptions behind its practical use (is it a part of automated design system or not).
3. Implementing the gathered knowledge to obtain the intelligent model, using a specific anatomical data set.
4. Verifying and validating the obtained model with another, realistic dataset, usually resulting in a number of iterative cycles of improvement of the created KBE solution.

These translate to mainly six stages of more strictly defined activities realized within the MOKA methodology. These six stages are:

1. *Identification* – which answers basic questions about the intelligent model.
2. *Justification* – which answers if it is a viable option to implement the intelligent model in a particular case.
3. *Knowledge acquisition* – focused on getting as much (unordered) knowledge coming from both tacit and explicit sources.
4. *Knowledge formalization* – aimed at making the knowledge organized, structured and ready to implement.

5. *Implementation* – building the model in CAD environment.
6. *Validation* – testing if the model works and if it generates useful data, to be used within the design process.

How to implement these stages in a typical scenario of building an intelligent CAD model for the machine industry (non-medical, non-anatomical one) is widely described in available literature, starting from MOKAs creators (Stokes 2001), through other notable implementations, such as (Skarka 2007) as well as publication of the team of the author (Górski et al. 2016, Zawadzki 2018). Here, only crucial things will be described, important from the viewpoint of anatomical models, not to fill the whole book with too much details.

Identification is the very first stage of development and basically it starts with a set of questions, which should be answered as carefully and thoughtfully as possible. These questions are:

1. What is the product and its main qualities and who is using it?
2. How is it designed/manufactured now and who does it?
3. What is the largest bottleneck in the design or manufacturing, that should be reduced?

At the stage of identification, it is most important to define the target – so the recipients of a given product – and the problems of the current design and manufacturing process of the product. In the case of anatomically customized medical devices, the recipient is usually a patient (sometimes – a doctor – if we consider anatomical surgical aids) and the bottleneck is usually a long time of waiting and high price, related to often non-digital methods of creation of these products. However, although the general answers may be similar for most classes of anatomically shaped devices, the details will greatly vary and they will have a substantial meaning in the further phases of intelligent model development. It must be answered, first and foremost:

a) what is the product made of and which of its parts (areas) are anatomically adjusted to the patient?
b) what are the customization and personalization possibilities and range?
c) how high anatomical individualization is required?
d) who has the most tacit knowledge about the product? What is the most important aspect of that knowledge, difficult to translate into explicit knowledge?
e) are explicit knowledge sources available for the product?

The more detailed answers will be given at this stage, the better for the viability of building useful intelligent models. The course of action in getting these answers can include using Design Thinking methodologies (Zupan et al. 2022), namely empathizing with the participants of the whole process of development of a given medical device. The examples of process of identification can be as following:

1. creating a sketch of an idea of the intelligent model and its potential usage,
2. defining potential stakeholders (patients of a specific group, doctors, technicians, physiotherapists, etc.) and their needs and problems related to the product,
3. realizing deep interviews with the interested parties, mostly about current state of the art, meanwhile discussing the new idea with them,
4. selecting the most prominent problems and the most important target group to focus on,
5. making a final decision about continuation of work and its possible directions.

The process might obviously be realized in a totally different manner – here is just an example straight of the author's own experience while building the AutoMedPrint system and other solutions. Regardless of how it is done, identification brings answers to some questions and often can result in ceasing further actions whatsoever, mostly by lack of capabilities to improve the situation (e.g., automation is too difficult or it is apparent already that it would be too labor-consuming) or by lack of a distinctly defined problem that the intelligent model would possibly solve.

If the identification stage ends with decision of continuation, there comes the justification stage. It is another rather analytical activity, where we ponder various scenarios of building and implementing a given KBE solution (here, an intelligent model of a medical device). The following activities should be realized:

1. Identification of potential sources of both medical and technical knowledge and its possible forms.
2. Assigning human resources to build the application – building a team of interdisciplinary competences.
3. Determining the hardware (manufacturing) and software (CAD) parts of the solution necessary for its construction in accordance with the constraints set at the identification level.
4. Marking the time frame for building and implementing the application.
5. Risk calculation, final decision to build and implement the model.

The main problems at this stage, characteristic for medical automation, require answering the basic technical and medical problems. It must be estimated how to automate (or if it is even possible) the collection and processing of anatomical data. Also, it must be answered how to include anatomical data in the CAD model of a product part/assembly. Going deeper and more technically, it must be also answered how to define classes, relations, and parameters in autogenerating models in relation to diverse cases of anatomy and pathophysiology. All this will translate into estimations of working hours of specialists that will be required to implement the model, as well as risk and potential unknowns (if there are more unknown, uncertain stages in the future, planned process of automated design, the risk goes higher). Here, various techniques of calculation and risk estimation might be brought to table, to obtain a highly believable prediction of how the process of implementation of a given intelligent model will go and how far it can be taken, with what results, both in terms of economic and medical success (bringing properly working devices to patients by a reduced price and in shorter time).

Finally, summarizing both the identification and the justification stages, there comes a decision. This decision is a complex one, requiring to answer not only a question whether to build an intelligent model, but also what type (level) and making a decision if the complete automation will be pursued, resulting in building also an automated design system (which has certain implications and is a complex endeavor), or not – resulting in building and implementing only the intelligent model. Figure 4.23 presents a very basic and rough decision tree that illustrates the mind processes that must be realized in order to decide the initial course of the operations (which might be later altered, depending on the results and requirements, like upgrading the model from level 2 to 3 or from 3 to 4).

After justification, there comes knowledge acquisition. In the case of design of a medical product, often worn by patient, customized anatomically on the basis of anthropometric measurements, this knowledge will be deeply interdisciplinary and varying greatly from source to source. That is why the acquisition process will be much different to the processes performed in production companies and it will require much more empathy and soft skills, as opposed to mostly hard, technical knowledge required when gathering knowledge related strictly to building of mechanical devices (not aimed to use in medicine).

In general, the following knowledge sources may be distinguished:

1. Available scientific and technical literature:
    - textbooks and popular science books,
    - clinical case studies,

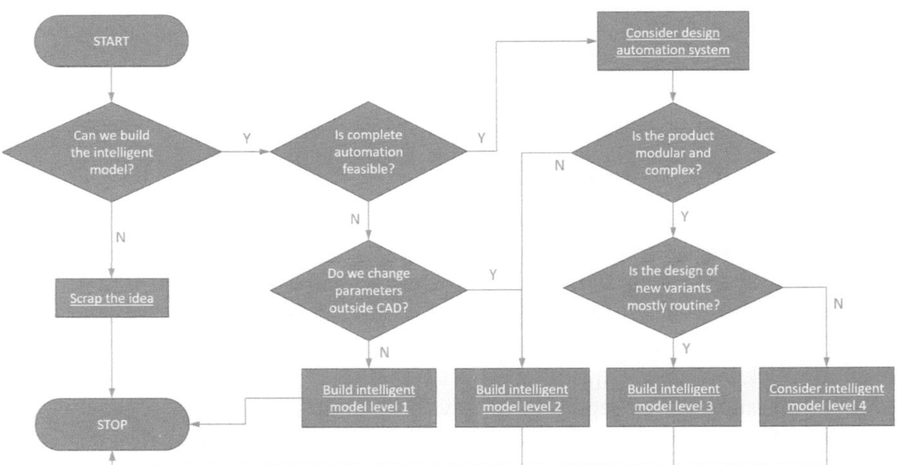

Figure 4.23 Justification stage basic block diagram – making decisions about the types of intelligent models.

- cohort studies,
- technological studies,
- standards and acts of law,
- interdisciplinary studies and others.

2. Current product documentation:
   - 3D models,
   - drawings,
   - schemes,
   - other technical documentation.
3. Current process documentation:
   - material and biomaterial characteristics,
   - manufacturing processes and their medical compliance,
   - assembly and post-processing instructions,
   - fitting and testing instructions and standards.
4. Current usage documentation:
   - instructions of use,
   - statistics of use and failures,
   - patient feedback.
5. Tacit knowledge, gathered by empathizing:
   - doctors,
   - design engineers,
   - process engineers and technicians,
   - patients,
   - caretakers (e.g., parents),
   - physiotherapists,
   - medical technicians,
   - other medical personnel.

The processes used to acquire knowledge in a technical environment are the domain of knowledge engineering and are widely described in literature, in addition to the items already mentioned, also in (Schreiber 2012, Kendal and Creen 2007). The medical knowledge acquisition processes are also well known (Lanzola et al. 1995). For this reason, these aspects will not be discussed in detail here. The end result of the knowledge acquisition process is a digital knowledge base (in various forms), consisting of files representing its various forms and sources. The knowledge base should contain an index so that it can be easily searched and retrieved specific Sunits, as well as to find out whether it is complete or requires supplementation. The informally recorded knowledge base should be verified in terms of completeness and clarity of the record, as well as redundancy of the data presented therein (Górski et al. 2019). Often, the informal model assumes use of notions such as MOKA's ICARE forms (Illustration, Constraints, Activities, Rules, and Entities) (Zawadzki 2018, Silesian University of Technology 2007). Then, the next stage should be implemented, which is crucial from the point of view of the implementation, i.e., formalization of knowledge.

Formalization of knowledge is a process in which the knowledge collected at the previous stage, previously informal, becomes structured, quantified, and fully ordered, capable of being understood (read, written, processed) by computer algorithms (Górski 2017). In the typical MOKA approach, a so-called formal knowledge model is being built, usually with the use of formal languages (UML or specially profiled MML). The formal model in MOKA is the structure of the knowledge base, the representation of which is intended to be understandable to humans and at the same time processed by the computer system. The structure of the database is permanently defined in terms of its main elements and consists of the product model and the process model. The product model is the basis for further information, which includes defined views of the product structure, which constitute separate namespaces and MML language packages. These are the following views: structural, functional, behavioral, technological, and representational. Views and models of these views allow for the classification of knowledge in the knowledge base (Silesian University of Technology 2007).

In building the intelligent models for medicine, the full formalization process with use of UML language variations and MOKA recommendations is not always necessary. Usually a simpler representation of the model's structure can be created, using UML diagrams, as presented in Fig. 4.24. All the parts, elements, assemblies,

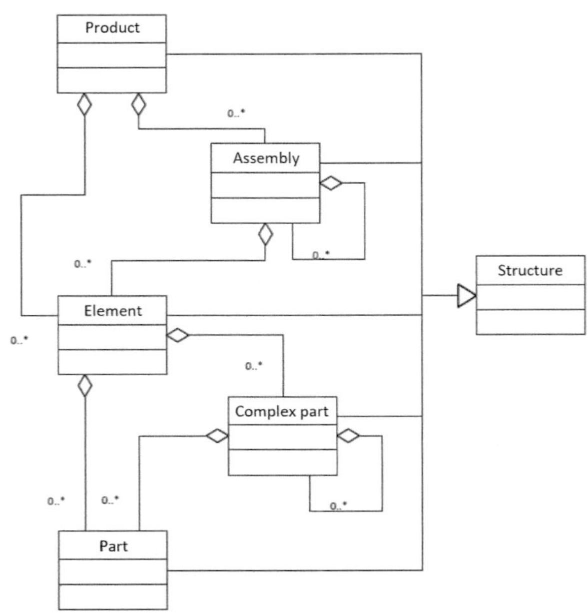

Figure 4.24 Formalization of product structure using UML (based on (Zawadzki 2016)).

and subassemblies are described and linked in the structure, helping to visualize mutual interconnections and dependencies, resulting straight out of the knowledge gathered in the previous stage.

The formalization also includes building full block diagrams of the model history tree, showing the full decision process taken by design engineers while preparing a given model. Usually, on the basis of many different processes implemented separately by different design engineers (who have different workflows in CAD systems, resulting from individual preferences), a single unified process is created and described. The full formal description of the CAD model is similar to the basic properties of intelligent models – it requires to unequivocally define the following:

1. *features of the model* – step-by-step operations needed to create a given shape (e.g., sketches, 3D operations, etc.),
2. *parameters* – numerical or logical values defining how the features are described (e.g., point coordinates, lengths, angles, geometrical constraints such as perpendicularity or tangency),
3. *relations* – mathematical formulas linking together parameters of various features (for instance, wall thickness divided by two can be used as hole diameter),
4. *rules* – conditional statements, used for selection of given set of features, parts and assemblies based on values of parameters, to form a final model structure (for instance, if length is higher than a certain value, divide the part into two solids).

If the model belongs to lower levels, some of these will not be necessary. However, even if it was decided at an earlier stage that the built model will be of level 1 or 2 (technically not requiring rules, with minimal amount of relations), it can still be made different at the formalization stage, where often it turns out of the acquired knowledge that additional logic is necessary in order to create a universal, fitting solution, potentially working for most recipients. If the model would belong to level 4, at this stage the future AI algorithm must be also fully defined and described formally, stating specific technologies used, sizes of training data and all the parameters and rules needed for the AI to be created and implemented within the model.

When the formalization stage is finished, the formal model is ready for implementation. The power of the formal model is that it is unequivocal and universally understood both by computer systems and by human users. That is why knowledge acquisition and formalization can be realized by a different specialist (or team) than the implementation itself and the formal model is then just a method of communication between a knowledge engineer and a CAD expert building a model in a selected software. The formal models are usually not suited for any specific, particular CAD – they can be implemented in any software that has the designated capabilities.

Implementation consists in the actual creation of an intelligent CAD model, based on all previous assumptions. It is a long, thoughtful, labor consuming and highly demanding process, very different in approach than regular CAD modeling. If we create a standard CAD model of an anatomically individualized product, we usually focus on fulfilling a specific purpose, e.g., giving a specific patient a product that will solve one particular problem. Building intelligent models requires much more than that – all the features must be thought of as universal and convertible for different geometries (coming from anthropometric measurements), while all the parameters must be somehow linked between each other and also available for modification outside the standard CAD operations (by forms or design tables). That is why it is important to carefully prepare the formal models, to avoid misinterpretation.

The intelligent CAD models can be created from scratch, or they can be based on existing, already verified designs, which will be "converted" from standard to intelligent models by careful description of existing parameters, remapping them so that they are linked (when necessary) by mathematical formulas and re-creating the connections in assembly to be governed by a specific set of rules. In the case of conversion (not building from scratch) it is very important to carefully go through the whole model and not leave any parameters unattended (i.e., hard-coded), because at a later stage of validation, such unlinked, undescribed parameters may generate heavy errors. An example might be a mounting hole that should gradually scale with the size of a prosthetic hand – if we leave a hard-coded dimension of that hole, it will be too large for small-sized hands and too small for large-sized hands. A rule should be established (possibly in form of mathematical function, or a switch conditional statement), linking a discrete amount of the hole diameters (3 or 4 sizes) to a specific diameter of a hand. Such situations can be encountered multiple (dozens) of times in a single model, depending on its complexity.

4.3.4 Automated design systems

Automated Design Systems (ADSs) are KBE systems belonging to the class of intelligent (expert) systems, in which expert knowledge is implemented in the form of a computer algorithm, enabling the product design process to be carried out without the need to engage qualified designers (or other specialists). As mentioned in Chapter 2, ADSs are a basic requirement if mass customization strategy is to be successfully implemented in a given company. They are also a way to achieve complete, or almost complete automation in the routine design processes, but their scope can get beyond that, influencing also production preparation and production itself (Zawadzki 2018), as well as enabling automation of some customer-producer relations (Scott 2018). However, building them is a highly demanding investment, taking much time and effort, with considerable risks on the way – so it must be very carefully considered.

The motivations for building design automation systems, as opposed to just singular implementations of intelligent CAD models to aid the work of design engineers, might be as following:

- "bottleneck" in design (relatively fast manufacturing, but slow design of individualized products),
- too many variants of products, mess in documentation (problems with accessing knowledge of historical product implementations),

- lack of standardized methodology, leading to problems with product functionality or re-creating functions already implemented in the past,
- no closed configuration range, leading to too much freedom for customers in choosing their variants,
- highly individualized products, lack of qualified design staff to perform complete, proper designs in an acceptable time (economically wise),
- need for interdisciplinary knowledge for designing certain cases, with simultaneous lack of possibility of continuous employment of a field specialist on a permanent basis.

These motivations can be roughly presented as four possible scenarios from the perspective of a given production enterprise:

1. "We want more!". The designers work well, the designs are correct and the products have high quality, but the whole process is too slow and too routine.
2. "We go for new!". The company wants to produce an entirely new product to expand its portfolio, but it lacks full know-how, or the traditional know-how is not applicable in given production conditions (e.g., company with a farm of 3D printers wants to produce orthoses – which traditionally are shaped through manual labor with thermoplastic materials in sheets).
3. "Customers want more!". Inflated customer requirements result in too many individual projects (taking much time and effort), which cannot be slashed down, due to high risk of downing the company's reputation.
4. "We want new quality!". Product functions and/or quality need to be improved, and daily routine design tasks take too much time – the design engineers do not have sufficient time for creative or innovative design.

Regardless of the motivation behind the building of a given ADS, decision about their implementation usually should be consulted with many divisions at a given company or institution, as these systems usually strive for complete automation, at least in certain cases. Such an approach introduces fundamental changes to the way a given company interacts with recipients of its products and will also alter the way the whole company works, especially at the earlier stages of product lifecycle (related to conceptualization, design, and production preparation). These systems are rarely used just internally, within a company – as motivation behind their creation is usually closely connected to demands or needs of the customers, these customers will be involved in the building and operation of ADSs, one way or another.

The structure, workflow and composition of a given ADS will be highly vary depending on the type of product, group of recipients, methods of communicating with them and methods of production. There are no two ADSs exactly identical in structure and usually building them requires a LEGO-like approach of taking various pieces and trying to get them together. In the core of any ADS lies an intelligent (autogenerating) model, or a group of models, but they are just the beginning and in themselves cannot be treated as even the simple substitutes of a complete system. Many other components are required in order to make the system work. An example

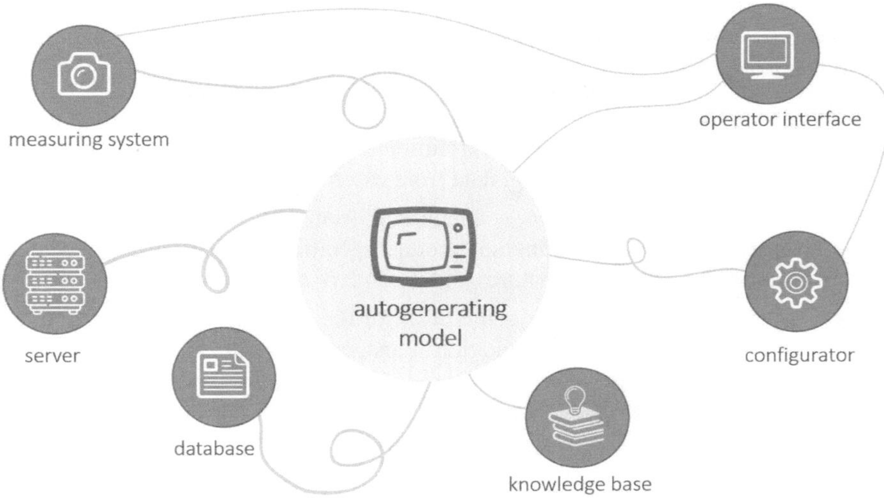

Figure 4.25 General structure of an automated design system for design of anatomically customized products.

of a basic structure of a system for design of anatomically customized devices is laid out in Fig. 4.25.

In the exemplary, general structure shown in Fig. 4.25, the autogenerating model is a central point of the system. The operator interface enables controlling both the measuring system – a source of anatomical individualization data – as well as the configurator, a source of customization data (process realized with or without the customer). The operator also initiates launching of the model itself. After it is launched, it is assumed that it automatically accesses the individualization and customization data, as well as makes use of cases stored in database and knowledge base during the process of creating a new design. The design is then stored both in local database and on a server. From the server, a manufacturing system may be allowed access to read it and go on with the production. This structure is basically true for the core implementation of the previously mentioned AutoMedPrint system developed by the author with the team in previous studies (Górski et al. 2020b).

The structure might have been composed entirely differently, using the same pieces – the server could be, for example, placed between the operator and the model. This assumes cloud operation and is possible in many modern CAD systems, or can be made possible by proper programming of mutual interfaces. Then, the configurator is also server-based and can be used to provide anthropometric measurements. Depending on the use case scenario, measurement can be realized in many different locations and its results could be simply uploaded, not requiring local connection between the measurement system and the intelligent model.

The basic components building the systems for automated design and their roles are as following:

- *autogenerating model* – the heart of the system, a model or set of models on a given level of intelligence (minimum level 2, according to previous classifications, enabling reading data from externally filled design tables),
- *configurator/ordering system* – data input from the customer, starting and queuing the "order", also containing product specification (in the case of medical products, mostly aesthetic features and secondary functions),
- *server or cloud* – support for some or all of the algorithms: CAD generation, configuration support, documentation management, queuing, etc., sometimes in between the operator and the model, in other scenarios used for storing and sending out the generated designs for production,
- *knowledge base* – a place where saved projects are stored in the case of complex logical relationships (verification/update for both new and old clients),
- *database* – a place where data for project implementation is stored (previous measurements, models, standard parts, materials, feature descriptions, etc.),
- *measurement system* – only present in the case of anatomical individualization, data entry point for shape individualization – it can be a 3D scanner, a photogrammetric system, a CT or MRI or any other anthropometric measurement system, with possibility to store the measurement in a digital form usable by the other components of the system.

It is noteworthy, that in the case of advanced manufacturing companies, utilizing integrated computer manufacturing systems and software such as various PLM solutions, as well as advanced CAx systems, these components can be built relatively quickly and integrated within company's processes without much hassle. Sometimes all it requires is introducing a proper intelligent CAD model and making it connected with the other software solutions used in the company, through proper data integration (i.e., automated file transfer between ordering system, database, server, knowledge base and the model itself). It is more difficult when working on a small scale, in a medium or small enterprise, where most of these solutions must be created anew, either as simple applications written on-site (or out sourced) or using low-cost or freeware software available. In the production of medical personalized devices it is usually the latter case, as the modern process of production of 3D printed products such as orthoses or prostheses is not pursued by big manufacturers, who stay mostly in the traditional way – it is rather implemented by smaller, innovative med-tech companies, trying to make a difference with their approach. That is why building an ADS in such conditions can be much of an effort.

An example of an ADS applied for medical production is the AutoMedPrint system, devised by the author and created together in a team at author's university, mentioned multiple times before. The system and how it works, including case studies of specific patients, is described in more detail in the Chapter 6.

4.4 CAD modeling of anatomical shapes

4.4.1 Basic definitions and concepts

When designing variant medical products with anatomical individualization, the main problems that can be encountered are as following:

1. How to take into account the patient's anatomy in a CAD design? (data source, knowledge of anatomical structures)
2. How to design new products efficiently without repeating the entire process from scratch for each patient?
3. How to standardize the product design process regardless of the type of medical case, while meeting the requirements and standards (MDR)?

These problems can be projected on the selection of an appropriate software and modeling method, as well as facilitate possible automation of design in further steps of development.

Concerning problem no. 1, the main question is what kind of data will be used for realization of the modeling activities. The following choices are usually available:

- no special data (free modeling), or measurements taken from anthropometric atlases,
- 2D data (images – photographs, X-ray scans, ultrasound scans, etc.),
- 3D triangle mesh (as a whole),
- measurements (dimensions) made on a triangle mesh,
- selected point coordinates, curves, or surfaces, extracted from a triangle mesh.

The first approach is rarely used when designing anatomically adjusted products for specific patients (involving customization or individualization). It can be utilized, when designing specific parts of devices that are otherwise individualized anatomically, e.g., taking predefined values of lengths of a human hand out of an atlas to create a mechanized prosthetic hand, or using arbitrarily assigned dimensions for some mechanical parts (such as distance tubes, couplings, springs, etc.).

In terms of the second approach, it usually involves uploading a 2D image of an appropriate scale into a CAD modeling environment and manually drawing basic curves of parametric sketches to represent the contents of a 2D image. This technique is well-known in realm of automotive design when converting artist sketches to technical designs (Kara and Shimada 2008) and it can work the same way in the medical realm. It can be used either to construct an anatomical model (3D representation of a body part) or directly a given medical device.

Subsequent approaches require having a 3D mesh consisting of triangles representing anatomical shapes (or voxels, alternatively). How such a mesh could be obtained and processed is shown in the previous chapter of this book. The way of conduct in the process of its utilization will largely depend on a given case, its uniqueness and number of predicted iterations of repeating it. The mesh-based approach (using a complete mesh as an input) is the easier, but less repetitive way, while extraction of points and creating parametrized curves out of them is more difficult, yet more prone to automation (and standardization). In both cases, manually made measurements directly on the surface of a mesh model can be helpful to complete the model.

In terms of the second and third problem, usually a parametrized approach, with step-by-step, normalized way of conduct for each processed patient is more effective in achieving these goals. However, the nature of a given design process and technological limitations may disallow achieving a fully repetitive process and it is often required to realize the same steps over and over, as there are no automation capabilities. However, if no automation is possible because of necessity of implementing manual activities by human designers, the process can still be standardized and made more repetitive (making it more effective and consistent) by utilizing the "wizard" approach, creating software tools that guide designer through the process, starting with a mesh (or another dataset) and enforcing certain design activities to be able to go into another step of design. Such an approach is usually implemented in dedicated software for design of a specific class of products (such as dental implants (Pereverzyev 2023) or hip endoprostheses (Guo et al. 2022)).

Further chapters present the two approaches typically implemented by the author together with the team to design given classes of anatomically adjusted products. These will be further explored in Chapters 6 and 7, using a number of case studies.

4.4.2 Mesh-based approach

The mesh-based approach is a general way of conduct in designing anatomically adjusted products, in which it is assumed that a triangle mesh is a foundation, upon which all the other geometrical features are built. In this general approach, the mesh – representing anatomical features of a given body part, set of tissues or internal organs – is used as a whole, inside a 3D modeling tool (of user's selection). The tools and methods of modeling may be actually selected by designer out of available ones in a specific software – it is important to realize that mesh-based approach does not mean that only mesh modeling techniques will be utilized – as it can be realized also with a fully parametrized solid or surface modeling. However, at the bottom of the model history tree, there will always be a basic mesh, of which all the geometrical features derive.

In general, several different pathways can be taken when decided that this approach will be pursued, depending on the tools available and the specific requirement for a model. These pathways, classified based on the author's experience, are shortly described below.

1. *Pure mesh pathway* – anatomical mesh model is processed in an appropriate software and no parametric modeling is implemented – the derivative geometry (e.g., – an orthosis model out of patient's hand mesh) is also created using mesh tools, most notably offsets, cutouts, mesh bridges and so on. There is no model history to be recorded and stored, just a set of files containing triangles (usually in STL format).

2. *Mesh-surface pathway* – after appropriate processing, the mesh model is imported into a software with parametric surface modeling capabilities. The software can be a non-CAD 3D modeling tool (such as 3DS Max or Blender) or a CAD. The model is created using curves forming various surfaces around the basic anatomical mesh and it may be later converted either into a solid or a mesh, for use in 3D printing processes. Some parametric representations may be saved and stored for future editing purposes.
3. *Mesh-solid pathway* – here, the mesh model is imported into a parametric CAD software, with capabilities of using mesh data to create new solid features (e.g., Autodesk Inventor with Mesh Enabler plugin or CATIA v5 with Digitized Shape Editor module). On the basis of a mesh, new solid features (extrusions, rotations, lofts, etc.) are created – the mesh can be converted to a "dumb" solid model or its topology can be recalculated to create a new, optimized surface model (with no triangles). Here, a full model history is available, with possible editing on every step.
4. *Hybrid pathway* – here, a mix can occur – some operations might be realized in a pure mesh approach and then a switch can be made into a parametric CAD modeling software.

Paths 1 and 2 can be realized using free software for mesh processing, as well as specialized commercial software with dedicated tools for designing of specific types of products. An example of such software is Mecuris platform – an online tool for designing leg and hand orthoses on the basis of 3D scan models, enabling also processing of the scans themselves, as well as correcting anatomical positioning of a leg or hand after marking a set of anatomically characteristic points, enabling the software to construct a "skeleton" for mesh deformation (Kour and Singh 2023).

To be able to realize paths 3 and 4 (partially path 2 as well, depending on the required tools), CAD software is required, and it should be a professional one, with a set of tools for mesh importing and processing, what is not always the case when working with standard, basic versions of software such as Inventor or CATIA. However, it guarantees repetitive, accurate result with possibility of easy editing of any of applied operations, just as in the case of parametric modeling.

The pathways with parametric modeling are necessary when the anatomically adjusted product has some "technical" geometries that need to be added to it in the same solid as the anatomically adjusted part. For instance, these can be mounting holes (of standardized size), threads, cases for electronical devices, bearings, grooves and other specified shapes, difficult to obtain using purely mesh operations or non-technical modeling systems. This approach is a must when designing mechatronic devices, containing electronical components – sensors and actuators, as these must be appropriately reflected in the construction.

In general, the mesh-based approach can be characterized as a viable strategy for short product series or one-time efforts for patients with unique anatomy. In the hands of skilled, experienced designers, creating a new variant of an already known product can take less than an hour – of course it only applies to certain product types, with less geometrical features (such as vacuum sockets for leg prostheses, faithfully representing anatomy of a stump (Wang et al. 2022)). Still, if such situation occurs and there is no need to create more sophisticated geometries, or account for high variability of the design itself (e.g., maintaining control over thickness, offsets, curvature, etc.), this might be the perfect way to go, especially that the patient's geometry is accurately represented in the model, with no deformations resulting from approximations.

However, this strategy, regardless of which specific pathway is selected, is flawed in the terms of universality, repetitiveness, and parametric representation of anatomical shapes. Automation capabilities are generally low, as the model for each subsequent patient needs to be practically created from scratch – previous features are usually difficult to recreate on a new mesh, so the whole process must be repeated manually. It is, of course, possible to automate mesh importing and processing, as well as creating a new model on its basis, but it is generally more difficult than the full parametric approach. For a skilled engineer, repeating the whole process is not a big of a problem, but to introduce mass customization capabilities for more complex products – this is not enough. That is why the parametric approach can be alternatively utilized when high need of automation of routine tasks arises.

4.4.3 Parametric approach

The parametric approach is not a natural, intuitive way of dealing with anatomical geometry, obtained via 3D scanning or medical imaging. It assumes indirect use of such geometry, to create a set of mathematically defined parameters, upon which parametric geometry is built and further operations are realized. Here, the mesh is not used in its whole – rather some values get extracted of it, usually in form of basic parameters – lengths, widths, diameters, surface areas, point coordinates and so on – enabling to create sketches and then 3D features, using wireframe, solid or surface modeling tools, depending on the particular shape and requirements towards accuracy of representation.

Similarly as in the case of mesh-based approach, there are many variants which can be selected on how to proceed with parametric models, that will depend largely on the desired shape and desired level of faithful anatomical representation. These pathways, classified by the author's experience, are shortly listed below.

1. *Simple parameters pathway* – here, geometry of the patient is used to extract the simplest of parameters. These are diameters, or widths and heights of section cuts, lengths between characteristic locations, basic angles and so on. This pathway assumes having a detailed, intricately constructed 3D model, which needs only basic parameters to reshape itself for a new patient.
2. *Spline curves pathway* – here, a number of section cuts must be realized and out of them a number of points are extracted. Of these points, spline curves are created and then extraction is realized between them (either as a solid or as a surface) – in such a way, a basic and first shape obtained in the history of a given 3D model is representation of anatomy, similar to the whole mesh – except that it is parametric and further features can safely derive from it, with easier replacement of data for subsequent patients.

3. *Hybrid modeling pathway* – in this approach, number of point coordinates and other characteristic values are extracted from the basic mesh. On their basis, a wireframe model – of the anatomy or the basic outline of a designed product – is created. This wireframe model is then a basis either for a surface model or for a solid model – either way, many sequences of operations must be realized to create a proper geometry, using all three basic parametric modeling methods.

As mentioned before, anything relying on parametric modeling will require CAD software – as such, parametric approach is not possible to realize without a proper CAD system. But there is more to that – to properly extract data for CAD models, a set of tools are required to work on meshes and get these measurements from them. One of such tools is MeshLab, that enables manual measurements, as well as creating and extracting section cuts and subsequent point coordinates. There are also many others – the point is to obtain proper coordinates of points and other values that can be put in the sketches in CAD model to build a parametric representation of anatomical shapes. If these values are stored within a proper design table stored externally (e.g., as Excel file), as mentioned before, CAD models fed with such a table by definition become intelligent ones.

It is therefore apparent that the parametric approach, albeit more cumbersome and problematic to properly realize than the mesh-based approach, naturally delves into automation capabilities, potentially removing the need for repetitive modeling and routine tasks. However, intelligent models with many data points that need to be extracted from anatomical meshes actually take longer to update for a new patient than creation of mesh-based models – in some cases. That is why it is always advised to create also a set of tools for efficient extraction of these parameters, preferably in an automated or semi-automated fashion, not to introduce another set of repetitive tasks.

Advantages of this approach are enormous, when done properly – the models are flexible, can be easily updated and changed, more parameters and features can be added as the knowledge gathering progresses and these features can be effortlessly controlled. There is an option to have just one model, constantly evolving, updated with data of new patients. Building such models is quite difficult and time consuming, but definitely possible – and routine tasks can be then eliminated almost completely, leaving more space for more creative enhancements. However, one must have in mind that it is only justifiable when there is a specific scale of production of given medical supplies combined with necessity of their customization.

Chapter 6 of this book presents descriptions of use cases pertaining to the parametric approach – namely, wrist hand orthosis (which follows spline pathway) and modular hand prosthesis (which follows the simple parameters pathway, combined with spline pathway for some parts). At the current state of the art, it seems like the only plausible way to get automation of designing anatomically shaped products going, except for the use of advanced AI algorithms.

Chapter 5

Additive Manufacturing of Anatomical Shapes

5.1 Overview of additive manufacturing technologies

5.1.1 Overview of technologies and their standardization

Additive manufacturing is an already established, various, wide group of manufacturing technologies. As mentioned in Chapter 2, it should be now considered on equal terms with the two other groups of technologies well-known for centuries (that is, subtractive and volumetric processing). It must be understood that these technologies, although based on the same, simple yet brilliant idea of layer-by-layer addition of material, are actually very different to each other. The industry has long been aware of this and conditions of use, prices, manufacturers, and areas of implementation are quite different. This chapter makes an attempt to show how the most important AMTs work out in general and in the context of building customized anatomical models.

The increasing popularity and firm establishment of 3D printing as a technology to stay is reflected in the standardization process. Quite a number of standards were prepared by the industry experts and are to be used as reference points when discussing technology foundations (Martinez-Garcia et al. 2021). We will start with the ISO/ASTM 52900 standard, that provides standardized terminology for Additive Manufacturing (AM) processes, classifying them into seven primary categories. Each category encompasses distinct methodologies that manipulate material into the desired three-dimensional object. Here, each process is briefly characterized (International Organization for Standardization 2021):

1. *Material Extrusion*: Material is drawn through a nozzle, heated, and extruded onto the build platform. The material solidifies immediately after extrusion as it cools. Fused Deposition Modeling (FDM) is a common technology within this category, primarily used with thermoplastic polymers.
2. *Vat Photopolymerization*: This process involves a vat of liquid photopolymer resin which is selectively cured by light-activated polymerization. A light source (laser or lamp) solidifies the patterned cross-sections of the material layer by

layer to form a solid part. Notable techniques within this category include Stereolithography (SLA).

3. *Powder Bed Fusion*: Energy (typically from a laser or electron beam) selectively fuses regions of a powder bed. The process includes various techniques like Selective Laser Sintering (SLS), Selective Laser Melting (SLM), and Electron Beam Melting (EBM). These methods can process polymers, metals, and composites.
4. *Binder Jetting*: This technique involves depositing a liquid binding agent onto layers of powdered material. The binder acts as an adhesive between powder particles. After each layer is bonded, the powder bed is lowered, and a new layer of powder is applied. This process is used for metals, sand, and ceramics.
5. *Sheet Lamination*: Layers of material in sheet form are bonded together using ultrasonic welding or adhesive, and the desired part shape is cut using a laser or mechanical cutting tools. Laminated Object Manufacturing (LOM) is an example of this process. This method is particularly suitable for rapid prototyping and building large structures.
6. *Material Jetting*: In this process, droplets of build material are selectively deposited through inkjet printer heads. Material is jetted onto the build platform and solidified, typically using light or heat. The process is analogous to 2D inkjet printing and allows for high-resolution and multi-material capability.
7. *Directed Energy Deposition*: A focused energy source (laser, electron beam, or plasma arc) is used to melt material as it is deposited by a nozzle onto a specific surface (Bibb et al. 2024). This process is commonly used for repairing parts, adding features to existing components, or building near-net shape parts from metals.

Each of these processes varies significantly in terms of the materials used, the mechanics of layer construction, the achievable precision and complexity of the parts produced, as well as the economic and temporal efficiency of the manufacturing process.

Beyond just categorizing and describing the main additive manufacturing processes, the ISO 52900 standard also includes definitions of terms related to the properties of materials, the features and geometry of parts, and the qualities of the final product, which aids in standardizing discussions and documentation in AM.

In addition to the main ISO/ASTM standard for AMTs, there are several other ISO standards specific to various aspects of additive manufacturing. Each of these standards targets a different aspect of the AM process, from material specifications to process and quality guidelines:

1. ISO/ASTM 52901 – Additive manufacturing – General principles – Requirements for purchased AM parts (International Organization for Standardization 2017).

   This standard specifies requirements and recommendations for the purchasing, supply, and preparation of design specifications for additive manufacturing parts. It is a guide for acquiring AM parts and establishing the criteria for production, acceptance, and delivery.

2. ISO/ASTM 52910 – Additive manufacturing – Design – Requirements, guidelines, and recommendations (International Organization for Standardization 2018).

   This standard provides guidance on designing parts to be manufactured by AM, including considerations for mechanical design, material selection, and the evaluation of structural and performance characteristics. It aims to optimize designs specifically for AM processes.

3. ISO 17295 — General principles: Part positioning, coordinates, and orientation (International Organization for Standardization 2023).

   This standard defines specifications for positioning and orientation of parts, regarding coordinate systems and test methodologies that are relevant to AM, essential for ensuring consistency and reliability in testing and measurement of samples and final products achieved using these technologies.

Apart from these general standards, there are also a number of standards pertaining to specific technology groups (processes). An example is a two-part standard ISO 52903, focused on extrusion technology (FDM), with Part 1 covering the process and equipment aspects and Part 2 covering material selection and validation, among other things. There are also separate standards for various material types (e.g., powders, resins, etc.).

5.1.2 Data formats and preparation

3D printing and additive manufacturing rely heavily on specific file formats to transfer data from design software to the machines that build the objects. These file formats contain important data: mostly about the shape, sometimes the color, texture, and material properties of the object to be printed. As opposed to the conventional manufacturing processes, there is no drafting involved (no flat drawings with dimensions), so the file exported from CAD is an only source of information for the process engineer dealing with the 3D print. As such, understanding these formats is important for effectively using 3D printing technology (Kumar et al. 2023).

The STL format originated as the file format used in SLA technology. Nowadays, the acronym is understood as standard triangulation language (or standard tessellation language) and it is the file format most commonly used in the 3D printing industry. This format focuses exclusively on the object's surface geometry, representing it through a series of triangular facets. Each triangle is defined by its normal and its vertices, which simplifies the complexity of the model but limits the detail to geometric information only. STL files may be written in ASCII (an example was shown earlier – see Chapter 3, Fig. 3.12), or encoded in binary, weighing much less in such case. While STL files are universally supported and straightforward to produce from almost any CAD software, they lack the capability to encode variable color, texture, or precise material attributes. Moreover, the simplicity of the format can lead to very large file sizes when dealing with high-resolution models.

Another popular format is OBJ (Object File), which extends the capabilities of STL by supporting additional details like texture and color. Unlike STL, OBJ files can include not just vertices and faces but also edges and texture vertices, and they can reference material files (.mtl) that specify how textures are mapped onto surfaces. This makes OBJ particularly useful not just in 3D printing but also in animation and interactive applications – it could be also used in Virtual Reality (VR) or Mixed Reality (MR), maintaining the same level of visual information. However, this format is more complex to handle, and not all 3D printers can fully utilize the advanced texturing and material features provided by OBJ files.

Advancing further, the AMF (Additive Manufacturing File) format was developed as an XML-based solution to overcome the limitations of STL. AMF can encapsulate detailed information about an object's geometry, color, and materials, supporting varied textures and even allowing for the specification of different material compositions within the object, including gradients. This capability makes AMF files highly efficient and smaller in size compared to STL, but they are more complex to work with and have not yet achieved the same level of widespread adoption in the 3D printing community.

The newest addition to 3D printing file formats is the 3MF (3D Manufacturing Format), created by the 3MF Consortium to address the shortcomings of both STL and OBJ formats (Godec et al. 2022). 3MF files are designed to provide a comprehensive package of model information, including geometry, color, texture, and other print-related attributes. They support multiple objects in a single file and manage the relationships and positioning of these objects effectively. 3MF files are intended to be fully integrated with modern CAD systems and have a strong support from major industry players. Despite its advanced capabilities and integration potential, adoption of the format is still not universal – especially by manufacturers of inexpensive, popular printers along with software used to make programs for them. Its complexity requires more sophisticated handling than the more established STL or OBJ formats and it is still difficult to easily convert from typical CAD or detailed 3D models into it using available software tools.

Each of these file formats plays an important role depending on the specific requirements of the user and the capabilities of the 3D printer in use. The STL remains a safe choice while exporting CAD data, as it is widely favored for its simplicity and broad compatibility. Formats like AMF and 3MF offer enriched functionalities that help with advanced applications and reflect the ongoing evolution of additive manufacturing technologies – however they are difficult to easily achieve and in the case of typical unicolor CAD geometries to be produced by standard FDM or SLA, they are probably not worth the effort. However, these formats will probably advance as the technologies itself and will play a role in defining the capabilities and limits of what can be achieved with 3D printing.

Obtaining the 3D data is a first step towards preparation of the printing process, where it is a task of a process engineer to ensure that the printed objects meet the desired specifications and quality. This involves a series of steps and considerations that can vary depending on the modeling technique used—be it solid modeling, surface modeling, or mesh modeling, often requiring introducing changes to the 3D models.

In solid modeling, as previously described, the creation of solid components in three-dimensional space provides detailed information about the volume and mass of the object. Most CAD software utilizes solid modeling because of its precision and utility in representing real-world objects accurately. The preparation for 3D printing in this context involves several critical steps. First – and usually the easiest – is ensuring watertightness. There should be no missing faces, holes, or gaps in the solid structure, as such defects can lead to failed prints or weak spots in the final object. Additionally, checking for thickness is vital. Parts of the model should not be below the minimum thickness supported by the printer to prevent easy breakage in post processing or basic use. Simplifying the model where possible can reduce print time and material usage without compromising the structural integrity. This involves removing unnecessary complexities that do not add functional value.

Surface modeling, on the other hand, is often used for aesthetic and non-functional items where the internal volume is not relevant. Therefore, preparation for 3D printing in surface modeling starts with closing all surfaces – as it is a rare occurrence to have a surface model which is watertight straight from the start. It is essential to ensure that all surface edges are joined together to form a closed volume. Most 3D printers cannot print models that do not define a fully enclosed volume and the software will either generate improper results (such as total or partial filling of model internal volume with support structure) or fail to create a proper program at all. Checking normal orientation is another critical step; all normals (the directions the surfaces face) must be consistent and pointing outward to avoid printing errors. Since surfaces have no inherent thickness, applying a shell operation to give the model a printable wall thickness is necessary.

Data coming from the mesh modeling, as a non-parametric form of storing of surface data, also requires some work prior to 3D printing. Repairing mesh errors is the first priority. Using mesh repair tools to fix issues such as holes, self-intersections, and non-manifold edges is crucial to creating a printable model. Ensuring watertightness is again critical; the mesh must form a sealed volume without any holes. Optimizing the mesh by reducing the number of polygons without losing detail helps ensure smooth printing and minimizes file size and processing time, especially when low-end computers are used for that task. Uniform scaling is necessary to maintain proportions, particularly if the mesh was imported from another source that may use different unit settings.

For all modeling techniques, there are several general requirements to consider. Orientation is key, both in terms of economical coefficients and mechanical properties, as well as accuracy and surface quality (Alanazi et al. 2024, Górski et al. 2015, Górski et al. 2020b, Beattie et al. 2021, Buj-Corral et al. 2019). Orienting the model to minimize the need for support structures can reduce printing time and improve the quality of crucial features, as well as increase strength. Adding support structures where overhangs or bridges exceed a certain angle helps prevent model collapse or deformation during printing. Using slicing software to preview the layers of the print is a valuable step. This can help identify potential issues before the print starts, allowing for adjustments to be made preemptively. Lastly, considering post-processing requirements is important. Adjusting the model's orientation or design to facilitate easier cleaning and finishing can significantly reduce post-processing work.

Keeping these basic requirements and steps in an ordered manner helps streamlining the preparation processes, efficiently obtaining better results. Each class of products, be it an orthosis, prosthesis, implant, or surgery aid – to stay within the realm of focus of this book – needs its own basic methodology to be discovered and described. Numerous available publications and instructions may help to get good process parameters (Lalegani Dezaki et al. 2021, Mohamed et al. 2015), although no instruction can surpass hands-on experience.

5.1.3 Overview of 3D printing software

The notion of "software" used in 3D printing processes in general applies to the programs that are used for preparation, launching and monitoring of execution of Numerical Control (NC) code on additive manufacturing machines. These programs, if classified according to industrial conventions, belong to the category of CAM – computer aided manufacturing, as their purpose is basically the same as the other CAM programs – to get the 3D model of a given part (parts, or assembly) as an input, process it and create an output in form of a manufacturing program, ready to be executed on a given machine with a specific set of tools. However, as most of the popular 3D printing processes are very different in implementation than the industrial manufacturing processes (such as milling or lathing), the notion of CAM is rarely used here – the programs are popularly called "slicers", by the most basic function that they realize. This software takes a digital 3D model and slices it into layers, generating a set of commands that dictate how the printer should construct each layer. Slicers are pivotal in additive manufacturing, affecting the print quality, efficiency, and feasibility of a project.

Splitting it up in a more detailed manner, the slicers are the programs that allow to realize the following set of activities:

- import a 3D model of a part or assembly, in one of accepted file formats,
- visualize the part and a working chamber (table) of a given 3D printing machine,
- allow the user to position and orient the part in the chamber, by translation, rotation, and scaling (a very important stage, influencing greatly technical and economical coefficients of the process (Beattie et al. 2021)),
- selecting the used material and adjusting the process parameters, specific to a given technology, most commonly – layer thickness, internal filling, wall thickness, printing velocities and temperatures, among others,
- configuring additional printed structures, if necessary (mainly supports, in technologies that require them),
- simulating and visualizing the process course of a given set of 3D models with selected parameters, estimating time of completion and volume of used material (here the actual "slicing" is performed, i.e., dividing 3D objects into layers and simulating their deposition),

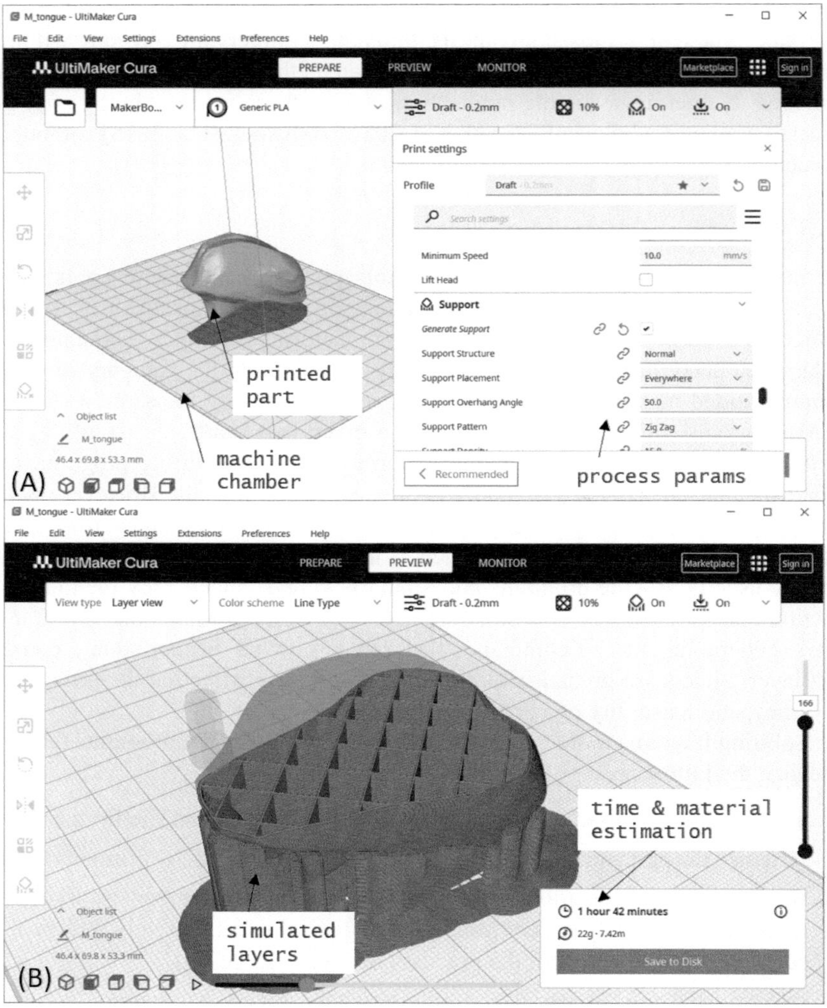

Figure 5.1 Working with slicer software, (A) imported and positioned model, setting up parameters; (B) process simulation and time estimation.

- exporting a created program to a code understandable by a given machine and saving it on an external data carrier (such as SD card or flash USB disk),
- optionally, connecting online to a given machine (via Ethernet, USB or wirelessly) and running the program on it directly, with possibility of monitoring its progress.

Some of these activities are illustrated in Fig. 5.1A and B, on examples of UltiMaker Cura software and an anatomical model of a tongue, processed for 3D printing using FDM technology.

The slicers can be categorized into three main groups, depending on their destination and availability (licensing):

A) *universal open source or freeware* – this category contains software that is generally free to use and can be applied for different machines and materials, often with machine and material presets created inside of a community of users of such software – examples of such software are UltiMaker Cura, Slic3r (with different versions), GrabCAD print and others; these are usually destined to work with various FDM machines, but not necessarily, GrabCAD being a notable exception, being able to work with PolyJet machines;
B) *universal commercial* – this software is also suitable for many different types of machines and materials, but it is licensed commercially with a specific license cost (usually per seat), usually offering technical support and customization for its users, as well as many advanced options, such as manual NC code editing, accessing advanced parameters, automation of generation of sophisticated support structures etc.; examples of such software is Simplify3D;
C) *dedicated (proprietary)* – this software usually comes in package with a given additive manufacturing machine and is developed by its producer; proprietary slicers are tailored to work seamlessly with their respective hardware in most cases, as the software already contains non-editable presets for given machine types and materials, as well as capability of connecting with a machine, monitoring its parameters in an online manner; frequently such machines cannot be operated with a different software than the one intended for them, a notable example here being the Zortrax machines, along with Z-Suite software made for them; some machines can be equipped with dedicated software, but still allow generating code of other software, with a notable example being MakerBot machines.

Slicers differ significantly when considering various additive manufacturing technologies. In extrusion-based printing (FDM), slicers emphasize settings like extrusion temperature, print speed, and bed adhesion, which are critical for controlling stringing, warping, and layer adhesion.

In stereolithography and similar processes, slicers manage photopolymerization with a specific light source. They prioritize parameters like exposure time, lift speed, and peel strength to manage the delicate balance between curing the resin and avoiding suction forces that can damage the print. Support structures are particularly important in SLA, as the uncured resin provides minimal support for overhanging features, especially in the inverse kinematics (upside down printing).

Considering the slicers working with powder-based technologies (selective sintering or melting), slicers focus on packing efficiency, layer height, and laser power to ensure consistent sintering and minimize powder wastage. Unlike FDM and SLA, SLS does not require support structures, but it does require careful thermal management to prevent warping and ensure uniform material properties.

Use of slicers, as opposed to CAM software used in subtractive manufacturing, is relatively simple, quick and can be easily taught even to non-technical personnel – such as doctors realizing the "hospital bed printing" concept. Of course, in advanced manufacturing cases or when problems with process stability or part properties occur, the expert knowledge is required. Similarly, setting up a new machine or material, as well as adjusting advanced, non-obvious parameters to perfect the surface quality or strength of a given class of parts requires lots of prior experience and knowledge, combined with careful testing via the trial and error method. But usually, for a combination of part class, machine and material, a set of more-or-less constant parameters can be developed and then applied by the users with less experience.

It is noteworthy, that some slicers – usually the open-source ones (such as Cura or Slic3r) can work in an automated mode. It stays within a realm of possibility to automate the preparation of programs for parts belonging to specific classes (e.g., prosthetic socket, hand orthosis part, etc.), leaving the operator with only one task of preparing and launching the machine – this has been achieved by the author's team while realizing the previously mentioned AutoMedPrint project. In the author's opinion, however, automation is not really necessary in low-cost FDM and SLA processes. Investment into programming of automated 3D printing, combined with the possibility of errors makes it not feasible economically, at least not on a current level of development. However, in the author's opinion, AI development will probably reach also the 3D printing branch, making it possible to create programs on a basis of simple text prompts in the future iterations of slicer software.

To sum up, it should be stated that the slicer software is integral to the success of 3D printing projects, providing essential functions and features while navigating the limitations inherent to each additive manufacturing technology. The differences in slicer requirements across FDM, SLA, and SLS highlight the need for specialized software tailored to the unique challenges of each process, ultimately enhancing the quality and reliability of 3D printed objects.

5.2 Polymer extrusion technologies – Fused Deposition Modeling

5.2.1 FDM technology – basic notions

As mentioned in Chapter 2, Fused Deposition Modeling (FDM) is a process that has been first invented by Mr. S. Crump, in late 80s of previous century (Crump 1991). The process, in itself, is not a groundbreaking one and did not require any "cosmic" technologies or fresh scientific discoveries to be created. Instead, concepts of polymer processing and subtractive manufacturing, well known and developed in the first half of XX century were used to conceive a process that is universal, robust, inexpensive in basic implementation and possible to implement in virtually any conditions, including the outer space (Musso et al. 2016).

The essence of the FDM lies in the extrusion of heated material. That is why, as mentioned, current standardizations do not stick to the original name of the technology, but rather focus on the aspect of layered extrusion. Different names for the process include Fused Filament Fabrication (FFF), as well as Layered Plastic Deposition (LPD), along with the MEX abbreviation (material extrusion). However, in this book we will keep the notion of FDM, in continuity of the author's previous scientific work started over a decade ago (Górski et al. 2013).

The FDM process in itself has a rather simple set of basic principles. A machine realizing the process must be equipped with a printing head, which is heated to an appropriate temperature, at which a given material starts to flow, being plasticized (it is not melting, however – if it melts and freely flows out of the print head, it means that the process parameters are wrongly set). Central to the printing head is a nozzle, which has a cone-shaped hole. On the one side, a thread of material (known as the filament) is pushed in, unwound from a roll by an electrical engine. On the other side, a thread with a smaller, precisely controlled diameter is pushed out, directly onto the machine table (a flat surface of specified dimensions, usually placed at the bottom of the machine chamber), or previously deposed material.

To create a given shape, a relative movement of the printing head against the table must occur. In the most typical scenario, it will be the print head moving around a flat plane (in two axes, usually denoted as X and Y, or X and Z), while the table will remain static as long as a single layer of material is being printed; after a given layer is finished, the table will move downwards to make space for another layer and the process of deposition will begin anew (Fig. 5.2).

This simple principle draws upon the well-known process of polymer extrusion, making the extrusion head movable akin to a numerically controlled milling machine. So in the end, FDM machines are just that – milling machines, but with the actual mill substituted with a head that extrudes plasticized material. There are some characteristic distinctions to the process. The extruded shape is subjected to gravity – it does not "freeze solid" the moment it leaves the nozzle, so in some cases, support structures are necessary to prevent some longer protruding geometries from falling down. If a machine has only one print head, these supporting structures need

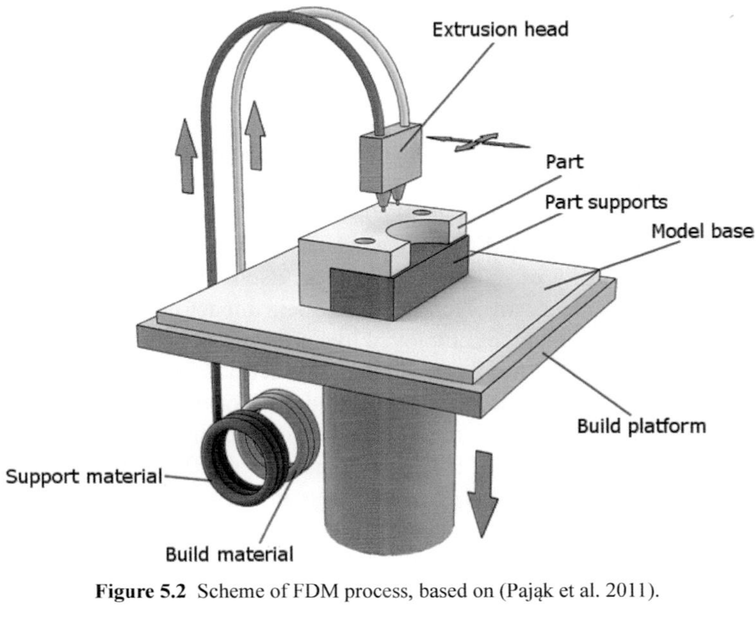

Figure 5.2 Scheme of FDM process, based on (Pająk et al. 2011).

to be created out of the same material that is used to build the main part, leaving mechanical separation as the only way of removing the excess material (which is a waste, and could be a huge one, by the way). Implementing another print head brings a possibility to make support structures out of a different material and apply chemical separation techniques (namely, dissolving the support in an agent that won't dissolve the main part).

Another quite important distinction of the FDM process is heating of other areas and parts than the nozzle itself. To maintain a proper, not too high temperature difference between the nozzle and the surface upon which the material is deposed, as required by certain materials (Wolszczak et al. 2018), the table must be also heated and this heat must be maintained throughout the process. Moreover, to avoid excessively large temperature gradients, in some cases the whole volume in which the printing is realized must be heated, or at least physically separated from the ambient temperature (insulated). That is why only the cheapest of machines do not have enclosed chambers, and the most advanced ones are fully closed, with a constant, appropriately high temperature maintained in the whole working space.

For a long time, FDM technology, similarly to other additive manufacturing techniques, was protected by patent claims and the machines, mostly produced by Stratasys company, were very expensive and not much accessible, costing roughly as much as a luxurious car. However, in the second decade of 21st century, the situation changed. There was a massive movement towards building cheap machines, what is completely achievable with constantly decreasing prices of electronic and electrical components of acceptable quality. Nowadays, prices of the least expensive machines are below prices of regular cellphones, so they are affordable from a viewpoint of hobbyists and households, leading to a possible revolution in consumer 3D printing, as mentioned briefly in Chapter 2.

5.2.2 Process characteristics and parameters

In Fused Deposition Modeling (FDM), a comprehensive understanding of the various process parameters is essential to control and optimize the properties of the printed products (Lalegani Dezaki et al. 2021). These parameters influence not only the mechanical properties and dimensional accuracy of the final product but also the efficiency and feasibility of the printing process itself. In this subchapter we will explore the key parameters involved in FDM, their impact on the properties of the printed objects, and the methods employed to control these parameters effectively.

The structure of FDM products is inherently defined by the layers comprising extruded filament, forming both the contour and infill of the object. The orientation of these layers and the angle of the tool paths significantly affect the mechanical properties and surface quality of the parts (Górski et al. 2015). At the interfaces of these layers and filaments, incomplete melting can occur, which can lead to weak points in the finished product. Ensuring optimal bonding between layers is crucial for achieving high-strength prints (Blaj and Oancea 2021).

Several key environmental parameters must be carefully controlled during the FDM process. The temperature within the build chamber, the pressure, humidity, and the temperature of the build plate all play important roles. A heated build chamber can reduce thermal stresses and warping, especially for materials like ABS, which have significant thermal shrinkage. Maintaining a stable and controlled environment ensures consistent layer adhesion and dimensional accuracy and also prevents instability of the printing process itself, as thermal shrinkage may lead to premature finishing of the process due to disjoint of a partial print from the table.

The orientation of the printed object on the build plate affects the strength and surface finish of the final product. Orienting the object to minimize the height (Z-axis) can reduce the number of layers, thereby decreasing the print time and improving the mechanical properties, as layers printed in the XY plane tend to bond better. However, the optimal orientation also depends on the geometry of the object and the need to minimize support structures.

Layer thickness is another critical parameter that influences the surface finish, mechanical properties, and print time. Thinner layers result in smoother surfaces and finer details but increase the print time. Conversely, thicker layers reduce print time but may compromise surface quality and detail resolution. A typical compromise involves using a layer thickness of 0.2–0.3 mm, which balances print quality and efficiency.

The infill pattern and density are also important factors. Monolithic infill provides higher strength but consumes more material and increases print time. Partial infill (originally named "sparse" in Stratasys software), such as a honeycomb or rectilinear pattern, reduces material usage and print time but at the expense of mechanical strength (Alvarez et al. 2016). The choice of infill pattern and density depends on the application's requirements for strength, weight, and material efficiency.

Other process parameters include retraction distance and speed, which prevent oozing and stringing of the filament when the print head moves between sections. Cooling fan speeds for the extruders must be optimized to solidify the extruded filament quickly without causing warping. The closure point of the contour, or the point where the extrusion begins and ends for each layer, can be managed to avoid visible seams on the surface of the print – or these seams can be removed mechanically or chemically in post processing.

The presence and geometry of basic support structures, such as rafts, brims, and skirts, help in preventing warping and improve bed adhesion. Rafts provide a solid foundation for the print, while brims increase the surface area of the initial layers to enhance adhesion. Skirts serve to prime the extruder and establish a smooth flow of filament before the main print begins, to prevent droplets of material sticking to the main surface of a printed part. Support structures are necessary for overhangs and bridges to prevent collapse during printing, and their density and pattern need to be optimized for easy removal and minimal material usage. Usually they are applied between the overhang angle between 45 and 60 degrees, but it can vary depending on the material and geometry of a product – expertise of a process engineer helps to make a proper decision. Most software slicers enable manual or automatic support structure planning, involving various strategies.

In terms of direct influence on the product characteristics, the orientation of the model, layer thickness, and infill method are primary factors. The orientation significantly impacts mechanical properties, accuracy, and surface quality. Layer

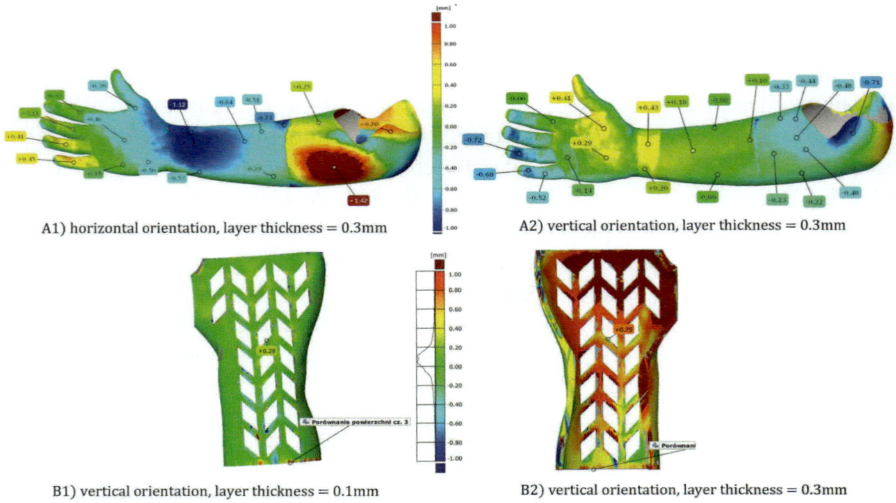

Figure 5.3 Influence of various parameters on dimensional accuracy of anatomical models produced by FDM technology: (A) orientation, hand prosthesis; (B) layer thickness, hand orthosis.

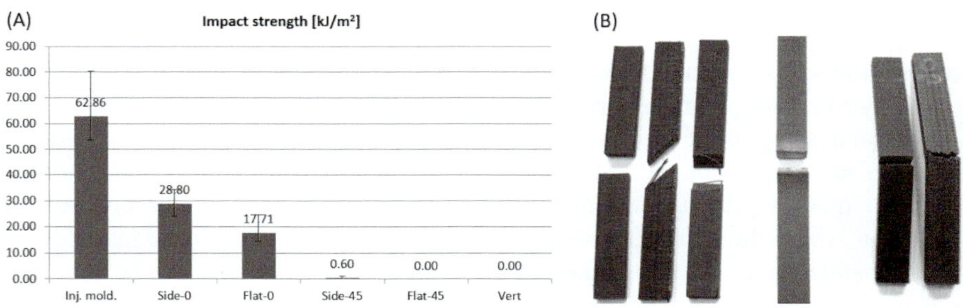

Figure 5.4 Influence of orientation on impact strength of samples produced by FDM technology of ABS material: (A) obtained values; (B) view of samples after the test (data taken from (Górski et al. 2014)).

thickness affects the resolution and surface finish, while the infill method dictates the material consumption and mechanical strength. Example of such influence is shown in Fig. 5.3, comparing accuracy of prosthetic and orthotic devices with different orientations and layer thicknesses (3D scanned printouts were compared to nominal CAD models and deviation maps were obtained).

Another example, greatly showing limitations of FDM's main principle of joining the threads of material by diffusion and adhesion, rather than complete fusion, is presented in Fig. 5.4. Results of Charpy's dynamic test are presented (realized by the author, using Stratasys Dimension BST 1200 machine and ABS material (Górski et al. 2014)), comparing 3D printed samples of different orientation with injection molded sample of exactly the same materials. The results are spectacularly bad for 3D printed samples, as the best FDM sample does not reach half of dynamic strength of injection molded sample. That translates to properties of products, such as prostheses or orthoses, in daily use – they are, the of author's experience, quite resilient in terms of static load, but tend to fail often when dynamic loads are applied.

Another important major limitations of FDM technology is the need for support structures, which can complicate post-processing and increase material waste. The size of the build volume also restricts the dimensions of the parts that can be printed. Moreover, the minimal wall thickness that can be achieved is typically between 0.1 to 1 mm, with recommended thicknesses above 1.5 mm for structural stability.

There are plenty of defects and problems that can happen during 3D printing using the FDM (extrusion) technology (Baechle-Clayton et al. 2022). The common defects in FDM prints include the staircase effect, seam lines, rounded edges, and thermal deformations. The staircase effect occurs due to the layered nature of the print, leading to visible steps on curved surfaces. Seam lines or weld lines appear where the extruder starts and stops on each layer. Rounded edges can result from improper cooling or filament deposition, while thermal deformations are caused by uneven cooling and material shrinkage. To mitigate these defects, careful control of the printing environment, proper orientation of the model, and optimization of the layer thickness and infill settings are essential. Additionally, the selection of appropriate support structures and their placement can reduce post-processing efforts and material wastage.

5.2.3 Materials used in FDM and their properties

The choice of material significantly influences the properties and performance of additively manufactured objects, regardless of the technology. The extrusion processes are no different here. FDM is based mostly on a considerable variety of polymers, each suited to different applications and offering unique properties. The most commonly used materials in FDM include standard, thermoplastic polymers like ABS and PLA, often used for general prototyping and functional parts, as well as more specialized materials like HDPE, PETG, TPU, and various engineering thermoplastics (Shanmugam et al. 2021). Typically, extrusion-based 3D printing uses materials in form of filament, unwound from a spool, with a constant diameter – 1.75 mm being one of the most common industry standards fit to most of the printing heads, with other possibilities also available, e.g., 3 mm. In the overwhelming majority of use cases, pure polymeric materials are used, but there are possibilities of using composite materials as well, basing on a thermoplastic as a matrix with added chunks, fragments or threads of a reinforcing material (Penumakala et al. 2021).

ABS, or acrylonitrile butadiene styrene, is one of the most popular materials used in FDM due to its strong mechanical properties. It was one of the first thermoplastics used in popular FDM machines manufactured by Stratasys, before the opening of the market. ABS is known for its toughness and impact resistance, which makes it suitable for creating durable parts. However, ABS requires a heated bed and an enclosed build chamber to mitigate issues related to thermal shrinkage, which is higher compared to other materials like PLA – although still not as high as many other polymers. ABS comes in various formulations, including transparent ABS and specialty blends, e.g., incorporating polycarbonate and glass fibers for enhanced performance (Kumar et al. 2023b). Despite its benefits, ABS can be challenging to print due to its tendency to warp and release fumes during printing, not being particularly healthy to the operators if not treated properly. Its properties post-printing – especially regarding dynamic ones – are also highly dependent on process parameters, most notably orientation, and they do not exceed characteristics of the same material processed traditionally (e.g., by injection molding), as noted in the author's earlier studies (Górski et al. 2013, Górski et al. 2014, Górski et al. 2015). ABS is also not particularly skin-friendly – although it can be subjected to sterilization and clinical use (Kuczko et al. 2018), its prolonged use in contact with the skin is not recommended.

PLA, or polylactic acid, is another widely used material, appreciated for its ease of use and biodegradability, as it is compostable in right conditions. Derived from renewable resources like corn starch, PLA is more environmentally friendly compared to other plastics and carbon footprint of products made of it – e.g., orthoses – is lower than their traditionally fabricated counterparts (Górski et al. 2020c). It has a higher rigidity and hardness than ABS but can be more brittle and less UV resistant. PLA's low thermal shrinkage makes it easier to print, often without the need for a heated bed and thermal insulation of the working chamber. The material produces a glossy surface finish and is well-suited for aesthetic applications, including models and prototypes. However, PLA is less suited for mechanical parts that need to withstand significant stress or high temperatures. The low temperature resistance may also prove to be a challenge when making useful daily products in the parts of the world where temperature frequently exceeds 40 degrees Celsius and the sun is very intense, e.g., Africa or Australia (building prosthetics for residents of these continents may require switching to another material).

PETG, or polyethylene terephthalate glycol-modified, combines the advantages of PLA and ABS. This copolymer based on a very popular PET plastic is easier to print than ABS and offers better mechanical properties and chemical resistance than PLA (Hsueh et al. 2021). PETG is favored for its flexibility, impact resistance, and transparency, making it suitable for producing clear and robust parts. It does not produce harmful fumes during printing and has good adhesion properties, reducing the risk of warping. In the case of biomedical applications, such as these described in the book, it is a solid second choice where PLA cannot be used due to its shortcomings and it is still required for the process to be relatively inexpensive and accessible. It can be sterilized and safely used with the skin (Yan et al. 2023). However, its ecological footprint remains generally negative (Junk and Vögele 2024).

Nylon, a family of synthetic polymers known as polyamides, offers excellent durability, flexibility, and resistance to wear and abrasion. Nylon is ideal for producing functional parts that require high strength and flexibility, such as gears, bearings, and tool handles – as well as prosthetic and orthotic product components. Unfortunately, nylon will not become a first choice of material due to its problems – first of all, it can be challenging to print due to its high melting point and tendency to absorb moisture, which can lead to print defects if not properly dried. It usually requires drying before printing, or heating the whole build chamber. Ensuring good adhesion to the build platform is also critical when printing with nylon, it favors build tables with a porous structure and cannot be easily processed on all machines, increasing a threshold of availability. However, it is relatively skin-friendly, less dangerous to environment than other typical thermoplastics, and has considerably acceptable properties for orthopedic and prosthetic devices used dynamically by patients (Górski et al. 2020b, Shahar et al. 2020).

If elastic, rubber-like properties are required from the FDM-printed parts, a basic choice is usually TPU. Thermoplastic polyurethane is an example of a flexible and durable material that resembles rubber in its properties (the other being TPE). TPU's high elasticity and resistance to abrasion, oils, and greases make it suitable for producing items like gaskets, seals, and wearable parts, as well as soft orthoses, prosthetic end effectors and limb phantoms for destructive and non-destructive tests (Górski et al. 2021). It can be challenging to print due to its flexibility, which requires careful calibration of print settings to avoid issues like stringing and poor layer adhesion.

In addition to these commonly used materials, FDM can also process a range of engineering thermoplastics that offer enhanced mechanical properties and thermal stability. Polycarbonate (PC) is known for its excellent impact resistance and transparency, though it is more prone to warping and requires precise temperature control. Polyphenylsulfone (PPSF) offers superior chemical and thermal resistance, making it suitable for high-temperature applications. Polyetheretherketone (PEEK) stands out for its exceptional mechanical strength, chemical resistance, and ability to withstand high temperatures, up to 250°C, making it ideal for aerospace and medical applications. Apart from PEEK, there are a number of other interesting "ultra-polymers" with good properties such as PEKK or PEAK. However, their quite good mechanical properties do not always justify their extremely high price comparing with typical thermoplastics such as PETG (Păcurar et al. 2024).

If using of polymers with enhanced properties is not a viable option, it can be also considered to use typical polymers – PETG, nylon, PLA – mixed with carbon or glass fiber, for better mechanical properties. Such a mixture increases the difficulty of processing, but helps obtaining products with better strength – although an optimal combination of material and processing parameters for high-demand medical products is yet to be found.

FDM also utilizes soluble support materials, such as PVA (Polyvinyl Alcohol) and HIPS (High Impact Polystyrene), which facilitate the printing of complex geometries. PVA is water-soluble, making it easy to remove post-printing, while HIPS dissolves in limonene, providing a clean and efficient support removal process. HIPS can be also used as a build material itself, although its properties for orthotic products are not particularly outstanding (Górski et al. 2020b). It is noteworthy, that in daily practice, soluble materials usually remain a nice addon to typical materials – their solubility may come in handy dealing with specific geometries (such as closed canals of RoboHand-like hand prostheses), but usually it is much simpler and less time-consuming to remove the support mechanically.

Proper storage and handling of FDM materials are a key factor to maintain their performance. Most filaments are supplied on spools and sometimes enclosed in cartridges with desiccants to protect against moisture, which can degrade the material properties, particularly for hygroscopic materials like nylon and polycarbonate. UV exposure, or lack thereof, is also crucial to maintain good mechanical properties of filaments, contributing to their brittleness, which may result in unstable manufacturing processes (breaking of filament fed into printing head mid-process).

To sum up, it should be stated that extrusion-based 3D printing has the most wide selection of materials of all popular additive manufacturing processes. The presented summary is not comprehensive by any means, as there are plenty of manufacturers, types, and sub-types, as well as admixtures and special ways of enhancing materials and their properties. As such, each medical use of FDM technology should begin with material consideration and careful checking of latest achievements in the field, as year by year scope and quality of existing material base is expanding.

5.2.4 Overview of medical applications of FDM

The extrusion 3D printing technologies gained quite interest in the medical field in the last decade, providing a cost-effective, versatile, and highly customizable approach to producing medical devices, especially personalized for patient's anatomy. The potential applications of FDM in medicine range from prosthetics and orthotics (known for over a decade (Palousek et al. 2014)) to surgical guides and educational models. By leveraging the unique capabilities of FDM, healthcare professionals can offer personalized solutions that improve patient outcomes and enhance the overall quality of care. Moreover, medical devices can be also printed in unusual places, such as patient's homes, schools, daycare facilities, etc., although use in hospitals is subjected to certain requirements (Otero et al. 2017).

One of the most significant applications of FDM in the medical field is in the creation of prosthetics. Traditional prosthetic devices can be prohibitively expensive, particularly for growing children who require frequent replacements. FDM offers a more affordable alternative, enabling the production of both mechanical and bionic prosthetics that can be customized to fit the unique anatomy of each patient (Neethan et al. 2020). For instance, 3D-printed prosthetics for children are not only cost-effective but also allow for rapid iteration and adaptation as the child grows. This customization extends to personalized myoelectric prosthetics, which can be tailored to the specific functional needs and preferences of the patient. There are also a number of open-source projects, developed by a growing community of hobbyist printers worldwide, such as RoboHand (Ten Kate et al. 2017).

Orthotics is another area where FDM technology can be successfully applied. Custom orthotic devices, such as Ankle-Foot Orthoses (AFOs) and Wrist-Hand Orthoses (WHOs), can be designed and produced using 3D scanning and printing techniques, as proven in earlier work (Górski et al. 2020b, Górski et al. 2022b). These devices provide critical support and alignment for patients with various musculoskeletal conditions, such as cerebral palsy or congenital paresis, as well as traumas such as bone fracture. The ability to create personalized orthotic devices that precisely match the patients' anatomy ensures better comfort, fit, and effectiveness. Additionally, 3D printing can significantly reduce the design and production time for these devices, allowing for quicker response times to the patient's needs.

FDM is also useful in veterinary medicine, where custom prosthetics for animals are becoming increasingly common. Simple examples include prosthetic legs for ducks and prosthetic wings for birds, demonstrating the technology's versatility and ability to improve the quality of life for injured or disabled animals (Carvalho 2024). There are also numerous successful cases of saving or improving lives of home pets, such as cats or dogs (Mendaza-DeCal et al. 2023). The customization offered by FDM ensures that each prosthetic device can be tailored to the animal's specific anatomy, facilitating better mobility and functionality, treating animals just like human patients and realizing the full procedure with them, starting with 3D scanning (often incapacitating them first), through CAD design and then printing.

Considering the area of surgical planning and intraoperative aids, FDM technology plays a crucial role by enabling the production of accurate anatomical models, of both hard and soft tissues. These models can be based on patient-specific imaging data, providing surgeons with a tangible reference for preoperative planning and intraoperative guidance (Banaszewski et al. 2018). For instance, models of organs with tumors, such as kidneys or livers, allow surgeons to plan and practice complex procedures, potentially reducing surgical times and improving precision (Jin et al. 2021, Żukowska et al. 2023). Similarly, custom surgical guides can be produced to fit the patient's unique anatomy, ensuring accurate placement of surgical instruments and implants. It is worth noting that FDM is normally not used for implant production, except for when special filaments are used, able to stimulate bone growth processes – these implants are then bioresorbable and do not stay permanently in the patient's body (Targońska et al. 2024).

Educational models for medical training and patient education are another valuable application of FDM. Detailed, patient-specific anatomical models can be used to teach medical students and professionals about complex anatomical structures and surgical procedures. These models provide a hands-on learning experience that is more effective than traditional 2D images or diagrams. Additionally, they can be used to explain medical conditions and treatment plans to patients, improving their understanding and engagement in their own care, as well as their psychological state (Nikitichev et al. 2018). Such applications are still rare, due to economic constraints, but as the "hospital bed printing" concept unfolds, such applications will be more and more common.

On a more general note, it is worth noting that the rapid prototyping capabilities of FDM enable the quick development and testing of new medical devices and treatments, just as it helps engineers iterate their designs in other, non-medical branches of industry. This iterative process allows for the refinement of designs based on real-world feedback, accelerating the innovation cycle in medical technology. For example, developers can quickly prototype and test new designs for prosthetic limbs, orthotic supports, or surgical tools, making necessary adjustments to improve functionality and patient comfort. FDM printers are generally clean, relatively quiet and harmless, not requiring specialized installations, so they can be used inside design offices with minimal safety and ergonomics constraints, streamlining iterative design optimization by rapid prototyping.

In general, successful use of FDM technology in medicine is dependent on the three main factors:

1. *material* – selection of biocompatible, appropriately strong, and resistant material, possibly also sterilizable lays a foundation for obtaining good quality medical products;
2. *process parameters* – as discussed, some controllable variables (layer thickness, orientation, etc.) can significantly influence the mechanical properties, surface finish, and dimensional accuracy of the printed parts, what is crucial especially when load-bearing applications are considered (orthotics or prosthetics);
3. *post processing* – it can further enhance the properties of FDM-printed medical devices and change their qualities to fit better the high requirements of medical world – e.g., applying biocompatible coatings can enhance the durability and safety of an orthotic device for the wearer, ensuring that it can withstand the rigors of daily use.

FDM technology, although inexpensive and simple, offers a powerful toolset for creating customized, cost-effective medical devices and aids that have a potential of significant improvement of patient care. Its applications in prosthetics, orthotics, surgical planning, veterinary medicine, and medical education demonstrate its versatility and impact. The continuous advancements in FDM materials and techniques promise to further expand its applications in the medical field, paving the way for even more innovative and effective healthcare solutions, especially considering attractive prices of the fabrication devices, making it possible to have them available directly at healthcare facilities.

5.3 UV technologies

5.3.1 Basic concepts of UV 3D printing processes

As shown in Chapter 2, UV printing (vat photopolymerization), commonly referred to by its popular subtype, stereolithography (SLA), was developed in the early 1980s by Chuck Hull (Hull 2015). This process, like Fused Deposition Modeling (FDM), did not stem from any revolutionary new scientific breakthroughs but rather evolved from existing chemical engineering and photochemistry principles well established by the mid-20th century. It leverages a method that is precise, versatile, relatively cost-effective in its basic form, and can be adapted for use under a variety of environmental conditions, just like the FDM.

The core of a basic process – vat photopolymerization involves the selective curing of a photosensitive resin. The technology is often referred to by different names such as Resin Printing (RP) or Liquid Resin Printing (LRP), with the abbreviation PPT (Photo Polymerization Technique) also in use. For continuity with prior work, this book will primarily use the term Stereolithography (SLA).

The SLA process operates on a straightforward concept. A machine equipped with a UV light source, typically a laser or a digital light projector (in such case the technology is usually called DLP) precisely hardens the liquid resin in a designated pattern. Central to the operation is the build platform, which initially dips into a vat of liquid resin. A light source projects onto the resin surface, hardening it in accordance with the cross-section of the desired object. Once a layer is complete, the build platform moves vertically, allowing fresh resin to flow beneath the just-cured layer, and the process repeats to build subsequent layers (Huang et al. 2020).

To achieve the object's intended geometry, the movement of the light source and the build platform must be accurately coordinated. Typically, the light source moves in two axes (X and Y), while the build platform's movement defines the Z-axis. The platform either ascends or descends depending on whether the machine builds from the top-down or bottom-up. The light source – if it is a laser beam – is usually directed through a moving mirror, otherwise known as the scanner, so the process of selective laser curing itself is sometimes called simply "scanning".

This method draws parallels with photolithography used in microfabrication, turning the UV light into a precisely moving tool, substituting physical cutting with photopolymer curing. SLA machines, just like FDM machines, share similarities with numerically controlled subtractive machines. Notable aspects of SLA include the need for supports to uphold complex structures and overhangs. If a model includes intricate or suspended features, support structures are printed alongside the primary object. These supports are usually generated from the same resin and later mechanically removed, often requiring additional finishing. However, as opposed to the extrusion-based processes, the liquid in a vat has considerable density and it can hold most of the protruding geometries, so that the supports are always openwork, with a lightweight structure, not wasting much material and requiring minimal workload to remove.

Another critical aspect of SLA is managing heat and exposure levels to avoid deformation or improper curing. Though the resin tank and build environment don't typically need heating like FDM platforms, precise control of light intensity and exposure time is crucial to achieving the right curing degree. More advanced machines offer tightly controlled environments to ensure consistent conditions. Anyhow, in the classic vat photopolymerization, full curing is not achieved in the process itself to avoid over-curing (leading to internal stress in the produced part), so that after the process a part must undergo post-curing, which usually involves putting it in a chamber with strong UV light sources, sometimes equipped with a rotary table (akin to a microwave). After such treatment, the material is usually fully cured and gains desired mechanical properties.

Historically, SLA, like many additive manufacturing processes, was governed by strict patent restrictions, primarily held by 3D Systems, a company founded by Chuck Hull. These machines were quite costly. However, in the 21st century, technological advances and the expiration of key patents have led to more affordable desktop versions of these machines, now costing obviously much more than the cheapest FDM machines, but often still less than many high-end smartphones, making them accessible to hobbyists and small businesses (less for households, due to many technological reasons, "dirtiness" of the process and potentially toxic vapors of liquid resin being some of them). This democratization of technology may potentially usher in a significant shift in custom manufacturing and design, as briefly mentioned in Chapter 2.

5.3.2 Photopolymerization-based processes and device types

Ultraviolet (UV) light-based additive manufacturing processes leverage the principle of photopolymerization to convert liquid photopolymer resins into solid parts. This category includes several distinct technologies such as Stereolithography (SLA), Digital Light Processing (DLP), Continuous Liquid Interface Production (CLIP), Daylight Polymer Printing (DPP), and Jetted Photopolymer (PolyJet). Each of these technologies utilizes UV light to cure specific areas of a resin material to build parts layer by layer (Huang et al. 2020). While sharing a common foundation, each process exhibits unique characteristics in terms of setup, speed, accuracy, material requirements, and application areas.

SLA, historically the earliest form of UV light-based 3D printing, uses a laser to cure photopolymer resin in a vat. A laser beam, directed by a so-called scanning system, traces the part's cross-section on the surface of the liquid resin, hardening it precisely where it hits. After each layer is cured, the build platform moves vertically to allow a new layer of resin to cover the surface, and the process repeats. SLA is known for producing parts with very high accuracy and great surface finishes, due to its low layer thickness (typically 0.05 mm or lower).

SLA process comes in two main variations, reflected in construction of a manufacturing machine – the standard and the inversed process. The standard process involves a vat containing an appropriate volume of resin to contain the whole printed part (i.e., height of resin level in the vat must exceed the height of the printed part). The build table has openwork construction and it moves downward the vat as subsequent layers are hardened, with the laser shining from above (laser beam pointing downwards). At the end of the process, the build chamber is again moved upwards and the parts can be removed. This process is mostly implemented in professional, industry-grade machines produced by prime manufacturers, such as 3D Systems.

The inversed SLA process has been implemented as a workaround to reduce machine size and cost, still enabling printouts with acceptable heights. Here, the vat contains a small volume of resin, just for a couple of layers. The laser is shining upwards, from below the vat. The build table is solid and it moves upwards as the print process progresses – so the part is effectively built upside down. At the end of the process, the table is at the full height, with the part hanging upside down, ready to be removed. Schematics of these two processes, underlining their differences, are presented in Fig. 5.5.

It is important to notice, that similarly as in the extrusion printing, support structures are necessary in SLA processes. In the standard process, due to natural support by the unbound, dense, liquid resin, the supports do not need to be massive and usually are required in the case oh heavy bridging or high overhang angles. The inverse process is more troublesome here, as there are practically no "natural" supports, so the support structures must be planned properly to avoid the part from

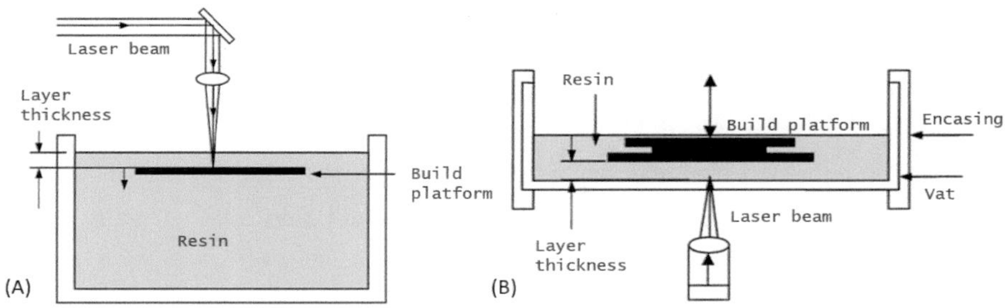

Figure 5.5 Basic types of SLA processes, (A) standard, (B) inverted.

falling apart and into the resin tank during printing. The support must be removed in post processing, and as it is made out of the same material, it has to be done mechanically – by cutting, polishing, etc.

The basic SLA process has certain drawbacks. The use of a single laser can lead to longer build times for larger parts, especially considering a high number of layers overall resulting from low layer thickness. The material options are also somewhat limited to those that can be effectively cured by the specific laser wavelength. That is why some alternatives have been coined out in recent years. One of them is DLP. DLP is similar to SLA in that it also uses a vat of photopolymer resin, but instead of a laser, it employs a digital projector screen to flash a single image of each layer all at once. This allows DLP to cure whole layers simultaneously, making it faster than SLA when printing large, complex parts. The resolution of DLP prints is generally high, though it depends on the projector's pixel size, which can be a limiting factor in achieving the finest details compared to the more focused approach of SLA's laser. DLP machines tend to be simpler and potentially cheaper to maintain than SLA systems because they contain fewer moving parts.

Developed by Carbon3D, CLIP is a further enhancement of the basic principles of DLP and another step forward in development of SLA-based techniques. It introduces a continuous printing process that eliminates the layer-by-layer approach and significantly speeds up the production process. CLIP uses a projector to cure the resin, but it does so continuously by employing a special transparent window that permits both light and oxygen to pass through. The presence of oxygen inhibits polymerization at the window interface, creating a "dead zone" that allows the resin to flow without curing immediately (Rolland 2016). This process results in much faster print speeds and smoother surface finishes, as – nominally – there is no such thing as "layer thickness" in this process. However, the technology requires specialized resins and equipment, which can increase operational costs.

DPP is a variant of DLP that uses natural light to cure the resin, instead of UV light. This makes DPP potentially more user-friendly and safer to operate in educational or low-resource settings, as the light sources are less hazardous and the technology can be simpler. The materials used in DPP are also different since they must be sensitive to the visible light spectrum, which can limit the range of properties and applications. While DPP offers an innovative approach to resin-based printing, the reduced control over curing due to variable ambient light conditions can affect part quality and accuracy (Pyteraf et al. 2022, Lojo et al. 2023).

All the variations of the SLA – be it the basic process, DLP or DPP – present a significant drawback related to handling of the material and the printed part. The post processing in all these processes involves the following steps:

- removing the part from the machine,
- cleaning the part from the leftovers of the unbound liquid resin – usually with a dedicated alcohol solution,
- removing the support structures and further cleaning of the part,
- post-curing – exposing part to UV light in a special chamber, to achieve full curing of the resin (Zhao et al. 2020).

Realizing the whole set of activities requires a lot of manual work, additional utensils and is definitely not clean. The fumes of resin may be harmful or even mildly toxic, and the smell of alcohol mixed with the resin is unpleasant, preventing the technology to be implemented directly in office spaces or, in other words, anywhere where people work or live. That is a drawback that is not present in the extrusion technologies, that can operate safely and cleanly and do not generate unpleasant fumes and do not require dealing with liquid agents.

These challenges are addressed by the hybrid technology of Jetted Photopolymer (JP, but popularly known as PolyJet), combining quality, low layer thickness and smoothness of the SLA-related processes (and using the same classes of materials) with flexible kinematics of extrusion-based processes, also enabling the process to be realized without manual dealing with liquid resin and other potentially harmful substances. PolyJet technology stands out by using a different approach where photopolymer resin is jetted, or sprayed, onto the build platform through multiple nozzles. This process allows for the simultaneous deposition of multiple materials, including different colors and material properties within a single print job. After deposition, the resin is cured by UV light immediately, not requiring additional post-curing in a separate chamber.

The kinematics of the PolyJet process are not strictly governed by the principles of the extrusion process, as no pressure and temperature need to be applied – the material (contained in a sealed cartridge) is in liquid state since the beginning of the process. As such, it is possible to implement kinematics based on a rotating table (Fig. 5.6A), where the material heads and UV lamp are immobile and placed in the "roof" of the working chamber. The process implies use of support structures just like in the extrusion processes, but conditions of room temperature make it possible to use easily removable, wax-like support materials, that easily go away under a stream of pressurized water (Fig. 5.6B).

All these features make PolyJet an excellent choice for creating complex geometries with intricate details and a smooth surface finish. The process can produce parts with diverse mechanical properties and aesthetics – many materials and colors can be mixed in the volume of a print, creating unique results, unachievable by any other process. The PolyJet machines and materials can be also certified for medical

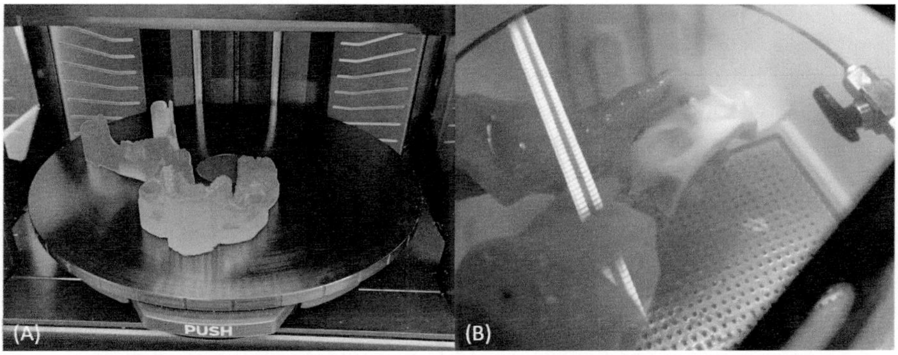

Figure 5.6 PolyJet process, Stratasys MediJet J5 machine: (A) working chamber with rotating build table, (B) cleaning of support by WaterJet.

uses (medical-grade resins can be used). However, the cost of PolyJet printers and their operational expenses can be high, making them less accessible for smaller businesses or individual users.

5.3.3 Examples of medical uses

The UV-based 3D printing technologies, particularly Stereolithography (SLA), Digital Light Processing (DLP), and Jetted Photopolymer (PolyJet), have garnered significant interest in the medical field over the last decade. These technologies offer precise, high-resolution printing capabilities, making them ideal for producing detailed medical devices and models tailored to individual patients' needs. Their applications span from custom prosthetics and orthotics to surgical guides and educational models, leveraging their unique capabilities to improve patient outcomes and enhance the quality of care.

One of the most significant medical applications of UV-based 3D printing is in the creation of highly detailed anatomical models for surgical planning and intraoperative guidance. SLA, DLP, and PolyJet technologies can produce models with exceptional accuracy, capturing intricate anatomical details from patient-specific imaging data (Żukowska et al. 2024a). These models provide surgeons with tangible references for preoperative planning, allowing them to visualize and practice complex procedures on replicas of patients' organs. For instance, models of organs affected by tumors, such as kidneys or livers, enable surgeons to plan precise surgical approaches, potentially reducing operation times and improving outcomes. Custom surgical guides can also be created to fit the patient's unique anatomy, ensuring the accurate placement of surgical instruments and implants.

UV-based 3D printing is also partially useful in producing custom prosthetics and orthotics. SLA, DLP, and PolyJet technologies offer a more affordable alternative by enabling the production of personalized prosthetics that fit the unique anatomy of each patient, enabling accuracy and good surface finish, when comparing with extrusion or sintering technologies. Custom orthotic devices, such as Ankle-Foot Orthoses (AFOs) and Wrist-Hand Orthoses (WHOs), can also be designed and produced using these technologies (Paterson et al. 2015), offering critical support and alignment for patients with various musculoskeletal conditions, although usually the small size of the chamber does not allow to produce full-scale orthopedic products (Chen et al. 2016). The precision of UV-based 3D printing ensures that these devices provide better comfort, fit, and effectiveness, significantly reducing the design and production time compared to traditional methods.

In veterinary medicine, UV-based 3D printing technologies are increasingly used to create custom prosthetics for animals, as well as surgical aids (Altwal et al. 2022). These technologies' ability to produce detailed and precise models ensures that each device can be tailored to the animal's specific anatomy, facilitating better mobility and functionality. Examples include models of animal teeth, bone models and wing prosthetics (Carvalho 2024), showcasing the versatility and impact of these technologies in improving the quality of life for injured or disabled animals.

Educational models for medical training and patient education are another valuable application of UV-based 3D printing. Detailed, patient-specific anatomical models produced using SLA, DLP, and PolyJet technologies provide an effective hands-on learning experience for medical students and professionals. These models enable a better understanding of complex anatomical structures and surgical procedures than traditional 2D images or diagrams. Additionally, they can be used to explain medical conditions and treatment plans to patients, improving their understanding and engagement in their own care. The high resolution and precision of UV-based 3D printing make these educational models exceptionally detailed and accurate, enhancing the overall quality of medical training and patient education.

Another critical application of UV-based 3D printing in the medical field is the development and testing of new medical devices. The rapid prototyping capabilities of SLA, DLP, and PolyJet technologies allow for quick development and iteration of new designs, enabling faster innovation cycles (Salmi 2021). For example, developers can prototype and test new designs for surgical tools, prosthetic limbs, or orthotic supports, making necessary adjustments to improve functionality and patient comfort. The precision and detail offered by these technologies ensure that prototypes closely match the final product look and feel, facilitating efficient design optimization – but it must be remembered that mechanical properties of UV-based 3D prints are usually not optimal and will be weaker than the final product.

The selection of biocompatible and sterilizable materials is crucial for the successful application of UV-based 3D printing in medicine. SLA, DLP, and PolyJet technologies can work with a range of materials that are safe for patient use, including resins that can be sterilized using standard medical sterilization methods (Bens et al. 2007). These materials ensure that printed devices meet the stringent hygiene standards required in healthcare settings, providing safe and effective solutions for patient care.

The process parameters in UV-based 3D printing, such as layer thickness, exposure time, and print orientation, significantly influence the mechanical properties, surface finish, and dimensional accuracy of the printed parts. For instance, thinner layers result in smoother surfaces and finer details, which are essential for medical devices that require high precision. The ability to control these parameters allows for the optimization of the printing process to achieve the desired properties of the printed objects, ensuring that they meet the specific requirements of medical applications.

Post-processing is another essential aspect of UV-based 3D printing in medicine. Post-processing steps such as cleaning, curing, and finishing can enhance the properties of the printed devices, making them suitable for medical use. For example, post-curing can improve the mechanical strength and stability of printed parts, while surface finishing techniques can enhance the aesthetic and functional qualities of prosthetics and orthotics. Ensuring proper post-processing is crucial for meeting the high standards of quality and safety required in the medical field.

In summary, UV-based 3D printing technologies like SLA, DLP, and PolyJet are useful tools for creating customized, high-resolution medical devices and models. Capabilities of these technologies can be leveraged in healthcare, as demonstrated in numerous literature examples, also shown in the later part of this book (Chapter 7). They mostly shine in creating personalized anatomical models with intricate details, that enhance patient care. Continuous advancements in UV-based 3D printing materials and techniques promise to expand their applications further, paving the way for even more innovative and effective healthcare solutions.

5.4 Sintering and powder-based processes

5.4.1 Basic concepts of powder-based additive manufacturing processes

The powder-based thermal additive manufacturing processes is an umbrella term, which can be applied to all techniques belonging to the AM group, where:

a) raw material has a form of a powder with uniformly-sized grains,
b) grains of powder are joined together by selective, directed temperature field, allowing for partial or complete fuse.

The basic technology in this whole group is Selective Laser Sintering (SLS), which was introduced in the mid-1980s, a development pioneered by Dr. Carl Deckard and his professor Joe Beaman at the University of Texas. This technology evolved from principles of powder metallurgy and thermodynamics, which were already well-understood disciplines by that time. However, only democratization of CO_2 lasers (i.e., great price reduction of commercial solutions) have made the additive powder sintering technology possible to be conceived (Davies 2020). SLS is recognized for its ability to fabricate parts that are durable, complex, and capable of being produced in various environmental settings, similar to other additive manufacturing technologies.

At the heart of SLS is the use of a high-power laser to sinter powdered material—typically plastic, metal, ceramic, or glass particles – layer by layer. The terminology around this technology may vary, with terms such as Laser Sintering (LS) and Powder Bed Fusion (PBF) being commonly used. Throughout this book, we will refer to it as Selective Laser Sintering (SLS) to maintain consistency with existing literature and standards (Kumar et al. 2021).

The SLS process relies on a simple yet sophisticated mechanism. It begins with a thin layer of powder being spread across the build platform. A laser then selectively sinters the powder, binding it together to form a solid structure corresponding to the slice of the digital model. After each layer is sintered, the build platform lowers, and a new layer of powder is applied on top of the previous one, repeating the cycle until the object is fully formed.

In SLS, the coordination of the laser and the platform's movement is crucial. The laser itself does not move but is directed by mirrors that adjust to aim the beam across the powder bed's surface in the X and Y directions. The build platform, meanwhile, moves vertically in the Z direction, with the precise lowering of the platform being essential for the thickness of each new powder layer. Another thing moving vertically in between layers is a platform of a material container – for each downward movement of the build platform, the material container moves upwards and additional portion of raw, unbound powder is mechanically distributed in the empty space of the build platform.

This technology parallels the sintering processes used in traditional manufacturing, where high heat is used to compact and form solid masses without melting the material to a liquid state. SLS machines, akin to their FDM counterparts, are essentially advanced automated systems that replace traditional labor-intensive techniques. An important feature of SLS is that it generally does not require support structures as the unsintered powder in the build chamber naturally supports overhangs and complex geometries, reducing waste and post-processing time (Kumar et al. 2021). That really is a game-changer as compared to other popular technologies – complete lack of need of support structures is one thing, already quite important – but similar capabilities can be achieved in UV techniques, specifically PolyJet, where removal of supports is not at all problematic, even in small spaces. However, what really can change the perception here, as compared to FDM and SLA technologies, is that the whole build chamber can be treated as a volumetric entity, meaning that different parts can be stacked one above another, fully optimizing the build space enabling production of multiple parts in a single batch, much more efficiently than in the mentioned technologies where build volume is rather thought in "2.5D" terms – arrangement is only done in a flat area, with height being considered as a limiting dimension.

Managing thermal properties is critical in SLS, as uneven temperatures can lead to warping or poor mechanical properties in the final product (Manshoori Yeganeh et al. 2019). Unlike SLA, SLS involves managing the heat of the entire powder bed to ensure proper sintering without overheating, which could degrade material properties. Usually the whole build chamber is preheated to a specific temperature (well above 100°C), to make it easier for the powder grains to melt. Another crucial aspect is maintaining of a proper atmosphere during sintering – inert gas must be used to prevent combustion of material and degrading of properties, allowable quantities of oxygen inside machine's chamber (depending on the material) should be very low, as presence of this gas influences properties of obtained products (Hu et al. 2018).

Historically, SLS technology was also subject to stringent patent controls, predominantly by DTM Corp. (Dimov et al. 2001). However, companies like EOS and many others slowly entered the market with their inventions. As with other 3D printing technologies, the expiration of these patents has led to a significant reduction in the costs of SLS machines. Today, they are increasingly accessible, not just to industrial users but also to smaller design studios and academic institutions, promising broader experimentation and more innovative applications in diverse fields. However, it is still one of the most expensive additive manufacturing processes and, as opposed to FDM and SLA, it cannot be really considered as a technology possible to implement in non-industrial conditions (e.g., at home). That is both due to price of purchase and maintenance and technical requirements – space, heating, requirement for inert gas supply, etc. make this technology impractical in home conditions, usually a special environment must be prepared for hosting such a machine.

Apart from the sintering or melting processes, there are a number of other proposals regarding selective joining of powders. Some of the solutions involve stream of electrons, others work with plasma arcs, there are also many hybrid technologies, with different kinematics than the original process. These are presented in the next subchapter.

SLS technology in its basic form is traditionally associated with use of polymers (remarkably, polyamide) or polymer-based composites. However, further work by the originators on use of metallic powders have led to creation of the whole branch of metal printing processes. Here we can distinguish direct sintering of metals (DMLS) and also Selective Laser Melting (SLM). Eventually, it has become possible to realize fully metallic solid printouts, using carefully crafted machines and proper materials, notably titanium and nickel alloys, as well as stainless steels and many other superalloys (Yap et al. 2015). As for now, selective sintering or melting is one of the basic methods of additive fabrication of metallic parts, with tons of practical applications (Sefene 2022, Yap et al. 2015). One of the most prominent fields of application of this group of technologies is medicine, especially in the field of customized implants. These applications are described more closely in the last subchapter of this section.

5.4.2 Overview of powder sintering processes and materials

The SLS technology is just one of the additive technologies based upon using heat to join grains of material in form of powder. However, there are many more technologies working around the same principle, but in a different manner. Among these technologies, several key methods have defined the advancement of both polymer and metal parts production. These methods include Selective Laser Melting (SLM), Laser Engineered Net Shaping (LENS), Electron Beam Melting (EBM), and Metal Jet Printing, among others, that due to limited space will not be presented here.

Selective Laser Melting (SLM), similar to SLS, utilizes a high-energy laser beam to fully melt and fuse metallic powders into solid parts. The process is characterized by its ability to produce components with high density and mechanical properties comparable to those produced by traditional manufacturing methods, but SLM produces parts in a slower way than SLS and most other additive fabrication methods. SLM is particularly beneficial for applications requiring strong, lightweight structures such as in aerospace and automotive industries (Hettesheimer et al. 2018).

The materials typically used in SLM include stainless steel, titanium, and cobalt chrome, with aluminum and copper alloys being more difficult, yet still possible to process, due to their distinctive properties. SLM cannot be used for magnesium alloys and has a very high requirement of inertial gas purity (Pauson 2021), in industrial practice often less than 0.01% of oxygen is acceptable. SLM machines are offered by many manufacturers, such as Renishaw.

Laser Engineered Net Shaping (LENS) is a form of directed energy deposition that involves blowing metal powder into a focused laser beam, which is directed onto the substrate where it melts and solidifies to form a part (Izadi et al. 2020). LENS can be used to repair or add material to existing components, making it an excellent choice for remanufacturing and enhancing part performance with gradient materials or complex features. The ability to process multiple materials and adjust properties at different locations within a single part makes LENS particularly useful for creating high-performance, functionally graded materials and components (Izadi et al. 2020). There are many kinematic variations of LENS process, based on printing heads placed inside multi-axial CNC machines, as well as on robotic arms. The technology also has many names, including Laser Powder Forming (LPF).

Electron Beam Melting (EBM) is similar to SLM, but uses an electron beam as the energy source instead of a laser. This process occurs in a vacuum and is well-suited for materials that are reactive at high temperatures, such as titanium alloys. EBM produces parts that are extremely dense and have excellent mechanical properties, making it especially relevant for the medical implant and aerospace sectors (Körner 2016). The parts produced by EBM are also characterized by a distinctive surface texture that can be beneficial for certain applications, such as osseointegration in implants. EBM is usually associated with a Swedish company Arcam.

Metal Jet printing is a relatively new technology that uses inkjet-like printing technology to deposit layers of metal powder followed by a binding agent, which is then post-processed to create dense metallic parts. This process is aimed at high-volume production of parts that are typically smaller and less complex than those produced by SLM or EBM but require fine details and excellent surface finish. Metal Jet Printing is gaining traction in the automotive and consumer goods sectors, where the ability to mass-produce customized metal parts is highly advantageous. The process was first put into practice by the HP company (Hendrixson 2022).

Each of these powder sintering technologies offers specific benefits and is chosen based on the requirements of the part being produced, including the type of material, the complexity of the geometry, the required mechanical properties, and the intended application. As the technology continues to evolve, the integration of these methods into traditional manufacturing processes is likely to grow, enabling more innovative designs and efficient production methods across various industries, including especially the medical industry, with the main application being fabrication of customized implants, as discussed in the next subchapter.

The powder processing additive manufacturing technologies utilize different materials and their available range is generally very wide (Tiwari et al. 2015), much wider than in the case of other technologies discussed in this chapter. These materials can broadly be classified into four main groups: polymers, metals, composites, and ceramics. The choice of material affects not only the printing process itself but also the mechanical properties, thermal stability, and biocompatibility of the finished products.

Polymers are among the most commonly used materials in SLS due to their ease of processing, versatility, and good balance of properties. The most frequently used polymer is nylon (polyamide), known for its strength, durability, and flexibility. Variants such as nylon 11 and nylon 12 are popular because of their reduced moisture absorption compared to other grades (Tiwari et al. 2015), which improves both processing stability and mechanical properties of the final product. These polymers are ideal for functional parts that require high impact resistance and flexibility, such as automotive components, consumer goods, and some medical devices, such as orthoses (Faustini et al. 2008).

Another significant polymer in SLS is polystyrene, used primarily for creating detailed prototypes or patterns for investment casting in the foundry industry (Yang et al. 2009). Additionally, thermoplastic polyurethane (TPU) is valued for its elasticity and is used to produce flexible and durable items (Pan et al. 2020), similarly as in the FDM technology.

Metals are predominantly used in SLM processes, though they are also used in SLS to a lesser extent (in such case, the name of DMLS is then used). The range of metals available for these technologies is extensive, enabling the production of highly durable and structurally sound components. The most common metals include:

- *titanium with alloys* – highly valued in aerospace and biomedical sectors for their strength, lightweight properties, and excellent biocompatibility, basic choice for anatomically customized implants (Leordean et al. 2015);
- *stainless steel* – commonly known for its robustness and corrosion resistance, stainless steel is used across various industries, including automotive, aerospace, and medical (Yap et al. 2015);
- *aluminum with alloys* – these are selected for their light weight and relatively high mechanical properties, making them suitable for aerospace and automotive applications, possibly also some medical (not inside human body due to poor biocompatibility).
- *cobalt chrome alloy* – this material is particularly resistant to wear and high temperatures, at the same time being highly biocompatible (Stamenković et al. 2019), making it ideal for high-stress applications in the medical and aerospace industries.

Each metal has its specific processing parameters and post-processing needs, which can include heat treatment and machining to achieve the desired surface finish and mechanical properties.

The least used are ceramic powders. Ceramics are usually associated with applications that require high thermal resistance, hardness, and chemical stability (Sing et al. 2017). Common ceramics include alumina, zirconia, and silicon carbide. These materials are often used in the manufacture of components subjected to high wear conditions, high temperatures, and corrosive environments – in medicine, mostly in implant department, especially in the dental branch (Chen et al. 2018).

Ceramics are very challenging to process additively, due to their high melting points and poor thermal conductivity, often requiring specific laser parameters and careful handling during the printing process (Chen et al. 2018). However, the development of ceramic-filled composite powders has helped to mitigate some of these challenges, enhancing the usability of ceramics in additive manufacturing (Kletetzka et al. 2023).

Composite materials in SLS and SLM usually consist of a polymer or metal matrix combined with reinforcements such as carbon fibers, glass fibers, or other fillers, including ceramics. Common connections include polyamide filled with short glass fibers, as well as polyamide with aluminum particles. In SLM, the use of Metal Matrix Composites (MMCs) is growing. These composites integrate metals with ceramic reinforcements to produce parts with superior wear resistance and mechanical strength, suitable for high-performance applications in industrial and engineering contexts (Kumar et al. 2022).

As additive manufacturing technologies evolve, the range of materials available continues to expand, offering new possibilities and improvements in terms of part quality and performance. The powder sintering technologies are – as a group – unique in the possibility to draw upon all four basic groups available – polymers, metals, ceramics, and composites. That is why these technologies are so desirable, but also makes them more costly and less available.

5.4.3 Medical applications of SLS

Selective Laser Sintering (SLS) and its close relative, Selective Laser Melting (SLM), as well as other powder sintering technologies, have found significant applications in the medical field, revolutionizing the way bespoke medical devices and implants are produced. Using these techniques it is possible to create highly customized and complex structures that are particularly beneficial in medicine (Mazzoli 2013), out of materials that are unavailable in other well-known additive manufacturing techniques (i.e., metals and metal composites).

In the realm of SLS and SLM, a high-power laser selectively fuses small particles of polymer, metal, ceramic, or composite materials into layers that form solid structures. This capability is particularly suited for the medical sector, where personalized solutions and intricate geometries are often required. Both techniques are employed to produce items ranging from personalized orthoses (Faustini et al. 2008), orthopedic implants (Leordean 2015) and dental fixtures (Stamenković et al. 2019) to prosthetics and even intricate components for surgical instruments (Mazzoli 2013, George et al. 2017).

One of the main advantages of using SLS and SLM in medical applications is the ability to produce patient-specific implants. These implants can be tailored to fit the unique anatomy of individual patients, which is derived from medical imaging data (Li et al. 2023). Design if these anatomically customized implants utilizes similar techniques of work with medical imaging and triangle meshes as already described in the previous chapter. This customization enhances the comfort and functionality of implants and reduces the risk of rejection (Li et al. 2021). Furthermore, the porous structures that can be created using these technologies facilitate better integration of implants with biological tissues (osseointegration) (Shaoki et al. 2016). This porosity can be finely adjusted to mimic the mechanical properties of bone, encouraging natural bone ingrowth and faster healing, as well as controlling the mass distribution, what might prove crucial especially in the maxillofacial region (Bormann et al. 2012).

SLS and SLM are also pivotal in the production of cutting guides and surgical tools that are specifically designed for each surgery, improving the accuracy and reducing the duration of surgical procedures (Krishna et al. 2011). These tools are designed based on patient-specific anatomical data, ensuring a perfect fit and precise surgical outcomes. This has a direct impact on reducing recovery time and enhancing the effectiveness of surgical interventions.

The already mentioned versatility of materials compatible with SLS and SLM extends their utility in medical applications. For example, biocompatible metals like titanium and cobalt-chrome are frequently used in SLM to produce sturdy, lightweight implants that are highly resistant to corrosion and wear (Leordean et al. 2015, Stamenković et al. 2019). Meanwhile, biodegradable and bio-compatible polymers used in SLS are ideal for temporary implants or scaffolds in regenerative medicine (Qu 2020, Kumar et al. 2016). Another critical aspect of these technologies in the medical field is the speed and efficiency they offer. Traditional methods of manufacturing custom medical devices can be time-consuming and costly. In contrast, SLS and SLM can significantly shorten the production cycle, allowing rapid prototyping and the ability to iterate designs quickly based on clinical feedback (Kumar 2011).

Today, the medical industry is increasingly embracing these techniques, not only for their efficiency and customization potential but also for the remarkable innovations they enable in patient care (Pacurar and Pacurar 2016). This shift is indicative of a broader trend toward personalized medicine, where treatments and healthcare solutions are tailored to individual patient needs. However, it is worth mentioning that due to the – still – high price of available solutions, their relative low widespread in medical industry, as well as multiple regulatory and formal hindrances (e.g., certification requirements (Booysen 2019)), process of obtaining of anatomically customized implants is much more costly and time-consuming than in the case of typical implants.

5.5 Vacuum casting

5.5.1 Principles of vacuum casting

Vacuum casting, or vacuum mold casting, has emerged as a versatile and cost-effective manufacturing technique that complements and expands upon the capabilities of various Additive Manufacturing (AM) processes. This method utilizes a vacuum to draw liquid material into a mold which is then allowed to set. Vacuum casting is particularly well-suited to creating high-quality prototypes and small series parts from polyurethane resins and other materials that mimic the characteristics of production parts (Górski et al. 2010b).

The workflow typically begins with an AM-generated model. This model is perfected and finished to the required surface quality since the vacuum casting molds will replicate every detail, including any flaws. The finished model is then encased in a two-part silicone mold. The mold creation process starts with the model being placed in a casting frame, where silicone is poured around it under a vacuum to avoid air bubbles and ensure a perfect mold. Once the silicone mold is cured, it is carefully split and the original AM model is removed, leaving a cavity that is a precise negative of the original model. This mold can then be used to cast multiple copies of the part. Resins that mimic various mechanical, thermal, and aesthetic properties of engineering plastics can be used in the vacuum casting process, making it extremely versatile (Wortmann and Frese 2022).

Additive manufacturing processes, such as Stereolithography (SLA), Selective Laser Sintering (SLS), and Fused Deposition Modeling (FDM), are commonly used to create the initial models or "masters" (Mustafa et al. 2020). The flexibility of AM to create complex geometries combined with the ability to replicate those shapes multiple times through vacuum casting creates a powerful toolset for rapid prototyping and small-scale production.

One of the key advantages of vacuum casting is its ability to produce high-fidelity, functional prototypes or end-use parts rapidly and cost-effectively (Wortmann and Frese 2022). For medical applications, this means that custom prosthetics, anatomical models for surgical preparation, and bespoke medical devices can be produced quickly and with high material quality, replicating the properties of more costly production materials.

The ability to use a wide range of materials and colors also allows manufacturers to simulate different material properties (such as transparency, flexibility, rigidity, and heat resistance) that are crucial for various medical devices and implants (Monteiro Ramos and Simoes 2009). This capability is particularly beneficial in the early stages of product development and clinical trials where functionality and aesthetics need to be tested without the high costs associated with traditional manufacturing methods.

For industrial applications, vacuum casting provides a rapid solution for the production of jigs, fixtures, and other manufacturing aids. Due to the accuracy and repeatability of the silicone molds, components that aid in manufacturing processes can be reproduced accurately ensuring consistent production quality. In the automotive sector, vacuum casting is utilized to produce durable, high-quality parts for use in prototypes and end-use components, such as dashboard pieces, handles, and other interior fixtures. The technology offers a significant reduction in lead times and cost, as well as weight (Vadiraj et al. 2019), especially for luxury or customized vehicle components that are produced in low volumes.

Similarly, in consumer goods, vacuum casting allows for the rapid iteration of designs for products such as electronic housings, wearable technology, and home appliances, facilitating a faster product development cycle and the ability to quickly respond to market demands.

Moreover, the materials used in vacuum casting, primarily polyurethanes, can be selected to meet specific requirements such as UV stability, flame retardancy, and high-impact resistance, which are essential for various industrial applications (Aggenbacht 2004, Wortmann and Frese 2022).

5.5.2 Scenarios of application in the rapid manufacturing processes

By default, the vacuum casting technology is based on molds made out of elastic, rubber-like materials, most frequently – additive (two ingredient, chemically hardened) silicones. Available literature shows many cases of uses of these materials for a mold (Wortmann and Frese 2022). The basic, most typical scenario of utilization of vacuum casting technology with these materials is as following:

1. Obtaining of a "master" model (pattern).

   This can be done, similarly as in the case of metal casting processes, with use of any suitable technology. As vacuum casting is often used together with 3D printing, this technology is a natural choice, but the models can also be fabricated out of wood, clay, hard foam, wax, or any other material.

2. Building a gating system and a molding box.

   Similarly as in the metal casting processes with disposable molds, a whole setup for material inflow into the mold cavity must be prepared. Usually it does not require sophisticated constructions as in the sand molds – it should consist of a simple, cylindrical (or conical) inflow hole and similarly shaped outflow hole at the other side of a model. The gating system can be made "ad hoc" out of predefined elements (plastic tubes), but can also be digitally designed and manufactured additively. The whole setup – the model and the geometrical shapes representing inflow and outflow – should be put in a molding box of suitable shape and size. The box does not need to be box-shaped – the shape can, and should be optimized to reduce raw silicone usage. There are no strict strength requirements for molding box, so it can be made of disposable, recyclable elements – such as cardboard – or out of modular elements (blocks with simple connecting shapes, akin to Lego bricks). The complete box should be sealed (e.g., with a sticky tape), to prevent liquid silicone leakage in the next step.

3. Filling the molding box with silicone in vacuum conditions.

   Here, the "vacuum" part comes in, as the material must be first mixed together, out of the base part and the chemical agent activating the resin curing process. Then, the material can be degassed in itself (in a separate container, put inside vacuum chamber), poured into the molding box and then degassed inside the box (again placed inside the vacuum chamber). Whether the material will be degassed once, twice or not at all depends on multiple factors – the lifetime of the silicone, material tendency to form air bubbles, mold size, quality requirements and many others and usually is decided on the basis of experience of a technician realizing it. Anyhow, it is generally favorable to perform double degassing, to avoid air bubbles potentially disturbing the shape and later causing improper model representation.

4. Demolding.

   After the silicone bounds and fully cures, the demolding can take place – it can be one hour or a full day, depending on the used material. Demolding does not have a specific scenario, as opposed to traditional metal casting methods. Here is when advantage of silicone molds comes in handy. It is elastic, so that the models themselves do not need to adhere to strict geometrical requirements (e.g., drafted walls) and the division of the mold is usually non-symmetric. What is more, the division surface is usually not a plane and there can be more than one surface – the mold can be divided into more than 2 parts. The process of demolding boils down to cutting the silicone block with a blade, forming a zig-zag line (for better alignment of mold parts later), caring not to damage the original model, if it is still usable. The model and the other geometrical elements must be removed, any leftovers as well. The mold is ready for casting immediately after.

5. Casting with target material – fabricating small series of a product.

   Now, when the mold is ready, it can be used for fabrication of a number of parts. Durability of typical silicones used for this type of process (such as XTX 45 (Wichniarek et al. 2014)) is usually of approximately one hundred parts, after which signs of degradation become visible and there can be problems with accurate shape representation.

The casting process course is dependent on the material. If it is a – typically used – chemically hardened resin (e.g., polyurethane or epoxy resin (Górski et al. 2010b)), then the process usually requires the vacuum chamber. The material is mixed together, degassed and poured into the mold in the condition of near vacuum. There are vacuum chambers designed for that type of process, equipped with actuators enabling stirring and pouring of the material. The mold must be sealed before casting, so usually its parts are tightly glued together with adhesive tape, to prevent any leakages. Vacuum helps to prevent short-run castings and smaller volumetric errors. However, in many cases pouring in vacuum can be skipped – especially if the material is degassed prior to casting or the volume of a part is small. Also, different materials require different approaches and some of them (e.g., wax or low-melting metals, which also can be used here) do not require degassing at all.

After the material is poured into mold, an appropriate time period must pass for it to fully bind, and then the demolding process is realized, usually by removing the adhesive tape, opening the mold and removing the casting trying not to damage it or the mold in the process (made easier by elastic properties of silicone).

Examples of a silicone mold, tools for its creation and casting using polyurethane resin are shown in Fig. 5.7 (based on descriptions and work realized under the author's supervision in (Jankowska 2015)). Notable examples can be also found in previous publications of the author with the team (Górski et al. 2010b), as well as in literature (Kroma et al. 2020). It is noteworthy that this process enables relatively low-cost production of a small series of products with customized shapes, so it can be frequently used and repeated in industrial conditions, both for prototyping and manufacturing of final products. By the author's experience, level of shape and

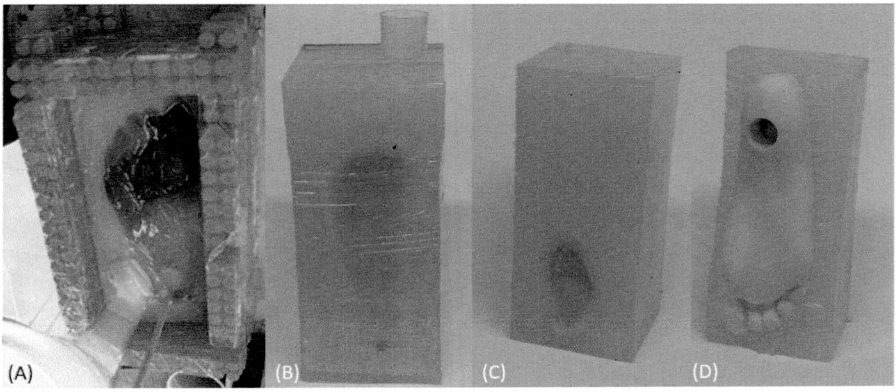

Figure 5.7 Vacuum Casting process for a foot prosthesis, (A) modularly built silicone mold, (B) mold prepared for casting; (C) mold after casting; (D) removing model from the mold.

dimensional accuracy is very high, i.e., the silicone very well represents the shape and surface quality of an original model. It is also noteworthy, that potential size of a silicone mold is not as limited as in the case of 3D printers' usual working chambers – quite large molds can be built, for quick casting of complex plastic parts.

The second type of casting process involving vacuum is basically a reverse of used materials. The mold – of a customized shape – is digitally designed and then 3D printed using any of available processes (FDM, SLA, etc.). Then, an elastic material – a silicone or something similar – is poured in, in conditions of vacuum or after degassing in vacuum – to obtain a part having rubber-like properties (Wichniarek et al. 2014). The requirement of existence of such a process is mostly due to lack of possibility of easy 3D printing with very soft silicone materials. Recently a number of additive manufacturing machines working directly with silicones were made commercially available, partially filling this gap (Liravi and Toyserkani 2018). Still, they are not as readily available, they can be costly and not all silicone materials are able to be handled by these machines. That is why the "reverse" process can be considered as a relatively easy alternative to direct silicone printing.

The stages of the process are as following, as based on previous work by the author with the team (Wichniarek et al. 2014, Żukowska et al. 2019):

1. Designing the part and the mold.

   Assuming that the original part already has a digitized form (either coming directly from a CAD model, or through reverse engineering measurements, as described in Chapter 3), the first step of the silicone vacuum casting process involves creating a digital shape of the mold. It should be designed according to general requirements and constraints toward molds for plastic materials, usually divided into two halves (with a set of geometrical features for aligning them together), with elements of gating system. Usually, this process starts with some Boolean algebraic operations – subtracting the original part's solid from a general mold shape (usually a box, but not necessarily). Then, the mold is divided into parts, gating system elements (holes) are added, along with features for alignment. Similarly as in the typical process, the material usage for a mold can be optimized through shape modifications.

2. Manufacturing the mold.

   At this stage, the designed mold is manufactured using one of available additive manufacturing technologies (again, not necessarily, but it is usually the quickest way). If surface quality is not crucial, simple extrusion printing (FDM) should be enough. If there are intricate shapes and requirements for better surface quality, UV technologies can be utilized, preferably PolyJet (for build chamber size, easiness, and tidiness of the process). The mold can be processed afterwards, to obtain the desired quality.

3. Casting.

   Similarly as in the typical process, here the material is poured into the closed and sealed mold. The material is usually silicone, requiring degassing in most cases, with the degassing possible to be realized before or after pouring the material into the mold. It is noteworthy that two or multi-material castings are possible. Some parts may be 3D printed beforehand and put into the mold before casting, therefore leading to obtaining of a part containing many structures made of different materials. Also, two or more different silicones can be used sequentially, first making a casting of one part and then making a casting of another part, with the first one inside.

After the material binds together, demolding takes place – it can be slightly more difficult, considering that the mold is usually rigid, but not as strong as metallic molds, conventionally used for plastic casting (e.g., in injection molding). However, elastic properties of the casting material should make it easier to remove the model. The process can be repeated many times, similarly as in the typical scenario. Examples of mold design and stages of fabrication are shown in the Fig. 5.8 (based on previous work, described in (Żukowska et al. 2022)).

It is noteworthy, that this "reversed" vacuum casting process can be realized using relatively low-complex setups – a 3D printer and a small vacuum chamber is all that is needed (except for the materials themselves). However, maximum sizes of molds are considerably smaller – as they need to be 3D printed, the size chamber limitation applies, as well as time of fabrication, which is usually considerably

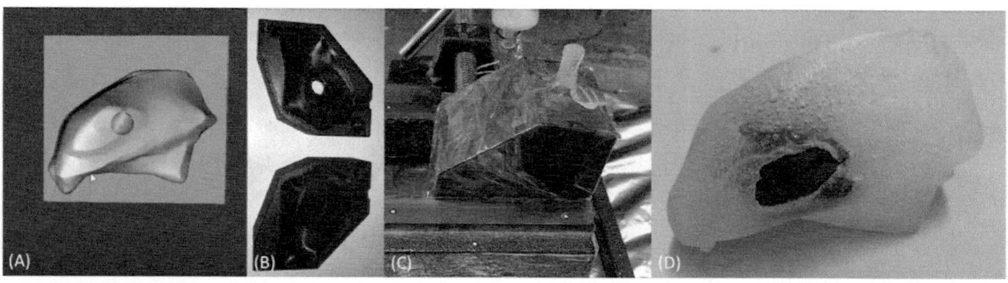

Figure 5.8 Vacuum Casting process for a tongue model, (A) mold design; (B) 3D printed mold; (C) mold filling; (D) finished part.

longer for a mold, than for a base model of the same shape. Still, it remains one of few feasible options to obtain customized, advanced shapes out of very soft silicone materials – a quality indispensable in a number of medical applications.

5.5.3 Examples of medical use of vacuum casting

Medical use of vacuum casting – in manufacturing of anatomically shaped devices – is especially concentrated on leveraging two important highlights of this technology, differentiating it from the regular additive manufacturing:

1. Possibility to manufacture a number of products of the same shape – quickly, cheaply and with acceptable repeatability and accuracy, possibly using different materials for properties and aesthetics customization between produced pieces.
2. Capability of building models out of very soft materials (silicones), also being transparent (or semi-transparent), as well as multi-material models involving these silicones and other 3D printed or cast parts.

These distinct characteristics favor a natural application of vacuum casting – it is building of multi-material anatomical models for surgical preparation, for the doctors being able to properly familiarize themselves with a specific medical case and train some procedures beforehand. A typical example is use of such models to simulate cutting out cancerous tumors with an appropriately large margin of healthy tissue to remove (to prevent cancer spreading (de Koning et al. 2021)). Such models can be also used for simulation of suturing. They are invaluable if vessels and nerves are included, together with transparency, as it gives the doctors a review of risks and potential errors that may happen during an actual operation.

Some examples of these models, out of prior experiences of the author and the team, are shown in Fig. 5.9. Such models can, potentially, be useful regarding almost all organs and areas inside human body. Figure 5.9A shows a model of a tongue, with cancerous tumor, in the process of simulated surgery. In this study, various types of materials were used to simulate the tumor, ranging from TPU processed by extrusion, to various types of silicones. This work is based on a previous thesis and a paper (Żukowska et al. 2022). Figure 5.9B shows an example of a kidney with a cancerous tumor, with kidney cortex being cast of silicone and all the other structures (tumor, veins, arteries) 3D printed out of rigid thermoplastic material and put inside a mold before casting (Żukowska et al. 2019).

Anyhow, such models are not limited to the kidney or tongue – literature analysis (Żukowska et al. 2023) shows that they can be used for other body parts as well, such as liver, heart, and parts of reproductive system.

Building these models requires applying the second, "reversed" process (with 3D printed rigid mold, as described in the previous subchapter). As they are useful in preparation for a surgery, it should be also noted that the combined process (design, 3D printing, casting) can be long and cumbersome, so that in many cases such models can be obtained more easily through use of technologies such as PolyJet, as shown in a case in Chapter 7 of this book. However, PolyJet has material limitations, which are virtually non-existent in the vacuum casting process (regarding the soft

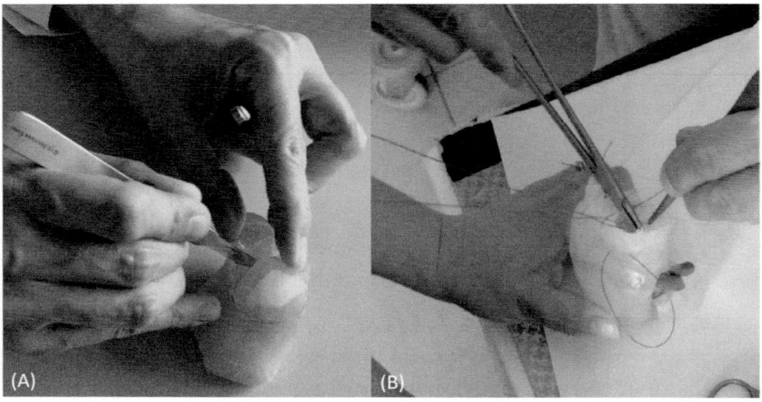

Figure 5.9 Vacuum Casting medical examples, (A) kidney; (B) tongue.

silicones). Anyhow, as the molds can be used many times to multiplicate the models (cheaply and without using the working time of 3D printers), they could be also used for different purposes. These are, mostly:

1. practical education of medical students and young doctors – the cases can be re-created for didactic purposes and training of surgical procedures;
2. improving psychical comfort of the patients and their families – building these models and showing them to patients before the operation can calm them down and make them believe in confidence and competence of the doctors, which may affect healing results (Hu et al. 2016);
3. for post-surgery purposes, e.g., to compare with 3D imaging and/or 3D printed model of a given organ of the same patient some time after the operation, to evaluate results of undertaken healing procedures.

There are, of course, other applications of vacuum casting in medicine, compatible with the already mentioned general applications, exploiting the benefits of this process in comparison both with conventional and additive manufacturing technologies. Examples – of the author's experience – include, but are not limited to:

- producing repetitive, not anatomically adjusted parts of low-cost customized prosthetics or orthotics,
- producing anatomically adjusted prosthetic parts out of elastic materials or using many materials, e.g., foams used for prosthetic feet,
- producing customized, not anatomically adjusted surgical tools (e.g., customized for a given doctor or group of patients, thus requiring relatively frequent fabrication),
- fabricating parts of prototypes of medical equipment before launching actual production, including both anatomically customized and non-anatomical shapes.

To sum up, it should be stated that vacuum casting, supported by models created via additive manufacturing, offers a unique combination of speed, cost efficiency, and material versatility, making it an invaluable part of the prototyping and small-scale production toolkit, especially in medicine and medical engineering. It bridges the gap between one-off prototype development and full-scale production, allowing for thorough testing and modification without the expense and time commitment of traditional manufacturing techniques, as well as use of a range of materials normally unavailable for popular additive manufacturing processes. This synergy not only accelerates product development cycles but also enables a level of customization and experimentation that is otherwise difficult to achieve.

5.6 Possibilities and limitations of available technologies

5.6.1 Additive manufacturing versus conventional manufacturing processes

As already mentioned, additive manufacturing is a group of technologies that is the youngest of the whole portfolio of fabrication techniques that we have developed as a technological civilization. Obviously, a general idea of starting with an empty "room" and then adding material in a layered fashion to obtain a desired shape is as old as humanity, probably. However, to do so in a mechanized way, using capabilities of machine tools was not available until the last two decades of the previous century.

Having that in mind, the most obvious fact about additive manufacturing is that the technology is purely dependent on computers and all the 3D printers are numerically controlled machines. As opposed to many other processes – drilling, lathing, foundry, forging, etc. – here, not having a digital 3D model of a part that we want to fabricate makes the very fabricating impossible. There are no "analog" 3D printers and, as some of them may be manually operated (for instance, for debugging purposes), it is not really an option to fabricate anything of value. This makes for a huge difference between 3D printing and previously mentioned technologies, and also can be a hindrance, as shown below:

1. To make a 3D print, we need a digital, "watertight" 3D model of a part that we want to fabricate.
2. To prepare a 3D printing program, we need a software that will be able to process such model and create an NC code for the machine.
3. To run the NC code, we need to transfer it to the machine, either offline (via USB connection, memory card, etc.) or online (via wireless connection or through Ethernet).
4. The machine requires to be powered up the whole time the fabrication process is realized – it cannot be really paused and resumed, at least not in most technologies.

As opposed to many technologies where "analog" approach still works – e.g., forging, casting, welding, etc. – the additive manufacturing requires extensive use of computers, which is dependent on commonly used operating systems, available software (and license for it, often based on subscription model these days), as well as firmware of the machines themselves. As such, maintaining production capability might be more troublesome and costly than in the case of traditional technologies.

Another difference between 3D printing and other, older technologies is a requirement for specific competences. The operation of a low-cost 3D printer is not troublesome – it is a rather quick process to learn, although the details and troubleshooting might require deep experience (in general, the "easy to learn, difficult to master" saying works perfectly here). However, to even begin, the aforementioned 3D model must be acquired first and creating these models requires a separate skillset in itself. For starters, CAD or other 3D modeling software knowledge is required; if medical models are considered, also knowing how to work with medical imagery might be essential. It means, that traditionally skilled technicians – i.e., operators, perfectly capable in working with manual tools and machines such as lathes, presses, injection molding machines, etc. – cannot realize the whole 3D printing process as they are not able to create 3D models themselves, requiring another person – a designer – to be involved. That is not the case in traditional technologies – many skilled technicians can "recreate" a given part without any kind of 3D model, just based on the physical part, drawings, or technical descriptions. That might pose organizational and logistical challenges, as launching a 3D printing facility requires employing people of diverse competences.

Discussing additive manufacturing, "paradigm shift" is a notion that is frequently used – and it applies not only to the whole layered deposition process, and not even to the whole 3D modeling problem, mentioned above. There is another huge difference in the additive process course as compared to other processes, manifesting itself after the process starts. It is, by nature, completely automated and uninterruptable. In traditional manufacturing processes, we may divide a single process into a series of activities, which might be manually controlled or automatically launched – such as a single movement of a machine tool over a processed part. These activities can be sequentially realized with pauses in between and the processed part can be removed out of a machine (e.g., out of a milling machine, or a robot welding cell) for the process to be resumed at another time. That is not the case in the 3D printing realm – once the layer deposition process is started, it cannot be stopped and resumed later. If the process is interrupted, what has already been fabricated is usually scrap, waste, or recycling material in the best case.

We cannot remove a semi-printed part and put it back into the machine to resume printing at a later date, for a number of reasons. These reasons are technology-dependent, but most of the widely popular additive manufacturing processes are like this – in FDM, the part is adhesively bound to a table and putting it back at the same location is nearly impossible. The same applies in SLS – putting the semi-finished part into the powder bed at exactly the same location is an impossible task. Even if we simply stop the process to resume it later, it still can end with failure, frequently due to thermodynamics – we will not obtain the same temperature gradient, enabling smooth connection between layers. All this, combined with quite lengthy times of layer deposition, causes the additive manufacturing processes not that elastic in terms of production planning as some of the other processes. However, the automated aspect also has its upside – operator's interventions and constant supervision are rarely required, so effectively the operator is free to do some other work at the moment the machine launches the program.

Before continuing with the more specific, technical details of additive manufacturing limitations, it is worth to bring forth another thing specific for the additively manufactured parts (for most technologies) – anisotropy of properties and their dependence on the process parameters. Orienting the part in a working chamber, resulting in a specific division into layers, has far more implications than any process parameter we might apply in the traditional fabrication technologies, such as machining. The layer division affects all the properties of the process and the part – cost and time, dimensional accuracy and strength, as well as surface quality (Górski et al. 2015, Vora et al. 2015). Moreover, the obtained parts are usually anisotropic in properties, e.g., having much different tensile strength in one direction than in the other (Maroti et al. 2018), or maintaining acceptable dimensional accuracy measured in one axis, while not keeping it in the tolerances field in the other (Górski et al. 2013, Vora et al. 2015). This influence is lower for the techniques with smaller layer thicknesses (such as SLA), but it still exists and is a factor that must be considered.

Overall, it is worth noting that the additive manufacturing technologies are simply different that the traditional ones, and mostly are not aiming to replace the older ones, but rather complement them and create entirely new fields and areas of application – with medicine being one of the strong examples. In the further subchapters, limitations of the current generation of additive manufacturing processes are discussed on a more technical level.

5.6.2 Limitations of additive manufacturing - overview

As demonstrated in further sections of this book, additive manufacturing brings significant advancements and transformative potential to the medical field, offering customization and rapid prototyping capabilities matched by no other technology. However, despite these advantages, there are some limitations of additive manufacturing when compared to traditional manufacturing technologies.

One of the most prominent limitations of additive manufacturing is the issue of material properties. Traditional manufacturing methods, such as injection molding and CNC machining, utilize a wide range of materials, including high-performance metals and medical-grade plastics, which are often superior in terms of mechanical strength, durability, and biocompatibility especially. In contrast, the materials currently available for 3D printing, particularly for processes like FDM, SLA, and SLS, are more limited. While advancements are being made, the range of materials that can be printed with the same reliability and performance as those used in traditional manufacturing remains relatively narrow (Iftekar et al. 2023). For medical applications, this limitation can be critical, as the materials must meet stringent regulatory standards for safety and efficacy. Basically, it must be stated that not many additive manufacturing processes are able to fabricate devices for direct use inside the patient's body (belonging to II and above class of medical devices, according to MDR regulations). It is actually only metal sintering and melting, plus some of the UV techniques (utilizing dental-grade resins) that make it possible to generate implant-like devices, as well as bone-growth stimulating filaments for the FDM technology, but these are still quite rare (Targońska et al. 2024).

Another significant limitation is the accuracy and surface finish of 3D-printed parts. Traditional manufacturing techniques – especially subtractive ones – can produce parts with extremely high precision and smooth surfaces, often required for medical devices that come into direct contact with the human body, not to allow any kind of microorganisms to settle in the pores of an uneven surface (requiring frequent sterilization). In comparison, 3D-printed parts, particularly those produced by FDM, can exhibit layer lines, surface roughness, and dimensional inaccuracies. These surface imperfections can be detrimental in medical applications where precision is crucial, such as in surgical instruments or implantable devices. Post-processing steps, such as sanding, polishing, or coating, are often necessary to improve the surface finish of 3D-printed parts, but these additional processes can increase production time and costs. Often they require manual labor, making the process troublesome in "battlefield" conditions (e.g., when printing in a hospital).

However, it is worth mentioning that there are cases when rough surface and porosity of additively manufactured anatomically adjusted devices comes in handy – especially when discussing implantable devices that are meant to be overgrown with human tissue. Additively manufactured bone implants, for example, should be porous to facilitate osseointegration. However, in most other cases, the imperfect surface quality is rather a drawback, than actual advantage.

The production speed of 3D printing is another area where traditional manufacturing holds an advantage. While 3D printing excels in rapid prototyping and producing complex geometries, it is generally slower than traditional methods when it comes to mass production. Traditional manufacturing processes, like injection molding, can produce thousands of identical parts in a fraction of the time it takes to print the same quantity using additive manufacturing, whereas the layer deposition process is usually lengthy and in some processes – such as extrusion – the "glass roof" has already been hit and not much more can be done to improve the speed, except changing the very principles of the process. This limitation makes 3D printing less suitable for large-scale production runs of standardized medical devices, where speed and efficiency are paramount. In general, 3D printing should not be considered as a viable, long-term option for manufacturing more than a couple of dozens of pieces of identical parts, except in the case no other manufacturing processes are available (e.g., in conditions of war or other catastrophes, where 3D printers, due to their portability, might easily be used even in unfriendly environments).

Cost is also a critical factor to consider. Although the initial setup cost for traditional manufacturing can be high, especially for custom molds or specialized machinery, the cost per unit decreases significantly with high-volume production. In contrast, the cost per unit for 3D printing remains relatively constant regardless of the production volume. This economic disparity makes traditional manufacturing more cost-effective for producing large quantities of medical devices, especially if their shape is not very intricate and – as a result – difficult to manufacture using traditional technologies. However, for low-volume or highly customized products,

3D printing can offer a more economical solution, considering the whole process (along with the design phase).

The scalability of 3D printing is another limitation when compared to traditional manufacturing. While additive manufacturing can produce highly customized and complex parts, scaling up production to meet large demands can be challenging. Traditional manufacturing processes are designed for scalability, allowing for the efficient production of large quantities of parts with consistent quality. In contrast, scaling up 3D printing often requires multiple machines – also known as 3D printer farms (Schmidt et al. 2022) and significant investment in infrastructure, which can be a barrier for widespread adoption in high-demand medical applications. Additive manufacturing is also not easy to fully automate using industrial robots – first of all, there is price disparity between 3D printers and robots (the latter being far more expensive) and secondly, it is much more difficult and economically unviable to program the robots to fully automatically realize all the stages of 3D printing of customized products (material loading and unloading, removing and post-processing parts, etc.).

Regulatory and certification challenges also pose significant hurdles for the adoption of 3D printing in the medical field. Medical devices must comply with rigorous standards set by regulatory bodies such as the American FDA or CE marking requirements in the European Union. Ensuring that 3D-printed devices meet these standards can be more complex than for traditionally manufactured devices, particularly when considering the variability in material properties and printing processes, as well as constant change in geometries when producing anatomically customized devices. Establishing Quality Management Systems (QMSs) and validation protocols for 3D-printed medical devices remains a challenge, that manufacturers must address to achieve regulatory compliance. That challenge has not yet been sufficiently described in available literature (Martinez-Marquez et al. 2019).

Additionally, the long-term durability and reliability of 3D-printed medical devices are areas of concern. Traditional manufacturing methods have a long history of producing reliable, long-lasting products. In contrast, the relatively recent adoption of 3D printing means that there is less data available on the long-term performance of 3D-printed medical devices, especially these used inside or in contact with the human body. Issues such as material degradation, wear, corrosion, and the potential for structural weaknesses due to layer adhesion (layer disjoint) problems need to be thoroughly studied and understood to ensure the safety and effectiveness of 3D-printed medical devices over time. There are plenty of studies available so far with promising results (Ali et al. 2020, Sheoran et al. 2020), but only the decades to come will show how reliable 3D printed devices really can be in the medical realm.

Despite these limitations, 3D printing continues to offer unique advantages that complement traditional manufacturing, particularly in the realm of customization and prototyping, as mentioned multiple times. The ability to quickly produce patient-specific devices, and to rapidly iterate designs based on clinical feedback, makes 3D printing an invaluable tool in personalized medicine. Furthermore, ongoing advancements in 3D printing technologies, materials, and processes are steadily addressing many of the current limitations, expanding the potential applications of additive manufacturing in the medical field.

As the field continues to evolve, the hybrid approach would probably be an optimal way to go. Integration of 3D printing with conventional methods promises to produce synergy for both technologies and enhance the capabilities and reach of medical manufacturing, paving the way for innovative and effective healthcare solutions, unavailable before emergence of additive manufacturing.

5.6.3 Strengths and weaknesses of additive manufacturing in medical use

As discussed in previous subchapters, additive manufacturing is a technology greatly differing from the conventional, traditional manufacturing methods. This provides some opportunities, some challenges and also considerable limitations, which were discussed on a general level. In this section, we will discuss it focusing on particular additive manufacturing processes, covered earlier in the chapter. Fused Deposition Modeling (FDM), Stereolithography (SLA), and Selective Laser Sintering (SLS) represent three of the most prominent additive manufacturing technologies, each with its own set of strengths, weaknesses, advantages, and limitations – so really they should be considered separately, when discussing technical details of medical manufacturing processes.

Fused Deposition Modeling (FDM) is widely recognized for its cost-effectiveness and versatility. One of the primary advantages of FDM is the relatively low cost of both the machines and materials. This technology supports a broad range of thermoplastic materials, including previously mentioned PLA, ABS, and PETG, which are affordable and readily available. The open-source nature of much of the ecosystem of the FDM/FFF allows for significant flexibility and customization to meet specific user needs. The ability to use different support materials that are easier to remove further enhances the versatility of FDM. Machines are simply designed and constructed, requiring low maintenance and repairs can be done without extensive skills. Additionally, FDM, in some of its implementations, is capable of producing large-scale prints and can simultaneously print with multiple materials – although usually no more than two. Still, it is beneficial for creating complex structures, especially in orthotics or prosthetics – such as fabricating orthoses with a hard, durable shell and internal filling made of soft, elastic material inside (Górski et al. 2019b). Shortly – FDM can be used to print almost anything, almost anywhere, at a very low cost.

However, FDM also has notable limitations. The process often results in parts with a rough surface finish and a visible layer or staircase effect, which can detract from the aesthetic and functional quality of the final product. Dimensional accuracy is typically around ± 0.2 mm (or worse), which may not be sufficient for high-precision applications, or precise mechanical connections in otherwise not as demanding parts (such as locking in two parts of a hand orthosis). The printing speed does not scale well with the number of parts being printed unless using specialized machines like Independent Dual Extruder (IDEX) printers (Gautreaux 2021). Creating small and intricate details can be challenging, and the mechanical properties of the printed parts are highly dependent on the print settings – especially orientation – and the quality of the material used. For certain materials with high shrinkage rates, such as ABS and PEEK, a heated enclosure is necessary to minimize warping and ensure layer adhesion. Moreover, the ability to recycle printed waste – fully possible from a mechanical point of view – is frequently limited material-wise, as the properties degrade after each cycle of melting and re-extrusion (Gomes et al. 2022), although there are some studies that prove that this may not be always the case (Vidakis et al. 2021).

In the medical field, FDM offers significant advantages, particularly in producing low-cost, customized prosthetics and orthotics, as mentioned multiple times and emphasized in the author's own work. These devices are tailored to the specific anatomical needs of patients, including children who require frequent replacements due to growth. The relatively low cost and ease of use make FDM a practical choice for printing medical devices in various settings, such as hospitals, clinics, and even patients' homes, realizing the "hospital bed printing" scenario. However, the primary disadvantage in medical applications is the rough surface finish and lower precision compared to other technologies, which may not be suitable for all medical devices requiring high accuracy and smooth surfaces. For instance, printing intricate bone structures in the maxillofacial area with use of FDM is not a viable option – as the author's experience indicates – the UV processes are far better for that purpose.

Stereolithography (SLA), including its variants, like LCD SLA and DLP, offers high precision and excellent surface finish, making it ideal for applications requiring fine details and smooth surfaces. SLA materials can be biocompatible and suitable for medical applications, especially regarding dental models (where resins identical to the dental ones could be used), surgical guides, and hearing aids. The ability to print very small features makes SLA ideal for producing detailed anatomical models for pre- and mid-surgical purposes, as well as intricate surgical instruments.

However, SLA also has many drawbacks. The raw materials, namely photopolymer resins, are typically more expensive than FDM filaments and also much more troublesome to handle. The resins are toxic in their uncured state, necessitating the use of protective equipment and proper ventilation during the printing process and after removing the part. Post-processing steps, such as washing in isopropyl alcohol and additional UV curing, are required to fully solidify and clean the printed parts, which adds to the overall production time, and, as already mentioned, making the process difficult to implement in many environments (including the hospital conditions). The working area of SLA printers is also usually much smaller than that of FDM printers, limiting the size of the parts that can be produced, practically eliminating the possibility to produce orthotic or prosthetic devices for limbs. Furthermore, the durability of SLA parts can be a concern, as they are often more brittle compared to those produced by FDM and SLS and SLA, in its typical form, allows printing only of a single material. However, implementation of PolyJet process helps overcoming most of these issues while maintaining the advantages of UV resins – at an increased cost, unfortunately, as already pointed out.

Selective Laser Sintering (SLS) offers a different set of advantages and limitations than the previously mentioned techniques and is, by many aspects, a whole different world. One of the key strengths of SLS – and the whole similar group of technologies based on joining powders with heat – is its ability to produce parts with excellent mechanical properties and relatively small anisotropy. The un sintered powder acts as a natural support structure, allowing for the production of complex geometries and intricate details without the need for additional support material, also allowing unique capability of realizing multi-part printing over the whole volume of build chamber (planning parts' positions not only on a plane on the build table). SLS also enables the reuse of leftover powder, frequently making it a more sustainable option compared to other additive manufacturing technologies.

However, SLS also has limitations. The surface finish of SLS parts is typically rougher than those produced by SLA, and the process can leave parts with a porous texture, especially in thin-walled sections. This might not be a problem in certain applications requiring porosity, however in most cases this means more work during the post processing. The machines and materials used in SLS are also considerably more expensive than those used in FDM and SLA, which can be a barrier to adoption for smaller businesses, individual users, or healthcare facilities. Additionally, SLS requires a slow and controlled cooling process to prevent thermal stresses and warping, which can extend the overall production time – already long, as the layer deposition process is usually much longer than in the FDM process (comparable to SLA processes). The necessity of post-processing stations and appropriate infrastructure further increases the complexity and cost of the SLS process, making it even more difficult to implement in variable, or "battlefield" conditions.

SLS offers substantial benefits for the medical domain, particularly in producing durable, complex, and lightweight prosthetics and orthotics, as well as implants ready to use inside a patient, when discussing printing out of biocompatible metallic powders (e.g., titanium or cobalt-chrome) (Dong et al. 2020). The ability to create intricate internal structures and lightweight designs without additional support materials is a significant advantage, especially for precise weight distribution control in bone implants and prosthetic devices. SLS is also useful for producing devices and instruments that require high strength and durability. However, the rough surface finish and higher cost of SLS can be disadvantages, especially for applications requiring smooth surfaces and affordability. A huge disadvantage is also cost and requirement of higher qualifications to program and operate SLS devices – this is a more high maintenance process and cannot be implemented anywhere easily.

Overall, each of these additive manufacturing technologies has its own set of strengths and weaknesses, making them suitable for different applications. FDM is cost-effective and versatile, able to be easily upscaled and widespread, but limited by its surface finish and precision. SLA (and the family of related processes) offers high precision and excellent surface quality but is more expensive and requires extensive post-processing. SLS (considered as the whole group of technologies) provides strong and durable parts with complex geometries with unique possibility of printing in biocompatible metal, but comes at a much higher cost and with a rougher surface finish, with more challenging implementation. In medical applications, the choice of technology depends on the specific requirements of the device being produced, balancing factors such as cost, precision, surface finish, and mechanical properties. Considering the devices presented in this book, all the mentioned technologies can be used, although each implementation requires careful consideration and experiments to find out the solution which will be ultimately the best for the healthcare recipients.

CHAPTER 6

Design and 3D Printing of Orthopedic and Prosthetic Devices

6.1 Individualized orthopedic devices: classification and design

6.1.1 Basic concepts and definitions

Orthopedic devices have been integral in the treatment of musculoskeletal disorders and injuries, as well as limb disabilities. The development of orthopedic devices has a rich history, with traditional methods forming the foundation for contemporary orthopedic technology. The literature review presented in this and the next subchapter explores the traditional, still maintained practices and techniques used in the creation of orthopedic devices, as well as the current methods employed in the production of orthoses and prostheses.

There are three main classes of orthopedic devices, two of which are frequently customized for the patient's anatomy and, as such, are of interest from the viewpoint of these books. These three classes are prostheses, orthoses, and assistive devices. Overview of these classes is presented in Fig. 6.1.

Orthoses are medical supplies that keep a selected part of a patient's body rigid and safe during healing or convalescence (Górski et al. 2020b). This is usually done by realizing two distinct functions: protection of a given area (by covering it in an appropriate material) and immobilizing a given joint, to ensure time and proper conditions for healing processes to occur. The orthoses can also be used for enforcement of a specific position and mutual orientation of various body parts (Górski et al. 2019b). They can be universal, relatively inexpensive mechanical devices or customized (much better in terms of healing function and comfort) products made for a specific patient based on their anatomic measurement (Górski et al. 2020b).

Prostheses, on the other hand, are a wider term pertaining to a notion of replacing a damaged, lost, or non-functional organ or body part with its synthetic counterpart, according to most definitions (Enderle and Bronzino 2011, Memon and Rizzo III

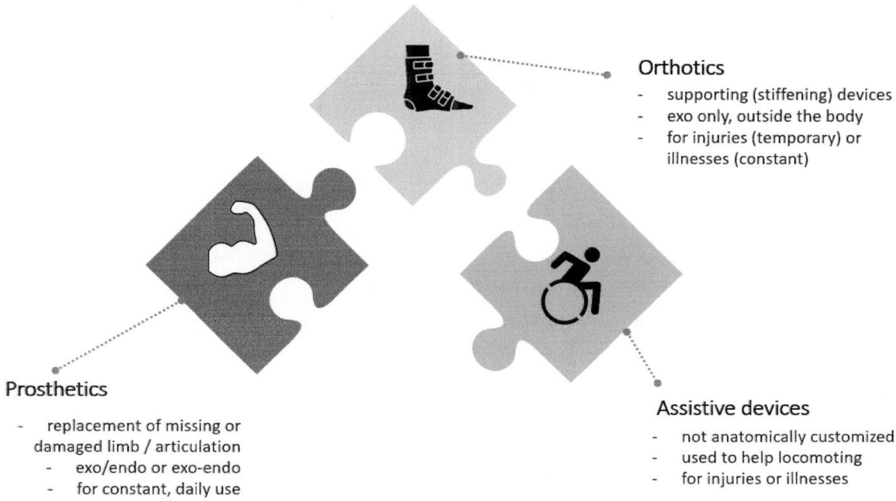

Figure 6.1 Overview of orthopedic devices.

2009). The concept is very wide and in its most broad definition it pertains also to artificial organs (Galetti et al. 1995), but in this book, we understand them as exoprostheses, i.e., prosthetics aimed to somewhat replace the missing external body part. For the purpose of the case studies considered here, the narrowing down will go even further, to comprise only devices replacing the missing limbs, or their parts. Treated as such, the limb prostheses can be classified to a similar category as the limb orthoses – devices aimed at improving pathological conditions and problems in the human movement system.

However, while orthoses are used mostly for treatment of specific conditions, and as such are treated as a temporary device, prostheses are usually needed by patients that otherwise could be completely healthy, just missing a body part. That is why in medical terms these two classes of devices function sometimes on a similar level, especially when narrowing the concept down to limbs (upper or lower), but at the same time can be treated completely differently – orthoses like a "cure", prostheses more like a "tool" or "toy" (depending on their purpose and age group of the recipients).

A general overview of conditions that may cause a need for use of orthosis, prosthesis or both is presented in a schematic in Fig. 6.2. This is by no means a comprehensive list – it includes only examples of conditions, also being a list of most frequent cases for which the author and his team developed and produced customized devices (some of which are presented in the further part of this chapter).

The limb orthoses can be defined as devices that can have one or more of the following functions:

- preventing mobility in a selected joint (one or more), to enable healing of a damaged bone or joint structure after an injury,
- adjusting the shape of the limb in its joints to correct pathological geometry, resulting from a long-term condition,

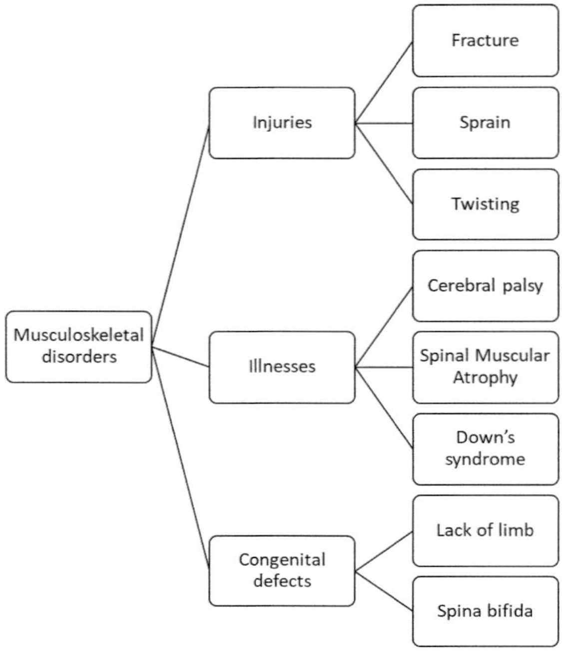

Figure 6.2 Overview of disorders necessitating the use of customized orthoses or prostheses.

- enabling proper and physiologically correct movement of a limb, partially or completely restoring its functions, in treatment of short-term injuries or long-term conditions – it can be a byproduct of either the first, or second, or both functions mentioned above and it can serve daily-life use or therapy (Jackson 2011).

The orthosis must cover and be adjusted anatomically to a selected part of the limb, so these devices are either made of flexible materials (easily adjusting to different geometries) or are custom-made for specific patients, based on their geometry. The orthoses are rarely a life-time use devices – for the short-term injuries their use rarely exceeds a dozen of weeks. For long-term diseases they are used longer, but usually, as the patient's state, abilities, and body dimensions change, they need to be replaced to adjust to it.

The limb prostheses can, similarly, be defined as devices having the following functions:

- replacing one or more functions of an original limb or its part that was lost by amputation or never developed (inborn defects),
- enabling to maintain proper body balance, preventing, or curing various types and degrees of scoliosis,

Similarly as in the case of orthoses, prostheses must be adjusted anatomically to represent selected features and dimension of a given part of the limb, which usually requires partial or total anatomical individualization (as well as customization).

Figure 6.3 Changing body size in the same patient, (A) hand orthosis; (B) upper limb prosthesis.

Usually the limb prostheses are long-time use devices in adult patients, although they still need to be replaced due to wear or change of body dimensions. The latter is especially true for child patients, where prosthesis exchange may be expected on a yearly basis (Ruggieri et al. 2013), but also for adult patients it can be true, especially considering the issue of changing body weight – some parts of prosthetics, such as vacuum sockets (Wernke et al. 2017) require the patient to not change their body weight significantly, as it may result in serious change of shape of the limb.

The problem of changing body size and functions for the same patient is illustrated for both types of devices in Fig. 6.3, showing problems of fit of corrective hand orthosis in a teenage boy case (6.3A) and change of size, as well as function for a growing child patient, using the prosthesis as an aid in bicycle ride (6.3B).

6.1.2 Limb prosthetics classification

The study and application of upper limb prosthetics have evolved significantly over the years, contributing immensely to the rehabilitation and improved quality of life for individuals with limb deficiencies. Here, we lay out basics of classification and structure of limb prosthetics, to provide an overview of these devices, necessary to comprehend the case studies presented in the further part of the chapter. For the sake of space and focus on the author's best experience, this overview will be limited to upper limbs.

Limb prosthetics are broadly categorized based on the cause of limb loss and the level of amputation. These categories provide a framework for understanding the diverse needs of patients and the corresponding prosthetic solutions.

Amputation, often resulting from severe injuries or diseases, necessitates the use of prosthetics for restoring functionality and appearance. The characteristics of amputations include traumatic injuries (e.g., car accidents, machinery accidents, burns) and diseases (e.g., severe diabetes, bacterial infections leading to sepsis, cancer). The stump is typically smooth and as long as possible, with surgical procedures tailored to accommodate future prosthetic fittings (Liu et al. 2015). Post-surgery, the stump requires rehabilitation and stimulation, and its shape may change over time. Patients often transition from temporary to permanent prostheses, adapting to changes in the stump and improving the fit and function over time (Dudkiewicz et al. 2004).

Congenital defects, arising from developmental anomalies or in-utero amputations, present unique challenges for prosthetic design and application. Congenital stumps – by the author's experience – are usually rough with atypical geometrical structures and may include residual fingers (Gold et al. 2011). Cosmetic surgeries may be necessary or optionally recommended to improve the stump's appearance. For children, prosthetics must be periodically replaced to accommodate growth, necessitating transitions from smaller to larger devices. Designing prosthetics for congenital defects is more challenging due to the irregular shapes and the need for precise customization to ensure proper fit and function. As it often requires working with small children, there is also an aspect of device familiarization and acceptance, as it must be willfully taken by the patient to obtain a full effect of its use.

Prosthetics are further classified geometrically, based on the level of amputation (or defect), each type serving different functional requirements and anatomical considerations. Finger prosthetics are primarily cosmetic but can include basic gripping functions, focusing on realistic appearance and lightweight design. Hand prosthetics include a functional wrist joint for basic hand movements, balancing aesthetics with mechanical functionality (Kargov et al. 2007). Forearm prosthetics (transradial prostheses) incorporate a functional elbow joint, enabling a range of forearm movements, ensuring seamless integration with the remaining upper arm, and providing robust yet flexible joints. Full arm prosthetics (transhumeral prostheses) frequently do not include an elbow joint, focusing on upper arm movements, emphasizing stability, and secure attachment to the shoulder region (Kumar et al. 2017). Main structural components of upper limb prostheses are presented in Fig. 6.4 – depending on the amputation level and needs, their amount and complexity may get appropriately reduced or increased.

The prosthetic socket is a critical component of any limb prosthesis (upper or lower), directly interfacing with the patient's residual limb. It must be anatomically adjusted and securely attached to prevent displacement. Vacuum sockets with liners utilize a vacuum to enhance fit and stability, often combined with a liner for comfort, providing a secure fit and reducing movement between the socket and the stump (Paternò et al. 2018). Open and closed (non-vacuum) sockets offer a simpler design without vacuum mechanisms, available in open or closed formats,

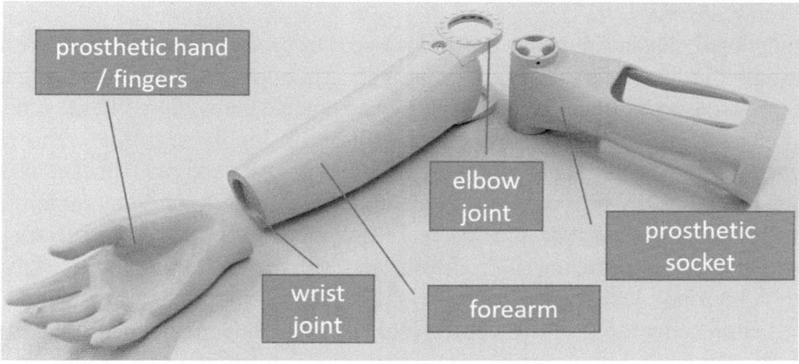

Figure 6.4 Main structural components of a full arm prosthesis.

making them easier to equip and remove, suitable for more temporary applications (not for prolonged continuous use). Compressive-Release (CRS) Sockets employ a compressive mechanism to secure the prosthesis during use and release it when needed, offering a balance between security and ease of use, simultaneously allowing the skin to breathe (Zawadzki et al. 2022).

Prosthetics can also be classified based on their functionality, ranging from passive devices to advanced electrically powered systems. Passive prostheses are primarily cosmetic, with a focus on aesthetic appearance without functional motor capabilities, often used as temporary prostheses or for patients prioritizing realistic visuals. Muscle-powered prostheses come in mechanical forms, enabling basic gripping functions without aesthetic enhancements, and hybrid "cosmetic+" forms that combine cosmetic appearance with movable joints, offering a blend of functionality and realism (VBionic 2020).

Electrically powered prostheses are equipped with motors or actuators powered by external sources. They use various control methods, including buttons or strings for simple control mechanisms for basic movements, muscle signals from myoelectric signals detected by armbands or electrodes to control ? prosthesis, gestures, or voice commands for advanced systems allowing control through specific gestures or voice commands, and brain signals using EMG or electrodes for cutting-edge technology that uses brain signals to achieve intuitive and precise control (Iqbal et al. 2018).

The evolution of upper limb prosthetics reflects advancements in materials science, engineering, and biomedical technology. Early prosthetic designs focused on basic functionality and appearance, with limited customization and adaptability. Modern prosthetics, however, leverage sophisticated technologies and materials to provide enhanced functionality, comfort, and aesthetic appeal.

Traditional materials for prosthetics included wood, metal, and leather, offering durability but limited flexibility and comfort. Current designs incorporate lightweight, durable materials such as carbon fiber, advanced polymers, and silicone, improving both comfort and performance. Customization has evolved from manual measurements and fittings, which often resulted in less precise fits, to advanced imaging technologies like 3D scanning and printing, enabling highly customized prosthetics tailored to the exact anatomical features of the patient (Zawadzki et al. 2022, Ten Kate et al. 2017).

Functional advancements have progressed from early mechanical prosthetics providing basic functions to modern designs incorporating intricate joint mechanisms and responsive control systems. Electromechanical innovations have integrated electronics and robotics into prosthetics, resulting in highly responsive and natural movements controlled by the user's muscle signals or brain activity (Iqbal et al. 2018).

Future trends in prosthetics include the development of biocompatible and bio-integrated materials that aim to create prosthetics seamlessly integrating with the body's tissues, reducing discomfort, and enhancing functionality. Advances in neural interface technology promise direct brain control over prosthetic limbs, offering unprecedented levels of control and dexterity. The incorporation of artificial intelligence and machine learning into prosthetics is expected to revolutionize their functionality, enabling adaptive and intuitive responses to the user's environment and activities (Triwiyanto et al. 2023).

6.1.3 Limb orthotics classification

This chapter explores the classification and applications of limb orthotics, providing a brief overview of their types, materials, and therapeutic uses. Orthotic devices of different types have a vast range of potential applications, advantages, and disadvantages. Detailed classifications and current state of the art can be found in numerous publications (Barrios-Muriel et al. 2020, Silveira et al. 2018, Huamanchahua et al. 2021).

The primary reasons for using limb orthoses include injuries such as fractures, dislocations, and sprains, as well as chronic diseases like cerebral palsy, rheumatoid arthritis, and congenital or acquired joint instability (Huamanchahua et al. 2021). Orthotic devices are classified based on the materials used in their construction and the body part they are designed to support. Basic types of orthoses, depending on the material, include soft fabric orthoses, stiffened fabric orthoses, and rigid thermoplastic orthoses, which are often adjusted (customized) to match the patient's anatomy. Customization can go beyond simple geometrical adjustment and include also mobility restraints, scheme, and schedule of use, as well as aesthetic properties (Popescu et al. 2024).

Orthotic devices can also be categorized based on the specific body part they are intended to support. The main categories include lower limb orthoses, upper limb orthoses, spinal orthoses, and cranial orthoses. Each category serves distinct therapeutic purposes, ranging from immobilization and correction to functional support and pain relief. Here, only limb orthoses will be considered, to narrow the possible cases down to the ones actually realized in practice and presented in the further part of the chapter.

The orthotic devices for limbs are diversified and sometimes very specifically constructed, with various functions and purposes intended. Lower limb orthoses, for example, are subdivided into foot orthoses, ankle-foot orthoses, and knee-ankle-foot orthoses, each with unique applications and benefits. Foot Orthoses (FOs), such as custom insoles or shoe inserts, address foot alignment and biomechanical issues.

They are commonly used for conditions like plantar fasciitis, diabetic foot ulcers, and flat feet. These devices improve foot posture, reduce pain, and enhance walking efficiency, although they may be limited in addressing complex deformities and may require frequent adjustments (Huamanchahua et al. 2021).

Ankle-Foot Orthoses (AFOs) extend from below the knee down to the foot, stabilizing the ankle and foot. They are used for conditions like drop foot, cerebral palsy, and stroke recovery. AFOs can be dynamic or static, improving gait and ankle stability. However, they may be visible under clothing and can cause discomfort or skin irritation. Knee-Ankle-Foot Orthoses (KAFOs), on the other hand, are encompassing the entire lower part of the leg, providing support to the knee, ankle, and foot. These orthoses are used for knee instability, polio, and severe muscle weakness. KAFOs offer comprehensive lower limb stabilization but are heavier and more restrictive, making them challenging to wear, posing ergonomic problems (Hasan and Hoque 2008).

Upper limb orthoses are aimed at supporting the hands, wrists, elbows, and shoulders. Wrist-Hand Orthoses (WHO) support the wrist, hand, and fingers, commonly used for carpal tunnel syndrome, arthritis, and post-surgery rehabilitation. They can be designed for immobilization or movement assistance, reducing pain and inflammation, although they may limit daily activities and require precise customization. Elbow orthoses stabilize or immobilize the elbow joint and are used for elbow fractures, ligament injuries, and tennis elbow. These orthoses promote healing by restricting detrimental movements but can limit arm functionality and cause discomfort (Choo 2022, Silveira et al. 2018).

Orthotic devices, whether for the lower limbs, upper limbs, spine, or head, share common goals of improving function, reducing pain, and enhancing the quality of life. The material used in these devices plays a crucial role in their effectiveness. Soft fabric orthoses are generally more comfortable and flexible, suitable for less severe conditions. Stiffened fabric orthoses offer a balance between support and comfort, making them suitable for moderate conditions. Rigid thermoplastic orthoses provide the highest level of support and are used in severe conditions where maximum stabilization is required. The most modern addition to the portfolio is openwork orthoses, produced by means of 3D printing, enabling greater comfort and ease of use by the patient (Howell 2019, Barrios-Muriel 2020).

The application of orthotic devices extends beyond mere physical support. They play a significant role in rehabilitation, enabling patients to regain function and independence. For instance, ankle-foot orthoses are vital in stroke rehabilitation (Carse et al. 2015), helping patients achieve a more natural gait, or in severe cases of congenital spina bifida, sometimes enabling the patient to walk (which would not be possible without the orthoses). Similarly, wrist-hand orthoses are crucial in post-surgery rehabilitation, aiding in the recovery of hand functions (Silveira et al. 2018).

The effectiveness of orthotic devices depends significantly on proper customization and fitting. Obviously, orthoses customized for a given patient work better then those off the shelf, providing better clinical results. Poorly fitted orthoses can cause discomfort, skin irritation, and even exacerbate the condition they are meant to treat (Swinnen and Keckhofs 2015). Advances in technology, such as 3D scanning and printing, have significantly improved the customization of orthotic devices, ensuring a better fit and increased patient comfort (Barrios-Muriel 2020). However, extensive reviews of effectiveness of therapy with orthoses created by 3D scanning and 3D printing (cohort studies) are not yet widely available. However, there are plenty of systematic reviews that help to shed some light on this technology (Schwartz and Schofield 2023).

Each type of orthotic device, depending on its design and material, offers a balance of advantages and disadvantages. While custom orthoses provide unmatched support tailored to an individual's specific needs, they often require a significant investment in terms of time, resources, and expertise to design and manufacture. Advances in 3D printing and digital scanning technologies are mitigating some of these challenges, making custom orthotic solutions more accessible and affordable. Moreover, ongoing research and development in materials science and biomechanics continue to enhance the effectiveness, comfort, and user acceptance of these indispensable aids.

6.2 Traditional and modern manufacturing of orthopedic devices

6.2.1 Review of traditional methods of production of orthopedic devices

The traditional methods of fabricating orthoses and prostheses for human limbs are quite similar for both these product classes, as regards to fabricating parts that are anatomically adjusted to certain areas of human limbs. The simplest way of manufacturing these was to use patient's anatomy as a basic "model" for molding, i.e., modeling the orthosis – or prosthesis – shape straight on the living human limb. The orthoses are usually devices of more simplistic nature, so the techniques of manufacturing them will be described as first.

Customized devices are essential in providing support and stability to individuals with hand-related medical conditions. Traditionally, the fabrication of customized, rigid orthotics, often referred to as splints, involved labor-intensive manual methods.

One of the earliest methods for treating orthopedic conditions involves casting and splinting. Plaster of Paris, first used in the early 19th century, allowed for the creation of custom-fit casts for bone fractures (Parmar et al. 2014). These casts provided stability, immobilization, and support during the healing process. While modern materials like fiberglass have largely replaced plaster of Paris, the fundamental concept of casting and splinting remains a cornerstone in orthopedic care.

The opposite of disposable plaster casts are rigid orthoses, made of more durable and stable materials. Orthopedic braces and external fixators have been in use for centuries. Metalwork and bracing techniques have evolved from rigid iron and steel constructions to lightweight materials like aluminum and carbon fiber. The Thomas splint, developed in the 19th century, is a classic example of an early orthopedic device designed to stabilize fractures (Robinson et al. 2009). Traditional orthopedic braces and external fixators aimed to maintain alignment and support bones during the healing process.

For softer orthotic devices, leather has been a traditional material for orthopedic shoe inserts and orthotic devices. Shoemakers and craftsmen in the past created custom-made orthotic insoles from leather to address gait abnormalities and foot-related conditions. These methods have transitioned to modern orthotics, which now use advanced materials like EVA foam and thermoplastics (Van Alsenoy et al. 2019). However, traditional leather orthotic practices were instrumental in understanding the biomechanics of the foot.

In the present day, when taking into account the traditional methods that are still in practical, clinical use, we might say that roughly two basic methodologies are used to create rigid, anatomically adjusted hand orthotics. The first one is plaster casting, which is applied mostly in the cases of injuries, requiring immediate treatment. The second one are lightweight, polymeric shells, used for injuries at a slightly later stage of healing, or for corrective purposes in hand therapy being a part of treatment of long-term diseases.

Plaster casting is a traditional method that has been widely used in the fabrication of orthotic, as well as prosthetic devices, including hand orthoses. This technique involves creating a plaster cast or mold to capture the patient's anatomical shape. A plaster cast is applied to the patient's hand and forearm to create a negative mold (Damiao et al. 2023). Plaster casting remains cost-effective compared to some other modern methods, making it accessible for patients in need of orthotic devices (Kim et al. 2006). Plaster casting also enables precise customization. It can capture intricate details of the patient's body, resulting in a well-fitted orthosis (Schlégl et al. 2022).

The biggest disadvantage of the plaster cast is that it can be a messy and invasive process for patients. It involves applying wet plaster to the body, which can be uncomfortable. After stabilizing and hardening, it cannot be safely removed without damaging it and it is impenetrable to light, generally heavy, bulky, and unpleasant for the skin, lacking any ventilation.

Recent studies have demonstrated modern applications of plaster casting, including rapid prototyping of die casting and the creation of orthotic devices. Plaster casting is still highly utilized in orthopedic immobilization materials due to its cost benefits and the ability to provide well-fitted orthoses (Ekanayake et al. 2023, Kim et al. 2006).

Plaster casting remains a relevant and cost-effective method in the fabrication of orthotic devices, with advantages in customization. However, its disadvantages include being time-consuming, uncomfortable, and not ergonomic for the patients. While modern applications exist, there is limited specific research on the strength of orthoses manufactured using plaster casting, as it is primarily chosen for its precision and fit.

The second method of building customized orthoses, a more contemporary one but still a manual one, is thermoplastic material shaping. In the process, polymeric or composite-polymeric materials are heated and manually molded to conform to the patient's hand (or leg, or other body part) shape. The process involves hands-on craftsmanship by orthotists and requires a high degree of skill (Damiao et al. 2023).

Thermoplastic materials allow for precise customization by softening the material and molding it to the patient's anatomy, resulting in a well-fitted orthosis.

Compared to plaster casting, this method offers rapid production, making it an attractive choice for patients in need of orthotic devices. Thermoplastic materials can be reheated and reshaped, making them versatile for adjustments and reuse. Orthoses made of these tend to be lighter and more comfortable for the patient.

However, thermoplastic materials may have limitations in terms of strength and durability, especially for complex orthotic designs. Effective shaping and fitting of thermoplastic materials require a certain level of skill and expertise. The problem of the patient's skin irritation and lack of ventilation also persists here, although holes can be made so that the air can flow in and out. While versatile, thermoplastic shaping may not be suitable for all orthotic applications, particularly those requiring high structural integrity and stability in higher temperatures – especially in hotter climates, or while dealing with heat sources, warping and distortion may occur.

In a study by Schlégl et al. (2022), the results of a three-point bending test showed that the flexural and shear strength of PLA (a type of thermoplastic) was three times greater compared to plaster casting. This indicates that thermoplastic orthoses can offer superior strength when compared to plaster-casted ones. However, the choice between these methods should consider factors such as patient needs, orthotic complexity, and material properties.

Thermoplastic material shaping offers advantages in customization, rapid fabrication, and reusability. However, it requires skill, and material limitations must be considered. Studies suggest that thermoplastic orthoses may exhibit greater strength compared to plaster-casted ones, but the choice depends on specific patient requirements.

As regards forming of customized shapes of prosthetic devices – mostly pertaining to anatomically fit prosthetic sockets – the general scheme of operation is based on the similar set of ideas, using the patient's own limb (the missing one, i.e., stump) as a pattern for making a mold – first a negative cast is made directly on the stump, then a positive cast – using the negative one, and on the basis of the positive cast, the prosthetic socket is then formed, using thermoplastic material shaping (Damiao et al. 2023).

Both these popular methods, as well as other traditional ones, have a common set of disadvantages: they lack the digital layer and the actual design stage. The patient is used as the mold, which can be uncomfortable, as well as logistically difficult – orthotic technicians and the patient must meet in one place with appropriate technical equipment for a set period of time. There is no project, nor detailed anthropometric measurement – when the same patient comes again, the production starts from the scratch and takes as long as the first time. The knowledge of used design solutions remains solely in the head (and hands) of the technician doing the job and is very difficult to carry on to other persons, as well as very distinguishing (almost as craftsmanship of pieces of art) and two different technicians can make different orthoses, with varying degree of therapeutic success.

To improve these processes and make them more suitable to 21st century methods of operation in the wake of 4th industrial revolution, which focuses on digitization, computer integration, smart factories and intelligent technologies, some other techniques need to be used for both measurement and forming of orthotic shapes. These techniques are described in the next chapter and in the following descriptions of case studies.

6.2.2 Overview of modern digital methods

Limb rehabilitation represents an integrative process aimed at reinstating functionality to patients with either acquired or congenital injuries. Employing an array of methodologies—including evaluation, diagnosis, tailored rehabilitation planning, patient education, pharmacotherapy, orthopedics, physiotherapy, functional training, psychological support, and monitoring of patient progress—this approach underscores the transformative potential of 3D technology in orthopedic and prosthetic manufacturing and delivery. Utilizing 3D scanning for precise patient data acquisition facilitates the customization of orthopedic fittings to meet individual needs. Moreover, CAD-based design processes enable the generation of 3D models crucial for adapting design solutions to patient specifications (Hale et al. 2020).

Notably, 3D printing technologies such as FDM, PolyJet, SLA, and DMLS are instrumental in manufacturing customized orthopedic and prosthetic equipment. These technologies, despite their distinct mechanisms, collectively contribute to reducing production and prototyping costs and timeframes, proving indispensable in Research and Development (R&D) and the deployment of novel technologies (Mavroidis et al. 2011, Palousek et al. 2014).

First examples of use of 3D scanning for development of customized hand orthotics can be found in existing literature around the beginning of second decade of XXI century (Kelly 2015). There are also many research papers focused on leg orthotics, as well as customized hand prosthetics and prosthetic sockets (Barrios-Muriel et al. 2020, Popescu et al. 2024, Damiao et al. 2023). A usual way of conduct is as follows:

1. Using 3D optical or laser scanning (stationary, manual or specialized automated scanners) to gather the anatomical measurements of a patient.
2. Processing the data using triangle mesh editing software, to extract the shapes necessary for building a given product.
3. Creating a model of a given device (orthosis, prosthetic socket, etc.), using mesh-based approach (no parametric modeling) or CAD modeling techniques. Optionally – subjecting it to simulations (strength calculations, virtual fitting, etc.).
4. Exporting the model into STL file or files (depending if it is needed to be divided into parts due to machine chamber size limitations).
5. Preparing NC programs for 3D printers and realizing the additive manufacturing process.
6. Post processing and assembly of obtained parts.
7. Tests with the patient and iterative improvement.

The technical details behind these general processes can be found in Chapters 3 to 5, while the more detailed implementations are shown in the further part of this chapter.

It is noteworthy, that the transition between the traditional and modern process does not always require radical makeover of the applied technology and required competences. Some elements of modern design process of orthopedic devices can rely on the traditional methods, thus slightly changing the approach. The most typical element of traditional process implemented in modern conditions can be making of a plaster cast of a limb and then 3D scanning it to gather the geometry and further process it (Damiao et al. 2023). It is also possible to realize the whole process of design digitally and then realize manufacturing without use of 3D printing, resorting to traditional forming methods, or only partially rely on 3D printing. It all depends on the available hardware and software resources, difficulty of work with a given group of patients and desired results in terms of technical and economical coefficients of manufacturing processes and products.

6.2.3 Challenges and premises for research

The modern, digital technologies for designing and manufacturing of anatomically customized orthopedic devices have many distinct advantages on a technical level – repeatability, storing the existing and easy improving of past designs, monitoring the patients, using modern materials and structures to make the products lightweight, just to name a few. However, there are some reasons behind the fact that the orthosis and prosthesis manufacturing industry is still largely based on manual labor and shaping of the anatomically adjusted parts of these devices on the basis of physical models, without involving any digital techniques whatsoever.

First and foremost, there is the "skillset issue", leading to lack of very distinct economic benefits and thus slower progress in replacing old techniques with new. The skillset issue can be summarized simply – going from old, manual processes into the new, modern ones does not make the process require less specialized qualifications and thus be more accessible. On the contrary – the manual skills of technicians being able to shape a prosthetic socket or a leg orthosis with their own hands are being replaced by the same skills of engineers, shaping these products in a digital domain. Of course, sculpting over digital set of data can be done very efficiently, but it still requires skills, experience, and time. Also, engineers are specialists with high paychecks, meaning that their time is very valuable. All this brings us to a situation, where the product obtained in a modern way is meeting all the requirements, with some extra features comparing to the traditionally made products, but it still takes a long time to design and manufacture and it costs the same (or even more) than the traditionally designed product.

This issue can be managed only by lowering the entry threshold into the process of design of orthotic and prosthetic devices based on 3D scans of the patient's limbs, to the point that the process is done in an entirely automated manner. It has already been discussed (in Chapter 2) how the mass customization works and how it can be related to medical products – the orthopedic devices are the most shining example of this. There are a number of challenges here, already mentioned in Chapter's 2 concluding remarks. The whole complex process of obtaining the patient's geometry and making an anatomically customized orthosis or prosthesis on its basis must be improved to introduce automatic, semi-automatic, or easy to learn solutions along the way, starting from the 3D scan itself – which should be easy enough for the patients to perform it themselves, maybe with a little assistance, through the design, which should utilize intelligent modeling (as described in Chapter 4), possibly AI-aided, to 3D printing (or other manufacturing technologies), where programs should be automatically, or very easily created and realized on accessible machines with popular materials. Meeting these assumptions will enable to obtain processes which are both effective (bringing in properly working devices, fulfilling medical purposes) and economically viable (shorter waiting time and/or lower costs than in the case of traditional solutions).

There are currently two main approaches to make it happen in the orthopedic industry, with many sub-approaches and hybrid methodologies. These main approaches are:

1. Building and implementing design software which relies on human operators to go through the whole design process – but with a number of simple choices and intelligent tools as well as friendly interfaces, making the design processes easy to learn and go through. Some notable examples are Mecuris (Spierings et al. 2023) and Spentys (Keller et al. 2021).
2. Building and implementing solutions that are completely automated, to the point of not requiring skilled operator at all – patient can be the operator of one part of the system, while medical personnel (nurse, technician, doctor) can operate the other part, making the whole design process as troublesome as a routine appointment at a specialized clinic, and taking as much time. This is an approach sought by the author and his team with the AutoMedPrint project (Górski et al. 2020b) and in some other work as well (Cirello et al. 2024).

These two approaches differ on a "philosophical" level. The first one assumes that there are always engineers in the process of developing orthopedic devices, just that they are engaged for a very short time, realizing simple tasks using software with friendly user interfaces, but still making important decisions along the way, based on their experiences and competences. There can be dedicated software for doing this job (as mentioned), or there can be know-hows and methodologies based on decision trees, whatever makes it work – and the challenges lie in preparing these methodologies to cover as much potential patient cases as possible, to avoid situation of designing things anew for each single patient.

The second approach assumes that engineer does not work with singular patient cases, but is involved in creation and maintenance of the system, while regular work is done by non-engineers trained in system's operation and thus realized in the metaphorical "battlefield" – at a hospital, school, doctor's office, physiotherapy facility or an actual battlefield. Challenges remain similar, but with the automation component added on top – way of doing things and making decisions in the design process by human engineers and doctors must be replicated in a computer system. Challenges of doing that are already described generally in Chapter 4, in the section focused on design automation. In general, bigger challenges are focused on making therapeutic devices for long-term diseases, where motion constraints of a given limb and a need to correct its posture are of key importance – these cases are difficult to automate, of the author's experience (Górski et al. 2024a). Repeatable, mechanical devices reflecting human's anatomy without need of correcting it digitally remain simpler and viable for automating – as shown in the subsequent specific case studies.

6.3 Design automation of orthopedic devices: AutoMedPrint system

6.3.1 Basic premises of automated design of orthopedic devices

The AutoMedPrint system facilitates the integrated design and manufacturing workflow for a specific subset of orthopedic devices, encompassing two varieties of limb orthoses—namely, the WHO orthosis and the AFO—and upper limb prosthetics, which include both cosmetic and functional (mechanical) subtypes. The methodology employed by this system leverages additive manufacturing technologies, commonly referred to as 3D printing, and initiates the process through an automated protocol that utilizes 3D scanning data of the patient's limb. Within the design phase, the system incorporates a patient-centric approach, allowing individuals to make critical decisions concerning the aesthetic aspects of the final product, including choices of materials, color schemes, finishes, and other design elements. The duration of this procedural phase ranges from a few to several tens of minutes, excluding the manufacturing period, which may extend up to several tens of hours. Notably, this streamlined process is engineered to proceed without the necessity for direct intervention or oversight by a qualified engineering professional, thereby enhancing efficiency and user engagement.

The general principle of operation of the system is presented in Fig. 6.5. The general contents of this chapter are results of research and development work carried out in the scope of AutoMedPrint project by the author leading a team of specialists. These results have not been publicly disclosed in full detail, apart from technical reports prepared by the author with the team along the course of the project. This, and subsequent subchapters contain descriptions that were originally presented in these reports, formed in a way to enable readers gain a full comprehension on how an automated design system dedicated for building customized prosthetic and orthotic devices can be built and operated. This work has been previously described in many partial publications, such as (Górski et al. 2020a, Górski et al. 2020b, Zawadzki et al. 2022, Górski et al. 2023b), and the system itself, as well as the scientific project on which it is based, is described on its website (AutoMedPrint 2021). The further descriptions are based on the development of the system, which was realized by the author (as a team leader and a main originator) and the team – detailed on the project website.

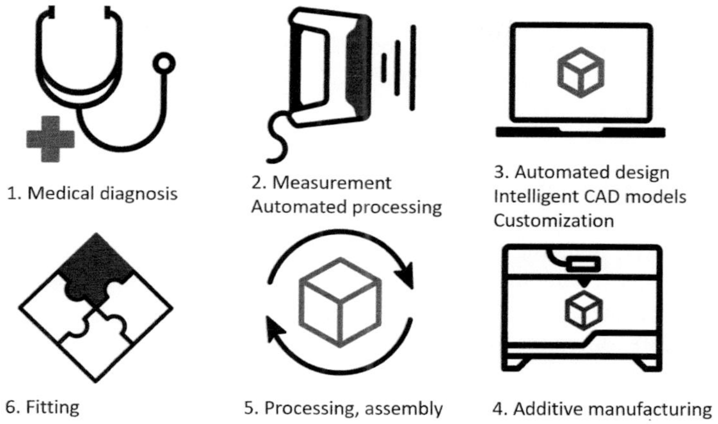

Figure 6.5 Main premises of work of proprietary AutoMedPrint technology – stages of work with the system.

The system makes it possible to realize the following activities:

- defining the basic features of the product required by a given recipient (product type and its purpose),
- entering the recipient's anthropometric data through the use of non-contact (non-contact) measurements, carried out semi-automatically - the operation of the measurement system initiated independently by the recipient without the participation of the operator, with the possible assistance of medical staff,
- completely automatic design of a new product individualized for a given recipient using a CAD system, based on collected anthropometric data,
- presentation and customization of the product by the recipient,
- completely automatic preparation of a manufacturing program for an additive manufacturing machine based on the developed CAD model,
- implementation of the manufacturing and assembly process without the recipient's participation, with support from technical staff.

Several users with different classes/functions take part in the process - the following should be distinguished:

1. The recipient of the system, i.e., the patient/customer (respectively: hospital, clinic, clinic, physiotherapy studio or company producing prosthetic/orthopedic supplies), who determines his need and enters his anthropometric data and receives the finished product based on them.
2. *Medical staff* – medical technician, physiotherapist or possibly doctor – supporting the recipient in defining the need and the data collection process. The presence of medical staff is not required in all cases.
3. *Technical staff* – 3D printing technician – carrying out typical assembly and finishing works related to the operation of an additive manufacturing device

(3D printer) and the preparation of a ready-to-use product and its delivery to the recipient along with instructions for use. In addition, the technician is responsible for keeping additive manufacturing machines ready for operation, i.e., supplying material, cleaning the head and working space, and removing any minor defects.

In addition, support staff should also be replaced, including: an IT specialist responsible for the current operation of the IT part of the system – maintenance and operation of computers and peripheral devices included in the system, administration of collected data, updating the system, etc. It should be emphasized that it is not necessary to involve qualified orthotists/prosthetists or biomedical or biomedical engineers in the operation of the system designers – their knowledge is contained in the automatically operating algorithms of individual system modules.

The method of working with the system can be schematically divided into several stages, creating a framework methodology.

1. **Preparatory stage** – involves preparing individual users using the system to start working. Especially:
    - *recipient (patient)* – is preparing for an anthropometric test – before using the system, the patient must know what product he needs, therefore a consultation with medical staff is necessary (as a result of the visit to the doctor, the patient receives a recommendation or referral), clothing that allows easy exposure of the limb should be used for which the product will be manufactured (arm or leg),
    - *medical staff* – consults with the patient on the need to produce a given product,
    - *technical staff* – regardless of the actions of the recipient and medical staff, prepares the machine (3D printer) for work, i.e., cleaning, loading support material, etc.
2. **Data collection stage** - involves collecting anthropometric data from the user. Especially:
    - *recipient (patient)* – independently or with the help of medical staff, runs the scanning assistant application on the touch screen; then takes a stationary position at the data collection station indicated by the scanning assistant. In the case of posture defects or body damage that prevents the patient from adopting the indicated posture, the patient uses the help of medical staff (e.g., physiotherapist, nurse); next, the 3D scanning process is started – independently by the patient (using a touch screen or voice control), or with the assistance of medical staff; scanning takes place, depending on the process, in several settings of the 3D scanner; the scanner can be moved semi-automatically (with electric support) or manually, independently by the patient or medical staff; after collecting the data and verifying their correctness, the patient receives information on the screen that the data has been collected successfully and can proceed to the next stage; if the same patient repeats the process of making a given orthosis/prosthesis (the 2 most common potential

situations: the need to make corrections to the shape and re-manufacture or destruction of the product, provided that the patient is sure that the anatomy will remain unchanged), this stage can be omitted and go straight to design,

- *medical staff* – supports the patient during the operation of the scanning assistant and the data acquisition process – the patient must keep the limb still during the entire strip projection process, which may take from a minute to several minutes; medical staff are supposed to support the patient in using the system and maintaining a still position (which is particularly important and problematic in the case of children).

3. **Design stage** – involves the automatic preparation of a design for an individualized orthosis or prosthesis based on anthropometric data collected in the previous stage and presenting the design to the recipient for adjustment and final acceptance. Especially:

    - *the recipient (patient)* – after waiting for the automatically generated design to be generated (it may take from several seconds to several minutes, depending on the product, the accuracy of the 3D scanner and the algorithms used, as well as the computing power of the computer at the workstation), can view their product in two basic ways, i.e., separately – as a 3D view of the generated prosthesis/orthosis – or using the so-called a virtual mirror, i.e., viewing a live image of your body generated by a camera located at the station, with a superimposed digital image representing a given prosthesis or orthosis; this helps the patient visualize its appearance after receiving and using a given product; the recipient can still visually verify the product at this stage (with the optional help of medical staff) and go back to the scanning stage if the measurements are incorrect; you can also use the configuration options and select additional features of your product, e.g., visual (stickers, colors, patterns, visual style) or secondary functional features (e.g., incorporating an additional hook or an accessory in the form of a bottle opener into the mechanical prosthesis); everything is done through interaction with a monitor/TV with a touch screen, after approving the project, the recipient can proceed to the next stage, i.e., ordering,
    - *medical staff* – supports the patient at the stage of selecting additional product features and verifies, based on visualization, the correctness of the automatically generated shape; if it is necessary to make corrections or another iteration for the same patient, uses advanced options that enable shape correction by modifying factors affecting, for example, the form of curves used to generate the external or internal shape of the orthosis adjacent to the body; medical staff use the same interface as the recipient, so theoretically the recipient can also select advanced options.

4. **Manufacturing stage** – involves starting the process of producing elements of a given medical device (prosthesis/orthosis) on additive manufacturing machines and assembling the finished product.

    - *technical staff* – supervises the correct course of the additive manufacturing process, depending on the system variant - either on-site (clinic, clinic, therapy room, etc.) or in a place intended for this purpose (prostheses/orthoses manufacturing plant) containing appropriate machinery; after the process is completed, the task of the technical staff is finishing processing (removing supports, manual smoothing) and assembling the products (assembling using screws or gluing) – simple operations that can be performed after short training, even by a low-skilled person, in a time not exceeding 2–3 working hours for selected products.

5. **Acceptance stage** – involves acceptance of the finished product by the customer with the support of medical staff, with possible immediate introduction of minor corrections through mechanical processing. It takes place at a certain time interval from the previous stages (maximum 3 days).
    - *recipient (patient)* – collects the product at a defined place (by default – in the place where the data was downloaded, a distributed variant or courier delivery is also possible), within a maximum of 72 hours from the start of production (according to the order queue), checking compliance of visual and functional features with the ordered design; if accepted, the product is put into use; if you want to make corrections, e.g., after a few days of use, the recipient reports this fact and returns to the system;
    - *medical staff* – (optional) supports the patient during collection – helps to assess the compliance of features, check whether the product fits the anatomy and makes a decision regarding the possible acceptance (or not) of the product,
    - *technical staff* – (optional) supports the patient during acceptance and accepts the product along with comments on corrections in the event of lack of acceptance during the first acceptance; alternatively (as a last resort), it makes physical corrections on site, using hand tools, if the joint assessment of the medical staff and the patient allows for the conclusion that such corrections will increase the quality of the product.
6. **The stage of introducing corrections** – optional, involves the patient providing comments regarding irregularities in the functioning of the orthosis/prosthesis after short-term use. If it occurs, it is carried out some time after the acceptance stage (several days or weeks). It can be implemented iteratively (many times until a ready-made, matching product is obtained), but each time corrections that are not a manufacturing error are introduced will involve costs for the patient (re-manufacturing the product).
    - After using the orthosis or prosthesis, the recipient (patient) reports within a few days the need for any corrections resulting from anatomical mismatch (incorrect measurements or changes in anatomy, e.g., swelling), design or manufacturing error, reappearing at the point of data collection; depending on the decision of the medical staff, there is a return to stage 2 (if it is necessary to download the data again), 3 (if using the advanced options of the automatic design module, the design can be updated to obtain the correct match), 4 (if a manufacturing error occurred that was not noticed upon receipt and, e.g., there is damage to the product that is not the result of incorrect use, it is enough to produce the same product again) or 5 (if only minor corrections made

manually by technical staff are required, e.g., grinding, bending or drilling a hole),

- *medical staff* – helps the patient decide on the place of the process to which he needs to return in order to obtain a properly functioning product, supports the patient at each stage of further work,
- *technical staff* – if only simple, manual corrections are required, they are made on site or after making an appointment with the patient, according to the instructions of the medical staff.

Work of the system is depicted in general terms in the block diagram in Fig. 6.6 (without taking into account certain details, e.g., manual corrections of the product or procedures when the patient uses the system for the second and subsequent time). The system and the recommended method of working with it enable iterative corrections to the product by the patient, with the support of medical and technical staff.

There are three basic classes of components in the system:

- *active components* – physical objects directly related to data processing in the system, powered by electricity, coupled with appropriate software downloading, processing, and transmitting data,
- *passive components* – physical objects related to data collection and processing processes, facilitating the collection of anthropometric data and preparation of additive manufacturing of selected products based on this data, not powered by electricity,
- *auxiliary components* – hardware or software not strictly related to the main process, but facilitating, making more attractive or accelerating the implementation of this process.

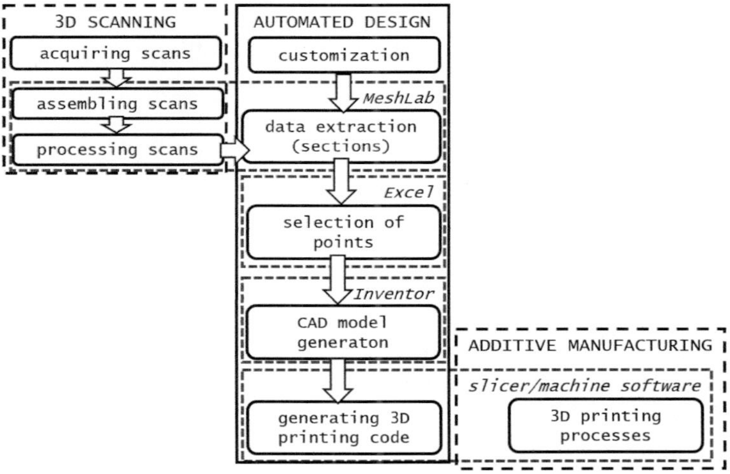

Figure 6.6 AutoMedPrint system simplified working scheme.

Active and auxiliary components can be divided into hardware and software. The most important active hardware components of the system can be listed as follows:

1. *3D scanner* – a device for carrying out anthropometric measurements.
2. *Computer* – integrates the software and hardware components of the system.
3. *Monitor/TV with touch screen* – used to display and operate the system user interface.
4. *Additive manufacturing device (3D printer)* – used to carry out the manufacturing process.

The active software components of the system are as follows:

1. *3D scanner software* – supports the collection of anthropometric data. Ready-made software, dedicated to the selected scanner, with a proprietary overlay (submodule) that automates mesh processing.
2. *CAD software + auto-generating models* – the most important part of the system, is used to automatically generate a prosthesis/orthosis design based on scanning data. The CAD system is a commercially available system, and the auto-generating models are an original overlay (submodule).
3. *Operator's panel* – used for basic work related to data entry and anthropometric measurements, as well as launching the auto-generating model. The author's own application, created in a 3D engine.
4. *Scanning assistant* – first submodule of the user (patient) interface, used to demonstrate the features of the 3D scanning station and teach the process to the scanned person.
5. *Virtual designer* – the second sub-module of the user interface, is used to visualize and adjust the prosthetic/orthotic design using touch screen, interactive 3D models and – optionally – VR goggles.
6. *3D printer software* – supports the manufacturing process. Ready-made software, dedicated to the selected 3D printer, with a proprietary automation overlay.

Passive components include:

- Supporting structure for the 3D scanner, enabling it to move relative to the patient's stationary limb. Optionally – mechanically supported (electric motors) and controlled from 3D scanner software, then it should be considered an active component.
- Frame and housing for a computer, TV and (optional) 3D printer.
- A position for the patient, enabling support/immobilization of the limb in a given position for 3D scanning.
- Hand tools facilitating the process of finishing and assembling a 3D printed product,
- Materials for the 3D printing process.

Auxiliary components include:

- Speakers and microphone enabling voice communication with interface modules.
- Applications supporting the patient's concentration during the data collection process (simple games, interactive or non-interactive entertainment videos, music player, etc.).
- Passive elements supporting the patient's concentration/distracting attention (colorful boards, moving play elements, etc.).
- Active elements enabling automation of the 3D printing process - remotely controlled electric buttons, electronic modules enabling control of the 3D printer from the outside, etc.

The above system components are grouped into four basic system modules: 3D scanning module, automatic design module, interface module and rapid manufacturing module.

The original concept of the AutoMedPrint system assumed the connection of all modules in a single location, i.e., the creation of fully independent instances of the system, where data collection, design and production take place in the same place, and some of the components are common to the modules (e.g., computer, frame, software). However, the results of conducted research on a prototype of the system, interviews, literature review and conclusions from the collected knowledge allow the formulation of alternative concepts of system operation, possible to implement under various conditions. Two independent scenarios for variants of the original system concept are assumed:

1. **Distributed design** – based on cloud data processing and client-server architecture, where anthropometric data are sent to an external server (common to a certain number of clients, i.e., local instances of the system), where there is CAD software with an auto-generating model. This approach is dictated mainly by economic considerations – the most expensive element of a single instance of the system is the CAD software license. In the case of a distributed architecture, there will be one server and one license for several dozen system instances (depending on the load – difficult to estimate at this point), which in this case may include software with more advanced capabilities. This approach is profitable, however, only when the actual number of active, implemented system instances exceeds a dozen or so. The disadvantage of this approach is the need for the system to operate online – a fast Internet connection is required, delays in operation are possible due to the queuing of tasks on the server, and additionally there is administrative (IT) staff and server infrastructure located in a separate place. This means that the entity implementing the system (product) resulting from the project will additionally have to maintain such architecture or build nodes common to several dozen individual recipients.
2. **Distributed manufacturing** – this concept assumes that the additive manufacturing machine is not part of the local system instance, and manufacturing takes place in a special facility (production plant), common to many, dozen or

several hundred system instances. This approach is also dictated by economic considerations and has similar advantages and disadvantages as in the case of distributed design. In such a case, technical staff (operating the 3D printer and assembling products) will no longer have to work at the place of data collection, but at the manufacturing plant. Reception can take place in three ways – at the place of data collection, at the manufacturing plant or by shipping the finished product to the customer. There are more advantages of this approach – the machinery park can include many machines with different capabilities, and its maintenance requires less expenditure (per one system instance). Additionally, such infrastructure can be considered for implementation in existing prosthetic plants, where the skills of prosthetists and orthotists could be used in a similar way as in the traditional process, while maintaining ultra-short design and manufacturing preparation time and low price, i.e., the advantages of the approach proposed in the system AutoMedPrint.

The system was developed with help of the virtual prototyping techniques. The initial design of the hardware (Fig. 6.7A) was used as a base for a virtual reality simulation. In an immersive environment, the process of 3D scanning of a patient was simulated to evaluate weak points and possible improvements. The first version of the scanning rig, after careful considerations in a team of interdisciplinary experts, was changed, mostly due to safety concerns, ergonomics, universality, and easiness of building a low-cost scanning station. Figure 6.7B shows the second version of the design, which – after certain improvements – was built, allowing to obtain a working

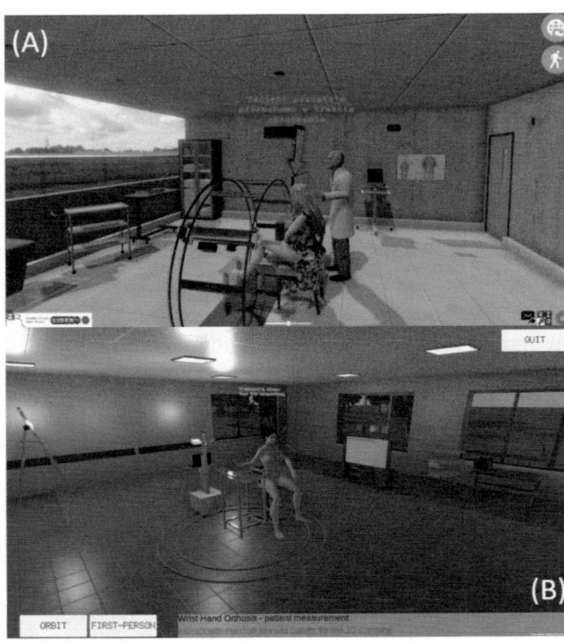

Figure 6.7 AutoMedPrint hardware – virtual prototyping: (A) initial version; (B) final version.

prototype of the system (this version was also used to build a VR training application for system's operators).

6.3.2 AutoMedPrint system hardware – measurement and manufacturing

The AutoMedPrint system consists – in general terms – of four proprietary software modules: scanning data processing module, operator's panel, user's interface and intelligent CAD models. These modules do not operate in isolation from the other modules – they work within one system and exchange data with each other, although some of them are more independent than the others. In the hardware layer, this translates into a total of 3 computer workstations:

1. *3D scanning and design station* – contains the operator panel and a 3D scanner with a computer and allows the operator to carry out the data acquisition process, as well as control other system elements. The user (patient) does not have access to this workstation and cannot see messages or the operator panel.
2. *User interface station* – a separate computer station, integrated with the operator panel via a network interface, containing a touch screen, a pair of VR goggles and applications that support the scanning process and product configuration. This station is operated by the user (patient), although they are not obliged to do so, and all necessary activities can be reduced to a minimum or performed by the operator.
3. *Rapid manufacturing station* – a station containing a computer with appropriate software and an additive manufacturing machine(s), along with the necessary instrumentation.

These modules were given a physical form, building the system's first prototype that was iteratively improved. It should be noted, that this prototype has been built with limited resources and as such it would require a different approach to forge it into a serially produced solution. However, it is innovative and unique, so for the benefit of the readers it will be described "as is", with a discussion of possible improvements.

The first module – station for automatic scanning of upper and lower limbs – consists of the following modules (Wichniarek et al. 2020):

- base with a circular track,
- a trolley with a computer and a 3D scanner,
- mechanical rig for scanning of upper limbs (both healthy and residual),
- mechanical rig for scanning of lower limbs.

The base of the entire stand has dimensions of approximately 3 × 2.5 m and consists of wooden boards. The track is screwed to the boards and consists of polyethylene pipes. There is also a central pole, permanently attached to the base, which is part of the lower limb scanning module. In addition, there are holes in the base with sleeves that support the frame for scanning the upper limb. Since the individual positions of the trolley are recognized using an optical system, the part of the base located between the pipes was painted white and code markers were placed on this surface in appropriate locations.

The trolley moves along the track through sets of wheels, powered by electrical engines. It is equipped with a 3D scanner and a computer that controls the operation of the scanner and the trolley's route. The trolley consists of a steel plate to which two aluminum poles are attached. On one of the poles, there is a platform with a computer and a power supply necessary to power the stepper motors and their drivers, while the other is equipped with a steel linear guide and other accessories responsible for moving the scanner between the upper and lower scanning positions. A linear carriage moves along the guide, to which a 3D scanner is attached using a photographic head, adjustable in two axes.

The trolley platform includes emergency stop sensors and supports to relieve the wheel sets when the station is not in use. An optical system sensor that recognizes the coded tags is attached to the bottom of the plate. Virtual prototype of the station and its photo are shown in Fig. 6.8.

The rig for scanning upper limbs and stumps consists of a chair and a frame with a mirror. The chair has been selected in such a way that a wide range of seat adjustments, independent adjustment of the footrest position and stable back support are possible. The more support points, the less chance of accidental movement of the scanned person during the measurement. The chair is independent of the frame and its setting is selected each time for the patient being scanned. The frame consists of square aluminum profiles screwed together, with appropriate construction elements to put the patient's limb on the rig, adjustable to the patient's size, ensuring safety and stability. Additionally, to ensure that the limb is positioned in a straight line, a line laser is mounted on the station. The line projected by the scanner makes it easier for the person being scanned to select the appropriate position.

The structure for scanning the lower limbs consists of a mast with an adjustable crossbar attached to it, with a swivel seat on both sides that makes it easier to take a

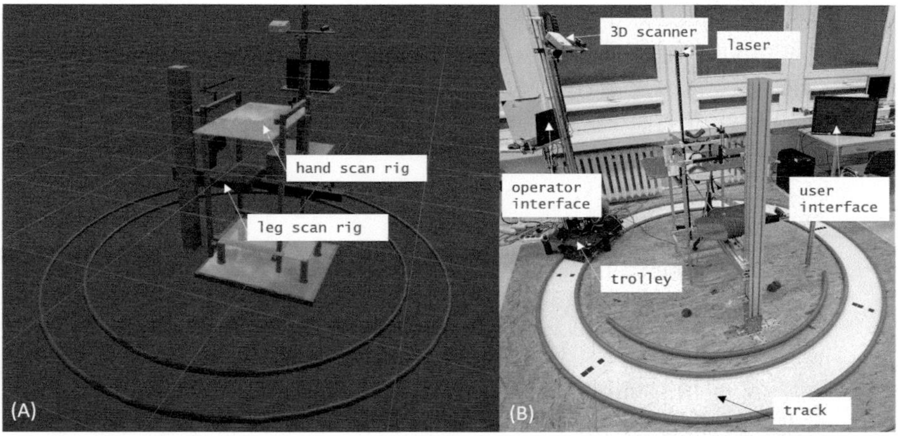

Figure 6.8 AutoMedPrint scanning station, (A) virtual prototype; (B) working prototype.

position at the station. Depending on the leg being scanned (right or left), the seat is placed on the appropriate side of the crossbar. An adjustable support for the second leg is attached to the crossbar. The scanned leg is placed in the appropriate position on a table with a glass top, and the table is placed on a platform with a mirror.

Starting work at the station begins with releasing the lock on the supports (screwing in their movable parts) so that the trolley touches the track with all its wheels and turning on the station's power supply using the button on the remote control located on the computer platform. After starting the computer and turning on the AutoMedPrint system operator panel, the scanner is set in the upper or lower position on the guide.

In order to perform measurements of a limb, the station should be adapted to a given patient. Adjustments involve setting the appropriate height of the chair and adjusting the positions of the elements, at which the patient's limb parts will be resting during the measurement.

The user (patient) interface station consists of a computer with installed software, a touch monitor on which applications are displayed and a virtual reality helmet thanks to which selected products can be configured in an immersive environment. The computer is equipped with an appropriately efficient graphics card to support numerous interactive applications and configurators with VR mode that make up the user interface. The user can work at the station using a touch monitor, keyboard, and mouse, as well as a helmet and VR controllers.

The rapid manufacturing module (Fig. 6.9) contains a computer controlling the module's operation, a device for additive manufacturing (by default, FlashForge CreatorPro machines were used) along with dedicated accessories, as well as cameras for monitoring and manual tools for post processing of the obtained parts. Data

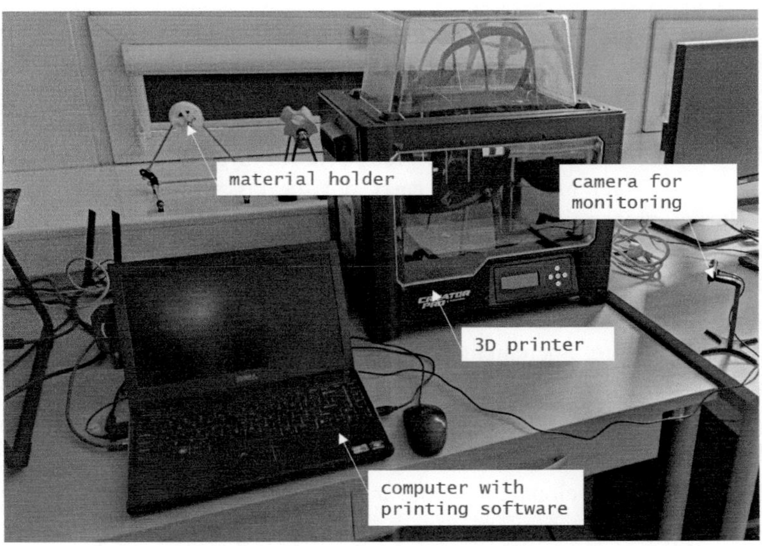

Figure 6.9 AutoMedPrint rapid manufacturing module.

regarding the product to be manufactured is transmitted remotely to the computer controlling the module's operation. Based on them, an NC code is automatically created for the selected additive manufacturing machine. This code is then sent to the machine via SD card. The manufacturing process, for safety reasons, must be started manually by the operator, but it can be controlled using cameras and, thanks to current relays, it can be stopped if a failure is detected. After the additive manufacturing process, the module operator can perform basic finishing of the product using the tools available in the tool box. After checking the suitability of the machine to produce the next part, the operator starts the next process from the waiting queue.

All the main 3 modules are independent on each other and can be exchanged. For example, for more "troublesome" patients (having problems with keeping their limbs static), the scanner on a track is not utilized but instead a manual scanner can be used (AutoMedPrint is equipped with EinScanPro scanner for that purpose) and the data is manually entered into appropriate software algorithms (described in further chapters). The rapid manufacturing module can also be replaced with any additive manufacturing system – a farm of 3D printers or any system of device, while the programs can be created manually on the basis of STL files exported from the intelligent models. The interface also is highly modular (as described in the next chapter) and does not have to be used in its whole. It brings a unique possibility to granularize the system and implement only its selected parts wherever and whenever need arises.

6.3.3 AutoMedPrint system software – design and user interface

The AutoMedPrint system contains many modules, connected to facilitate its main purpose: the automated design and manufacturing of orthopedic devices through 3D printing technology. This subchapter shows the structure and functionality of these components, offering a view into the system's modular architecture, which supports workflow from patient scanning to the final production of orthotic and prosthetic devices. Most of the software modules were programmed by the author, aided by the other members of the AutoMedPrint team.

The software that makes up the AutoMedPrint system consists of a number of smaller and larger modules that carry out the tasks of scanning, design and manufacturing preparation, linked together by means of diverse data exchange algorithms. Among the modules used, the following types of software components should be mentioned:

- Ready-made, unmodified commercial or free (open source) applications: David scanner software, incremental manufacturing process preparation software (Slic3r),
- Commercial or free software with the ability to automate by creating macros: MeshLab, Microsoft Excel, Autodesk Inventor,
- 3D application development engine and the modules built with it – Unity 3D,
- Other applications built by the programming method using development environments (Visual Studio) – CAM module, scan download module, etc.

The 3D scanning module is used to implement the process of collecting and processing anthropometric data from patients using the system's capabilities. The module consists of the following programs:

- operator panel (AutoMedPrint Operator Panel),
- programs to control the movement of the scanner along the track and vertically,
- single scan acquiring algorithm,
- scan processing algorithm (MeshLab).

The operator panel application – AutoMedPrint Operator Panel – is the central point of operation for the entire system, accessible from the 3D scanning workstation operator. The AOP application manages patient data acquisition, triggers automated 3D scanning and data processing and automated design and manufacturing procedures, and runs other user interface applications. The operator panel is a multiscreen 2D application containing simple user interface elements for entering data and running predefined functionality or other programs, either locally or remotely (on the computer containing the user interface applications or on a remote station for automated design or fabrication). The various application screens are discussed and shown below (Fig. 6.10).

1. Startup screen (product selection).
2. Patient and product data:
    - patient data: name, gender, date of birth, contact e-mail,
    - product data: the patient's height at the time and the side for which the device is designed (left/right),
    - writing data to disk and clearing the data.
3. Scanning:
    - a button to perform a single scan (start the scan download machine),
    - launch button for the software that controls the movement of the scanner,
    - button to delete all measurements taken in a given session.
4. Data processing and design:
    - *scan processing* (MeshLab macros): merging and matching, limb cleaning and reconstruction, point extraction from reconstruction,
    - *design*: remote launch of the configurator for a given product, launch of the relevant Excel sheet for filtering and completing design data, launch of the autogenerating CAD model loading design data automatically,
    - *preview*: enable preview of reconstruction (visual verification of correctness of scanning and data processing) and product (visual verification of correctness of design),
    - *data*: copy data to patient folder, open folder, delete temporary data.
5. Project acceptance.

The programs controlling the travel of the scanner along the track and its movement between the top and bottom positions on the linear guide were written in the

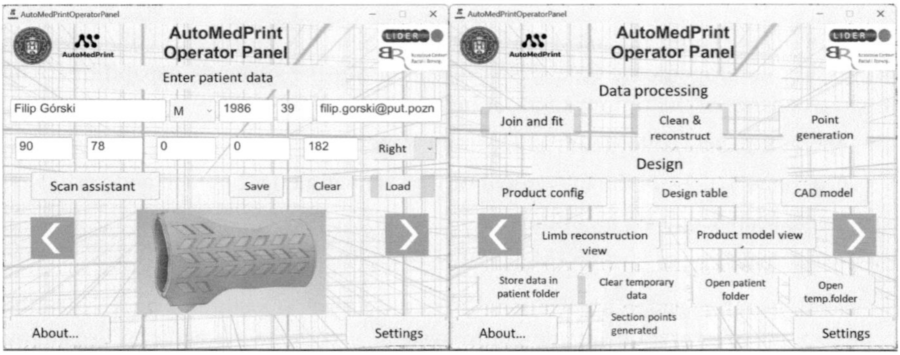

Figure 6.10 AutoMedPrint – operator panel screens (patient and product data, data processing and design).

C-based Arduino language. Because of the need for commands and communication, windows forms applications were created to easily read and ask the current position.

The first application controls the passage of the cart on the track. After connecting to the Arduino module, the application allows to select one of four predefined positions – A, B, C or D – and move to the set position. It is also possible to select the previous and next positions. The STOP function stops the platform, once the direction or position is selected again, the movement can continue. The application window displays information about the current position and the status of the sensors, as well as position errors and the absence of a set position.

The second application controls the movement of the scanner along a linear guide. It allows to select one of two basing options (depending on the limb being measured). In the case of the "Zeroing" option in the "Hand" field, the scanner will be based using the upper extremity, while in the case of the "Zeroing" option in the "Leg" field using the lower extremity. After basing, the user can select the scanning position (Position 1 – top position/Position 2 – bottom position, mirror image scanning). Using the "Adjust" button, the user can change the distances of position 1 or 2 from the zeroing point. Using the slider, the feed rate can also be changed, and the STOP button is used to stop the movement of the scanner. A console view is also available in the application window to check the status of the connection. Both applications are shown in Fig. 6.11.

The scan processing machine in MeshLab software takes advantage of the ability to create and group macros in this software and run them remotely in batches, by using the so-called MeshLab Server, i.e., software that executes macros exclusively from a console (command line) application, without a graphical user interface. The whole system consists of two main components:

1. Module for merging and matching scans:
   - 6 macros in the native .mlx format (one macro per each possible scan position), including translation, rotation, mirroring, and automatic cleaning of scans from junk data,
   - a program designed to perform best-fit (best-fit) operations on cleaned and positioned scans, using MeshLab's graphical interface,

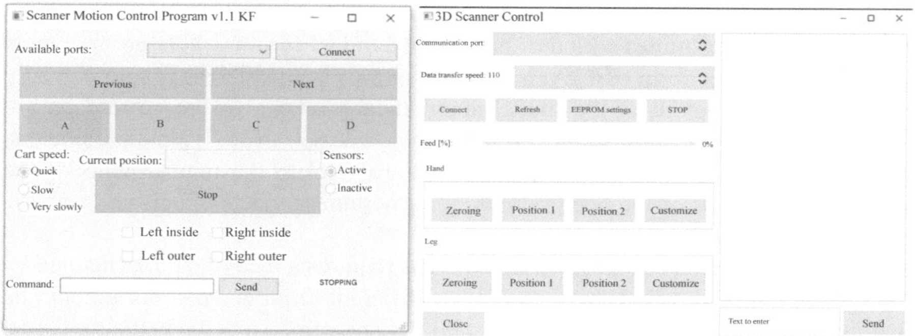

Figure 6.11 AutoMedPrint – scanner motion control programs.

- batch files in .bat format, containing MeshLab command line calls for each of the 6 macros, then assembling them into a single file and merging them into a single mesh (using a separate macro), distinguishing between limbs (one file for the arm, the other for the leg),
- MS Excel spreadsheet used to calibrate the workstation – the calibration data placed in it is used to automatically generate MeshLab macros that allow correct positioning of scans.

2. Cleaning and reconstruction module:
   - macros in .mlx format, containing translation and rotation operations to place the limb in the correct coordinate system, and then clean up by selecting and removing triangles above the limit of useful data,
   - macro in .mlx format used to implement Poisson reconstruction, i.e., to create a model of a smooth, closed shell in place of concatenated and cleaned scans,
   - batch files in .bat format, containing MeshLab command line calls to lay out the resulting limb model in layout, cleaning and reconstruction, differentiated for side (left/right) and limb (hand/leg).

The automated design module is described more extensively in Chapter 4 – in terms of basics – and in further part of the chapter, regarding specific products. In general, it consists of three basic submodules:

1. Intelligent CAD models, fed by Excel sheets containing anthropometric data generated using automated algorithms.
2. An automated algorithm created in the MeshLab environment for making cross sections through a reconstructed limb model and exporting point coordinates of these sections to text files.
3. An automated Excel spreadsheet implementing the process of extraction, selection and ordering of points derived from limb cross-sections and preparation of the actual Excel spreadsheet feeding the autogenerated models.

Intelligent CAD models form the basis of the operation of the design automation module in the system under construction. The autogenerating model of an object of a given type constitutes a kind of base/source model for the whole group. On its basis, the geometric form of a given variant of a prosthesis/orthosis is automatically prepared. The structure of the model automatically adapts to the entered data – thanks to appropriate scripts and saved forcing/exclusion of individual parameters, the program adapts the overall structure of the model to the requirements of the new variant. Details of the operation of autogenerating models of selected orthotic products are presented later in this chapter.

Correct operation of autogenerated models requires a feed of data, generated by the scanning module. Each base model has different requirements, but a common feature is the way the data is integrated into the program. An autogenerating model of a given type contains a set of key data necessary to describe the new variant. In the work on the AutoMedPrint system, as mentioned above, Excel spreadsheets were used, implementing the process of extraction, selection and ordering of points derived from cross-sections of the limb, which in effect feeds the autogenerating models. The following section synthetically describes the various sub-modules that allow the implementation of the automatic design process using autogenerated models.

The MeshLab algorithm functions on the same principle as the scan merging and cleaning automatons described in the previous section. Its essence is a macro-command used to implement the Planar Section filter, i.e., an operation that allows the extraction of a part of a common plane with a specific position and a closed shell (triangle mesh), in this case representing the patient's limb. Sectioning the limb model with a plane as a result allows to obtain a set of (usually) several hundred points.

The macro command comes in 3 different versions: for the full (non-amputated) upper limb, for the upper limb stump and for the lower limb. Each version has a different number of generated cross sections. The locations of the most important characteristic planes, between which the cross sections are made, are determined by the user graphically, using an application (an example of such an application is shown later in the chapter, when describing the parametric model of the WHO orthosis).

The final output of the macro-commands are design tables in MS Excel, which directly feed the autogenerated models created in Inventor. More details about how all these algorithms work and what are the results – in form of actual models for specific patients – are presented in the subsequent chapters, detailing intelligent models of wrist hand orthosis (6.4) and hand prosthesis (6.6).

6.3.4 Verification and validation of AutoMedPrint system

The AutoMedPrint system was and is thoroughly validated and evaluated, while continuously being improved with subsequent iterations of patients, products, algorithms and software modules. An approach of continuous improvement was taken, with certain assumptions taken along the way. These assumptions are briefly summarized below.

1. Work with the patients from day 0 – it was assumed that the system will be highly patient-oriented. As such, usually a given product model started out as a one-time, practically validated solution to a problem of a specific patient, introduced into the system as an intelligent model alongside with algorithms for its automated creation and data processing. As such, it was not a problem to validate the products – their concept and basic functions were validated before they were part of the system. Some selected cases are presented in the further part of the chapter.
2. Division of patients into "test patients" and "regular patients". The test patients belong to a group that works with the AutoMedPrint team on a frequent basis – a span of many years, spent on testing and long-term use of orthopedic products generated by the system, often as the first users of a given product type or subtype. Most of the products or their modules have their own test patients assigned to them, although the test patients also can change, due to lack of need of use of a product anymore, or change in dimensions (in the case of younger, still growing patients). The regular patients are frequently one-off – a single product is made for them, usually in an automated manner, as much as possible.
3. Work with medical specialists and physiotherapists – again, new methods of design, scanning, data processing and product concepts are always verified and validated by certified orthopedists or physiotherapists. They also take part in fitting and testing process of regular patients to assess the products.
4. Quantity evaluation studies – from time to time, the system is evaluated by a study realized on a larger group of healthy volunteers (usually students, or young professionals), using survey methods, observations, and interviews to find out potential ways of improvement (Górski et al. 2022a).
5. Quality management and risk assessment studies – recently, a quality management system was conceived for the AutoMedPrint system, as well as methods of risk assessment for the products generated by it, as described in a previous publication (Górski et al. 2022c). Future studies will include obtaining full conformity with all the medical device regulations present in the European Union and other regions of the world.

Some examples of validation through work with the test and regular patients are shown in the further part of the chapter for various types of products, showing how new solutions are conceived and tested, as well as how the already verified ones are tried on with new patients. These examples can be also found in previous publications (Górski et al. 2023b, Górski et al. 2024c).

6.4 Case study #1 – openwork wrist hand orthosis

6.4.1 Case and problem definition

The openwork wrist hand orthosis is a product that can have a therapeutic or corrective function in both short-term injuries and long-term conditions, requiring constant or periodic use. It is a new, modern solution aimed to replace standard solutions such as plaster of Paris, or polymeric shells, that due to typical manufacturing processes are constrained in a way of geometry – they are monolithic, both in the macroscale (no holes) and in the smaller scale (high density of material). This causes lots of problems with comfort, hygiene, but also in terms of therapeutic functions – the orthoses could be simply too large and heavy to be effective in the process of treatment of hand conditions.

The simplest division of orthoses is based on the degree of stiffening. There are three types of orthoses: rigid, semi-rigid and flexible. Most often, stabilizers are lined with a material that allows air to pass through and allows for quick removal of moisture. There are airbags between the materials, which increases the stabilization of the injured limb. Many stabilizers have Velcro fasteners, which significantly improve wearing comfort. The most important advantages of orthoses include the ability to put them on easily, monitor the condition and hygiene of the skin, and relatively low weight of the orthosis with good stabilization.

A response to the known set of problems with traditional orthoses is an openwork orthosis. The original version of the design (dubbed "Cortex Exoskeletal") first proposed in 2012 at the University of New Zealand by a team of scientists led by Jake Evill (Kelly et al. 2015). The only feasible and cost-acceptable way of producing such shapes is by means of 3D printing (additive manufacturing), as such it has only been available for a short time as a viable concept. The openwork orthosis was originally made of polyamide. Compared to traditional plaster casts and monolithic orthoses, the stabilizer is lighter, thinner, and more comfortable to use. Due to the openwork structure, the skin breathes and skin inflammation does not occur. The main disadvantage of the orthosis is the time-consuming nature of developing a digital model – developing it manually is pointless, as the time needed to prepare the orthosis must be short. In the case of wrist injuries, such as a fracture of the scaphoid bone of the wrist, the treatment time is 3–6 weeks, and the orthosis is applied approximately 1 week after the injury at the earliest. It is therefore fully justified to automate the preparation and production of this type of orthoses, which was achieved during the creation of the AutoMedPrint system. Here, in this chapter, various approaches to design of such orthoses will be presented, along with selected considerations of manufacturing and properties of obtained products.

The openwork orthoses are a relatively new product on the market and they must be prepared using 3D scanning and modeling techniques. This process is not yet unified (at the detailed level), which means that each entity preparing the design of an openwork orthosis uses their own approach, depending on the available equipment and software. However, the main stages of the project always include the scanning process and then 3D modeling – using the scanned mesh as a reference.

6.4.2 Mesh-based design approach

The approach described in this chapter is based on direct use of 3D scanned and processed mesh of the human limb – so-called "shell" approach. It is also an approach that can be named as a hybrid one, as it involves use of both non-parametric geometry (mesh) and parametric CAD operations of solid modeling. However, the way of conduct is linear, as the mesh is a starting point for the solid, that comes later in the process. This approach can be frequently found in many previous elaborations by other authors (Cirello et al. 2024, Keller et al. 2021, Kelly et al. 2015). It can be evaluated as an "ad hoc", intuitive solution – as it usually brings correct, patient-tailored results, does not require special, dedicated hardware and software and can be easily repeated by almost any mechanical or biomedical engineer with solid modeling CAD skills.

Here, we will consider a description of a process realized in a team of Poznan University of Technology in preliminary stages of development of automated system for production of orthoses. The process presented in this chapter was already described in a previous work – initially in a diploma thesis by Burdzinska, supervised by the author (Burdzińska 2015) and later in a common publication of the authors (Burdzińska et al. 2015). The further contents of this subchapter that pertain to a specific case study of a wrist orthosis are based on the studies described in these publications (originally published in Polish).

The first stage of the project was to scan the patient's diseased hand in the forearm in order to prepare its initial digital representation. For the purposes of this study, a 22-year-old female patient with no structural changes in the skeleton was selected and it was assumed that she suffered a fracture of the scaphoid bone of the wrist as a result of a fall on a straight arm (Burdzińska et al. 2015).

As described in previous chapters, contactless, generic scanners using structured light were used for scanning. Thanks to their high degree of mobility, they allow to collect a large number of point clouds during one light beam pass. It is important for scanning processes to be carried out in a darkened room – then the process runs with fewer disruptions. The patient's hand should be covered with an anti-reflective coating (chalk) to matt the skin. This makes measurements faster (elimination of light reflection). The scanning technician should select the appropriate working field for the scanner (the size of the area for which data is collected in one shot) depending on the type of equipment and the size of the patient's hand (different for a child and different for an adult). When scanning a limb, special markers – reference points are not used (it is difficult to maintain them on the injured limb), therefore it is recommended to use algorithms that automatically combine scanned images (this is usually one of the options available in the device software).

In the presented study, GOM ATOS I scanner was used, with a working field of 500 × 400 × 400 mm. During the scan, the patient should not move the limb, otherwise there may be a necessity of repeating the scan, or obtaining incorrect data. The measurement time, assuming a cooperating patient, is approximately 10 minutes. It can be realized without any special supporting construction.

The scanning results in the form of an initial mesh model require further processing, because the scanner collects too much data. The surplus data must be

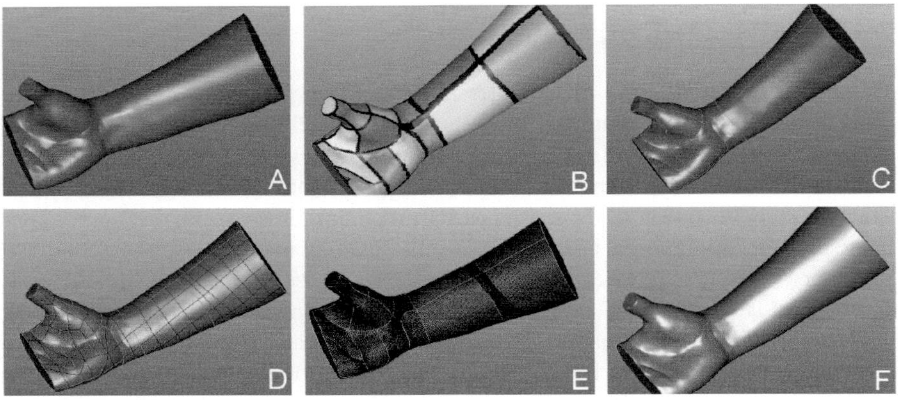

Figure 6.12 Stages of generating solid model from upper limb 3D scan, (A) – initial mesh; (B) – dividing into regions; (C) – generating basic patches; (D) – generating and manual adjustments of patches; (E) – densing the mesh; (F) – final solid; based on previous work (Burdzińska 2015).

filtered (to remove elements of the surroundings, close holes in the surfaces, etc.) to prepare the mesh model for further work on the design of the orthosis. Some of this work can be carried out in the scanner software (if its functions allow it), in special software for processing mesh models or in a CAD program (if it has such functions). Here, freeware GOM Inspect software was used to process the mesh – its results are presented in Fig. 6.12A.

The next stage of data preparation is converting the mesh model to a solid. This could be done automatically, using CAD software, but for the best results in terms of model topology it is advisable to perform the process in stages, shown in Fig. 6.12, based on (Burdzińska 2015). First, the curvature of the entire object is examined and divided into regions (Fig. 6.12B). The resulting contours are the basis for further development of the system of patches forming the surface of the solid (Fig. 6.12C). The next stage is to prepare a network of quadrangular patches covering the entire model (Fig. 6.12D). Thanks to this process, it is possible to eliminate potential errors in the surface by manually modifying them in selected places and then densifying the mesh (Fig. 6.12E). The last step is to combine the surfaces based on previously entered data and convert the model into a solid form, which is the basis for further design work (Fig. 6.12F).

Further work is realized in a CAD system – in this case, CATIA v5 software was utilized. The course of work was as described below.

1. Loading the hand model and determining its position relative to the center of the coordinate system. It is recommended to set the model's base plane as XY to prepare the model for easier export into additive manufacturing systems.
2. Execution of the Offset command on the outer surface of hand model at a selected distance – the offset value should be described depending on the scale of the injury, because this parameter determines the "looseness" between the skin of the hand and the orthosis. It was assumed 5 mm in the presented case,

but it can vary between 0–6 mm (of the authors' experience), depending on the actual injury case and also planned finishing of the orthosis (lining, foams, etc.).

3. Execution of the Offset command on the outer surface of a hand model at a selected distance of looseness increased by demanded thickness of the orthosis – in this case, thickness was assumed 6 mm, so the distance was 11 mm. Of the author's experience, thickness of openwork additively manufactured hand orthoses can vary between 3–6 mm, depending on applied material and requirements towards orthosis strength.
4. Connection of offset surfaces (closure), to obtain a solid model, also making a proper cutout in the thumb region.
5. Adding rounded or chamfered edges.
6. Marking the dividing line (usually along the axis of the hand, considering both manufacturing technology issues and assembly possibilities on the patient's hand) and division of solid into two parts.
7. Adding assembly features – sketching and extrusion functions with material subtraction. The method of implementation depends on the type of mounting, but there is freedom in the scope of CAD program functions used and skills of a given engineer.
8. Development of the shape of the openwork – subjective, depending on the designer and the adopted design; built on the external surface of the model, using sketching on the planes of the main system (with their rotation around the longitudinal axis of the model) or additional planes (e.g., tangent to the surface at a randomly selected point), using the material subtraction function; possible use of the solid subtraction function – Boolean algebra.

Model of the orthosis realized in such an approach, with more important stages marked, is presented in Fig. 6.13.

The resulting orthosis was additively manufactured out of ABS material, using Stratasys Dimension 1200 machine, with default parameters (0,33 mm layer thickness and monolithic infill). The resulting orthosis was tested for functionality, fit and comfort by a healthy volunteer and these tests ended successfully – the wrist was properly protected, but the functionality of the hand remained intact and there were no problems with daily activities or hygiene.

To sum up, this approach – as compared with the fully parametric ones – has an advantage of being fully tailored to a given patient, not considering a larger group of patients and not making any compromises to make the model "universal". The design approach presented in this chapter is a one-time procedure, repeated every time for a new patient. With correct assumptions (of medical and technological viewpoint), results will be appropriate. However, time of generation of such a model is usually quite long. In the given case it was 21 working hours of CAD modeling, which could probably be reduced significantly, but not below 1–2 working days, in the author's opinion. As such, this mesh-based approach is a fully acceptable strategy of generating orthoses from time to time, when direct need arises and there are no other suitable tools other than a 3D scanner, a CAD system and an engineer

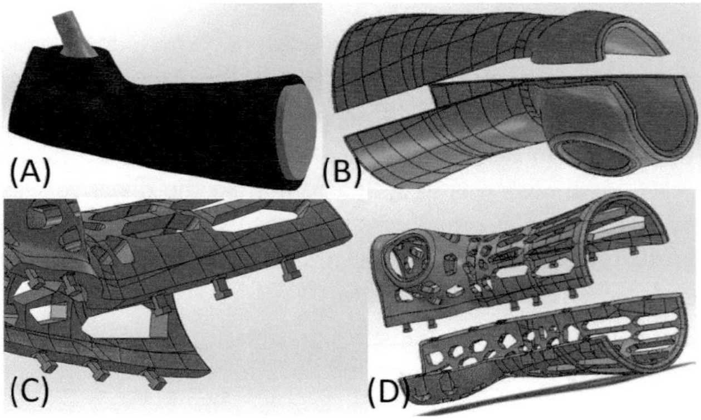

Figure 6.13 Modeling of the orthosis based on hand mesh, (A) – initial solid with thumb cutout; (B) – solid division and rounded edges; (C) – adding assembly features and openwork; (D) – final solids; partially based on (Burdzińska et al. 2015).

with a certain set of skills. However, for repeatability and larger amounts of patients (i.e., mass customization), a different, more universal and less time-consuming approach is needed, as described in the next chapter.

6.4.3 Parametric CAD design approach with complete design automation

The approach presented in this chapter is a result of a methodology developed to obtain a fully parametric individualized shape of the wrist hand orthosis, having in mind automation of design and quick changes of shape in response to different sets of anatomical data. As a result of previously realized project – AutoMedPrint, these results and their subsequent iterations were partially described in previous publications, namely (Górski et al. 2020a) and (Górski et al. 2020b). This subchapter and the following studies and cases descriptions are based on AutoMedPrint project reports, as well as these publications and it describes the second main approach – the "spline" approach.

The concept of parametric representation of individualized anatomic shape is based not directly on a mesh – a result of the 3D scanning process – but rather on sets of point coordinates extracted from that mesh. These points are used to draw parametrically described curves (in this particular case – closed spline curves). Based on these curves, a multi-section solid extrusion operation is performed in 3D CAD. This generates the three main bodies:

- the anatomical shape, representing more or less faithfully the scanned limb itself,
- the external boundary body, which is created by expanding the closed splines (offset operation) by a value of $l+th$, where l is looseness (offset from limb surface to inner orthosis surface) and th is thickness of the orthosis shell,
- the internal boundary body, same as the external one – however the offset is only by the l value and the operation is of Boolean subtraction (removal of material).

In practice, it is enough to draw the basic spline curves based on point coordinates, offset them twice (by *l* and *l+th* values) and then create a single multi-section solid extrusion based on the two offsets, to obtain a shell of the newly created orthosis.

During the design of the orthosis, there were plenty of considerations and iterations, which will be briefly presented. First of all, it has been decided to divide the orthosis into the forearm part and the wrist part. Originally, during early iterations, it was assumed that the forearm part splines can be less complex, having a lower number of points – while from the wrist upwards, geometry should be more detailed, presenting the sections of the wrist and hand's palm with thumb area, where the orthosis must be closely fit. A number of experiments by trial and error were realized to determine a number of points and planes for comfortable, safe and yet not too lengthy process of re-calculating the orthosis geometry after changing the dataset from patient to patient. Initially, in a version of orthosis described in (Górski et al. 2020a) 8 points for the forearm and 12 points for the wrist were used. Later, the orthosis was re-designed from the scratch, using 11 planes (6 for the wrist upwards, 5 for the forearm), with 16 points at each plane selected automatically out of the whole point set, generated automatically by MeshLab while realizing section planes (Fig. 6.14B) and stored in an MS Excel spreadsheet.

The point coordinates, placed in Excel design table, were used to create 11 sketches with basic splines, placed on the planes. Locations of four most important planes (start, end, wrist, thumb) are decided by the user in a simple 3D application – AutoMedPrint Limb Calibrator – created in Unity engine by the author

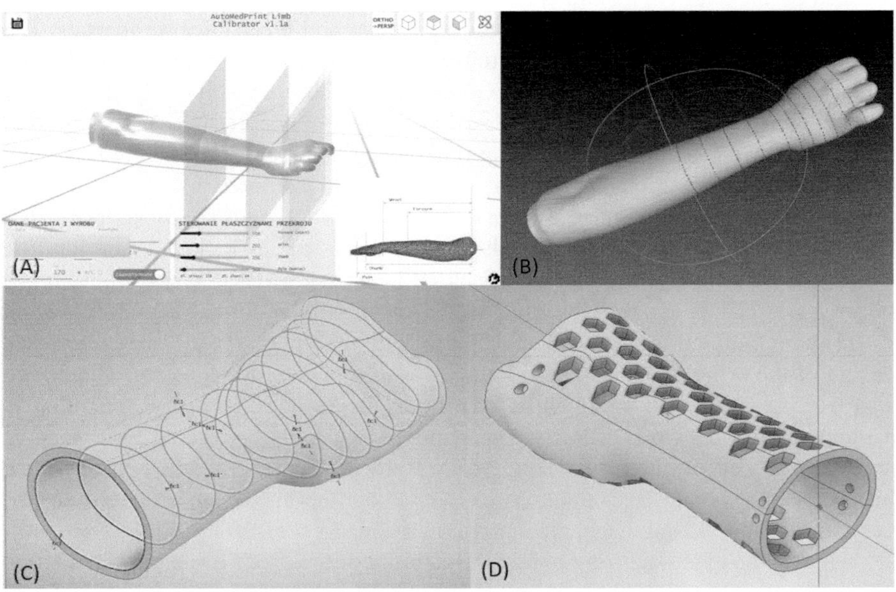

Figure 6.14 Intelligent model of the wrist hand orthosis in the AutoMedPrint system: (A) selecting the planes; (B) sectioning the limb mesh; (C) initial orthosis solid; (D) final orthosis model.

(Fig. 6.14A). Adding offset and thickness to these sketches, initial orthosis solid is created (Fig. 6.14C). Then, are a number of operations are realized – cutting thumb hole, splitting the orthosis into parts (2 or 4, depending on the size), adding openwork cutout shape and assembly features, as well as additional features (e.g., chamfers), until a final model is obtained (Fig. 6.14D). The whole process is automated and requires simple refreshing of the Inventor file and subsequent exporting of orthosis parts into STL format directly for 3D printing or for verification.

The created intelligent model belongs to level 2 of automation, according to the classifications presented in Chapter 4, as it contains a link to an external data source so that the whole geometry can be changed without touching CAD tools. There are some conditional statements: division into 2 or 4 parts depending on the 3D printer space, defined in Excel; mirror image of orthosis and type of openwork (hexagonal or other shape of holes). The project is an ongoing constant improvement – recent iterations include separate Excel spreadsheet for the openwork structure, additional features for therapeutic orthoses (cutouts), improved assembly features and a set of algorithms for automated generation of cutting patterns for EVA foam lining, to name just a few. The concept itself proven to be a viable one and a number of orthoses were made by the author and his team in the scope of AutoMedPrint project and its evaluation (Górski et al. 2022a). Some examples (patient cases) are presented in the further part of this chapter.

6.4.4 Additive manufacturing of wrist hand orthoses and their properties

This subchapter explores the methodologies, results, and conclusions of studies focused on the additive manufacturing of wrist-hand orthoses, emphasizing the key aspects of process stability, fit, patient assessment, and mechanical properties. The contents of this chapter are based on studies performed in scope of the AutoMedPrint project, wider described in the paper (Górski et al. 2020b), as well as other papers by the author and the team.

The research aimed to establish a reliable set of parameters for the rapid and cost-effective production of customized wrist-hand orthoses using popular 3D printers. This involved almost 100 manufacturing processes of orthosis parts in a controlled environment. The materials used in these studies included Acrylonitrile Butadiene Styrene (ABS), polylactic acid (PLA), nylon (PA12), and high-impact polystyrene (HIPS), processed using Fused Deposition Modeling (FDM) technology.

The manufacturing process for these orthoses began with the automated design of the orthotic device based on a 3D scan of a patient's forearm. The design was executed using an intelligent CAD model in Autodesk Inventor, resulting in a lightweight, openwork orthosis consisting of two parts connected by a snap-fit mechanism. The orthoses were printed in various orientations and with different infill percentages and layer thicknesses to evaluate the impact on process stability, economic viability, and mechanical properties (Górski et al. 2020b).

During the manufacturing phase, the stability of the process was assessed based on whether the 3D printing could be left unsupervised without resulting in significant errors. The most stable processes were achieved with ABS and PLA materials when the orthoses were printed horizontally. Vertical printing, although often resulting in higher surface quality, posed more stability issues, particularly with nylon and HIPS materials. Overall, ABS demonstrated the best process stability, with the fewest instances of detachment from the build platform or other errors requiring intervention.

The economic aspect of the manufacturing process was evaluated by measuring the time and cost required to produce the orthoses. The study found that most orthoses could be produced at a cost lower than the equivalent mass-produced items available in medical shops, with acceptable manufacturing times under 16 hours. The most cost-effective material was ABS, followed by PLA. Nylon, while offering superior mechanical properties, was more expensive and required longer production times due to its lower extrusion speed and higher need for support material during horizontal printing.

Fit accuracy and surface quality were critical parameters assessed in the study. All manufactured orthoses were tested for fit on a patient, and their surface quality was evaluated on a scale from 1 to 3. The orthoses printed in vertical orientation generally received higher surface quality scores, especially those made with PLA in the accuracy-oriented strategy, which utilized a finer layer thickness. However, vertical orientation sometimes led to reduced mechanical strength, as the bonding between layers was weaker compared to horizontally printed orthoses (Górski et al. 2020b).

The mechanical properties of the orthoses were tested using a quasi three-point bending test on a universal testing machine (see Fig. 6.15). The study aimed for a minimum load-bearing capacity of 300 N, simulating a static load from an object weighing approximately 30 kg. PLA and nylon orthoses showed the highest strength values, with PLA orthoses often exceeding the strength of those made from ABS and HIPS. However, the strength of PLA orthoses was significantly affected by the printing orientation, with vertically printed PLA orthoses showing inconsistent strength due to the weaker interlayer bonds (Górski et al. 2020b).

Nylon orthoses displayed exceptional durability and resistance to deformation, even under higher loads. This material's flexibility and toughness make it an ideal

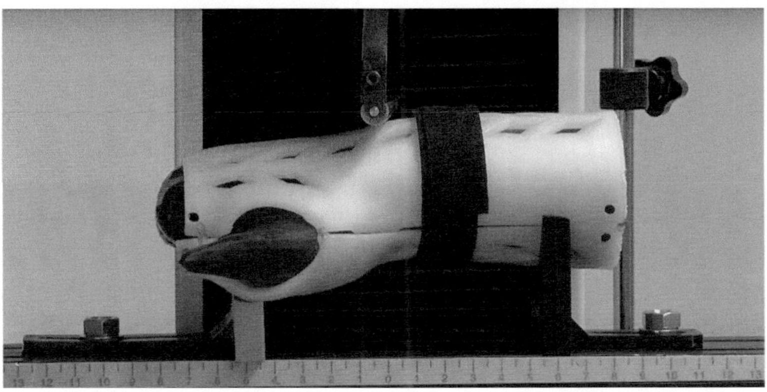

Figure 6.15 Testing static bending strength of a 3D printed hand orthosis.

choice for dynamic applications where the orthosis might experience varying loads. Despite these advantages, the higher cost and longer production times of nylon orthoses might limit their widespread adoption in clinical settings where cost efficiency is very important and in the "battlefield" conditions, nylon-capable printers are not always there.

In terms of economic and technical coefficients, the study concluded that the most balanced material for additive manufacturing of wrist-hand orthoses is PLA, especially when using the economic-oriented strategy (15% infill and 0.3 mm layer thickness) in a vertical orientation for better surface quality and patient acceptance. Nylon, while superior in mechanical properties, should be considered for specific use cases where its benefits justify the higher cost and production time – especially in cases where dynamic applications are considered.

The study also highlighted the importance of choosing the right orientation for printing, with horizontal orientation preferred for achieving maximum strength and vertical orientation for better surface quality. Ultimately, the choice of material, strategy, and orientation should be tailored to the specific requirements of the patient and the clinical context, ensuring a balance between cost, strength, and aesthetic quality. Summary of the obtained numerical results is presented in Fig. 6.16.

These studies, conducted as early as 2020, were a major breakthrough in proving that even using inexpensive, extrusion 3D printers along with the cheapest and most available PLA material, strong, durable and comfortable orthoses can be still obtained, in a time less than one working day. The 3D printers are constantly evolving and the times of manufacturing are now much shorter. Also, studies were conducted on more advanced materials, including the composite ones (Rybarczyk et al. 2024), leading potentially to developing devices capable to work in any conditions and under heavy loads, not compromising the patient's health even when doing advanced tasks with the orthoses equipped. It also led to another very important conclusion – automation of 3D printing program preparation is not really necessary, when there is a know-how on how to prepare these programs to meet certain requirements (depending on the

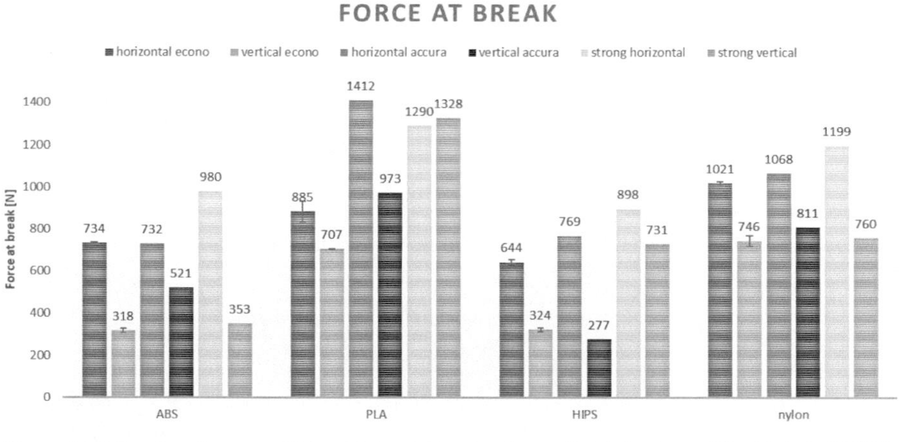

Figure 6.16 Results of testing of 3D printed wrist hand orthoses (data taken from (Górski et al. 2020b)).

assumed strategy). Having this know-how and a very basic training in 3D printing (a skill that can be developed by practically anyone, not necessary to have a technical background) it is viable to have human operators realize 3D prints of automatically generated geometries. In the further part, some examples of patients are shown, where practical, long-term use was achieved, based on these initial studies.

6.4.5 Patient case #1 – trauma healing

The first, and one of the defining cases in proving capability of design automation of anatomically adjusted orthoses was a one where lack of automation would not allow to create an orthopedic product in the limited time frame. The patient was a 9-year old boy. On the last day of school and a week before traveling abroad for vacation, Jan broke his left arm in two places while on the playground. The doctors decided to perform a surgery and insert Kirschner wires, which are an auxiliary and temporary element in stabilizing the fracture that has been used for more than a century now (Franssen et al. 2010). As the patient was an extremely mobile boy, and it was decided not to put a plaster cast on, his hand had to be immobilized. As the patient was travelling very soon, a matter of days remained to perform a full procedure of design, prototyping, manufacturing and validation. He was equipped with a thermoplastic orthosis, heavily limiting his movement and preventing him from water play and general sports activities – that is why it was decided to create a lightweight, mostly waterproof orthosis.

As the injury was serious (two bones were broken), it was decided to use a two-part orthosis for better stabilization. The parametrized WHO model was used to prevent wrist motion and protect the forearm, while a second part was created separately in form of a 90 degrees bent tube-like shape (adjusted anatomically), starting where the WHO ended, and ending approximately in a half of the bicep muscle. It was decided to split the orthoses into several parts, easily detachable, so that in the later phase of healing the elbow part could be completely removed and the patient could be left only with the forearm and wrist part, with possible elbow bending.

The patient was 3D scanned using the AutoMedPrint automated scanning station and rig for hand placement. The fractured arm was not scanned, it was decided to go for measuring of a healthy hand and applying appropriate looseness values to compensate for muscle swelling in the injured limb. Some additional scans for experimental purposes were realized manually, of both limbs (with doctors' and caretakers acceptance), but were not used in the process of design – only for testing if dimensions of the two limbs match (they did). For the manufacturing, it was decided to utilize low-cost FDM technology, creating two orthoses: one of PLA material and one of PA12 (nylon) for more dynamic use, especially in water. The first, test orthoses were made also of PLA material and with variable offset values, to test which one would be the best fit. For PLA printouts, low-cost FlashForge Creator Pro machine was used and for the nylon prints, Zortrax M200 machine was used (all printouts were realized with 30% infill and 0.25 mm layer thickness).

On day 1 of the project, the team met up with the patient and the scan was realized (Fig. 6.17A). The first version of the orthosis (the WHO part) was designed

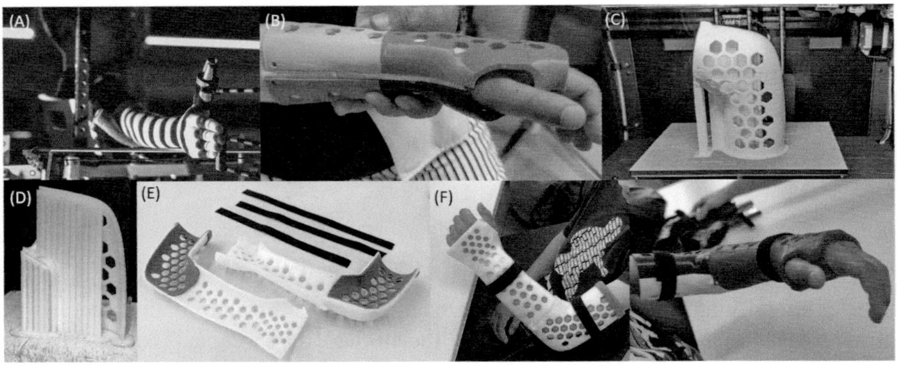

Figure 6.17 Trauma orthosis – case study, (A) patient scanning, day 1; (B) first try-on, day 2; (C) 3D printing – PLA; (D) 3D printing – nylon; (E) post processed orthosis with EVA foam lining; (F) testing by the patient.

almost immediately after scanning. Data extraction from the arm scan was carried out using automated MeshLab and MS Excel macros within the AutoMedPrint system, as described in the parametric model section (11 planes, splines with 16 points each). The orthosis was 3D printed in different variants and a test try-on was made on day 2 (Fig. 6.17B). In the meantime, the elbow part was designed – the model was not available before, so it was created in a way similar to the rest of orthosis, but with manual algorithms of point extraction out of the scan. After the try-on, some corrections were introduced (notably, looseness value was slightly increased, from 3 to 4 mm) and the whole orthosis was printed again (Fig. 6.17C–D). It was lined up with the EVA foam (Fig. 6.17E) and successfully tried out by the patient in both versions (Fig. 6.17F).

The case was proven to be a highly successful example of rapid development of an orthotic product. Automated design was used for proper generation of an orthosis for a real patient case and it was proven that in one day, a functional 3D printed orthosis can be generated. It also proved a possibility to use mirror image of a limb to create a design of an orthopedic product for another limb, which is now frequently utilized by the AutoMedPrint team led by the author. Cost of a single orthosis was estimated as being below US$20, even including post processing and basic design time (not counting manual development of the elbow part model). It would be fully feasible – in the author's opinion – to introduce the technology in healthcare facilities, to take care about at least some injured patients semi-automatically, equipping them with lightweight, modern, anatomically customized healing devices.

6.4.6 Patient case #2 – correction and rehabilitation

The patient case presented in this subchapter was a subject of work of the AutoMedPrint team and is described in previous publications – the basic case is described in the 2024 chapter (Górski et al. 2024a), while the further development of orthoses for therapy with aid of Virtual Reality is presented in a 2023 journal publication (Górski et al. 2023b). This chapter contains lots of citations of these works, but also some new insights from the designer's perspective.

The research presented in this study was focused on a case of a 13-year-old (initially) boy suffering from congenital paresis, a condition requiring specialized therapy. Congenital paresis involves partial or complete muscle weakness present from birth, caused by developmental irregularities or damage to the nervous system during fetal development. This condition can affect any body part, thereby limiting movement and functionality in the impacted areas. Prompt diagnosis and treatment are vital to enhance mobility and improve life quality for those with congenital paresis.

The patient demonstrated limited mobility across several upper limb joints. Specifically, in the shoulder joint, extension, abduction, and external rotation were restricted to about 80 degrees for extension and abduction, and roughly 45 degrees for external rotation. The elbow joint showed minimal flexion movement. The hand, positioned in pronation at the distal radioulnar joint, exhibited limited supination movement to approximately 40 degrees without any radial or ulnar deviation. Furthermore, the wrist joint, in an ulnar position, lacked active distal radiocarpal joint extension, which restricted hand movement. These limitations substantially affect the patient's daily activity capabilities, necessitating customized rehabilitation approaches to improve specific impairments and enhance functional mobility (Górski et al. 2023b).

The therapeutic orthosis designed for this case drew on 3D anatomical scans of both the patient's arms. Typically, the AutoMedPrint system employs one of two strategies for designing hand orthoses (or leg orthoses, for that matter):

1. A direct approach where the injured or unhealthy limb is scanned in the correct anatomical position to serve as the sole basis for the orthosis design. This method is feasible only if the limb's condition permits scanning without further damage.
2. An indirect method involves scanning the non-injured, healthy limb and creating a mirror image to design the orthosis for the injured side, incorporating appropriate adjustments for any swelling. This technique is used when the injured arm cannot be scanned and both limbs are symmetrical.

However, in this specific instance of a problem – a non-symmetrical set of limbs resulting from a paresis – neither strategy was suitable due to the limbs' asymmetry, the unhealthy limb's lack of proper muscle control, and the need for correction. Therefore, a third, hybrid approach was developed, characterized by:

1. Scanning both limbs: the healthy left arm in an anatomically correct position using a David SLS-3 scanner at the scanning station (Fig. 6.18A), and the right arm manually using an EinScan Pro 3D scanner in a wrist position comfortable for the patient.
2. Splitting both scanned meshes at the wrist, with the healthy left hand digitally "transplanted" (and mirrored) to the right forearm and adjusted based on the right hand's main dimensions.

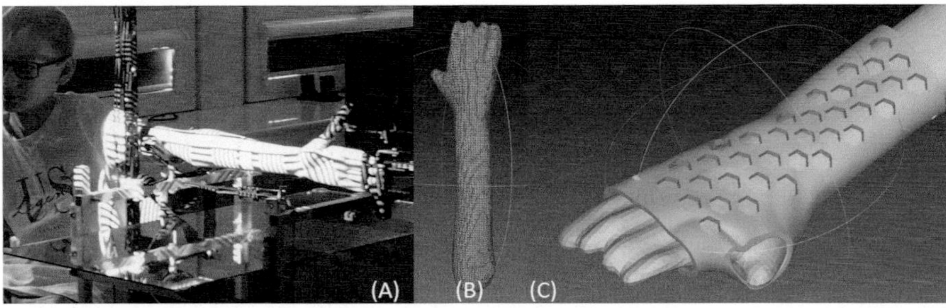

Figure 6.18 Therapeutic orthosis – design, (A) patient scanning; (B) "model" arm mesh; (C) virtual collision check.

3. Smoothing the connection on the resulting "model" arm, which then served as the foundation model for orthosis generation (Fig. 6.18B).

Data extraction from the "model" arm was carried out using automated macros, similarly to the earlier case study, using AutoMedPrint system capabilities. Additional settings included a 3 mm offset between the arm and the orthosis (to account for the lining) and maintaining the orthosis thickness at the standard 4 mm for durability and strength.

Subsequently, an advanced CAD model of the wrist-hand orthosis, previously designed in Autodesk Inventor, was adapted for the therapeutic orthosis. The model incorporated cutouts in the palm and thumb areas to facilitate grasping exercises and other daily activities while wearing the orthosis. This model was digitally aligned with the scan to ensure there were no potential collisions (Fig. 6.18C).

The orthosis was produced on FlashForge Creator Pro 3D printers using FFF (FDM) technology. The printing process followed standard AutoMedPrint system procedures, with a layer thickness of 0.25 mm and 30% infill, utilizing PLA material and taking about 4 hours per part. Post-processing involved thorough grinding, and the orthosis's internal parts were lined with foam. The patient tested several iterations of the orthosis, ultimately using it for rehabilitation over 9 months, periodically (mostly for exercises). After this period, marked improvements in motion functions, muscle development, and wrist position were observed, although the orthosis became unusable due to the patient's growth. Consequently, a new orthosis was 3D printed. The process was repeated few more times, and the span of therapy of this patient is now well over 2 years. The application of the three first orthoses for the patient are documented in Fig. 6.19.

The orthosis use was largely successful, allowing the patient to improve in functioning. After gathering feedback from the patient and working with another number of patients with similar conditions, it was decided to try making the therapy more compelling and interesting, by introducing gamification in virtual reality. The development work and initial studies on that concept can be found in an open access publication by the author with a team (Górski et al. 2023b). Also, a couple of other patients were treated using the same methodology, with generally positive results.

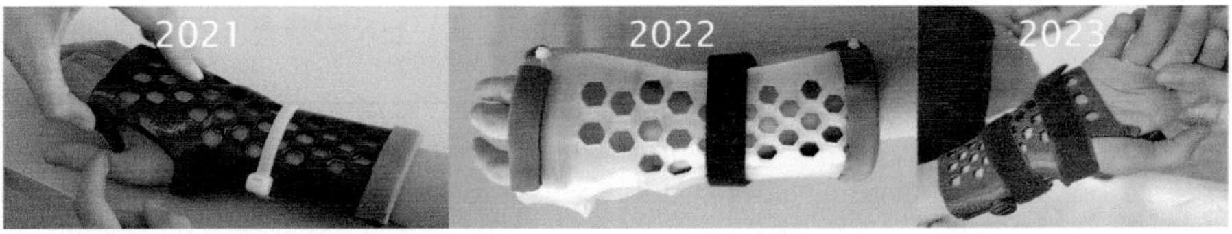

Figure 6.19 Therapeutic orthosis – long-term use, different (improved) models for the same patient.

The study yields several key observations and conclusions:

1. It has been demonstrated that designing a therapeutic orthosis automatically, based on a semi-automatically generated hybrid 3D model that integrates characteristics of both a healthy and a deformed limb, is feasible.
2. 3D-printed orthopedic devices can be manufactured to be lightweight, which is advantageous for therapeutic applications, especially when muscle tissue is significantly underdeveloped.
3. Minor skin irritation in the wrist area was noted, resulting from the corrective shape of the orthosis exerting pressure on the side of the wrist joint. This issue was addressed by incorporating a thicker foam layer to mitigate skin damage and discomfort.
4. The production of the orthosis is fast (within 2–3 working days), and its costs are relatively low compared to professional orthotic solutions available on the market.

The process of semi-automatically designing and 3D printing a cost-effective, lightweight corrective orthosis for congenital arm paresis has been successfully implemented. This orthosis has provided the patient with a viable rehabilitation tool, especially after unsuccessful attempts with commercially available orthoses. Furthermore, this approach has shown to be effective over an extended period, with the orthosis being successfully redesigned to accommodate the patient's limb growth almost every year.

The primary challenge in this process is the data processing involved in merging two disparate limb models, which requires significant expertise and knowledge in mesh data manipulation. There is potential for automating this step to enhance the efficiency of producing such orthoses. Nonetheless, the feasibility of automation greatly depends on the scanning conditions, as patients with paresis exhibit varying levels of mobility and capability during the 3D scanning process, often necessitating manual adjustments to the mesh of the affected limb.

6.5 Case study #2 – leg orthosis

6.5.1 Case and problem definition

The orthotic devices for lower limbs – shortly, leg orthoses – come in many variants, shapes, and applications, as described and classified in 6.1. It is very important to make them fully individualized, both anatomically (by shape) and functionally, by proper features, adapted to patient's specific needs. Here, similarly as in the case of hand orthoses, making devices for simply keeping the limb rigid for healing of a relatively short-time trauma (such as ankle twist, bone fracture, ligament damage, etc.) is much easier than in the case of necessity for geometry correction. The latter case occurs when the patient suffers from one of the long-term diseases, such as cerebral palsy, severe spina bifida, muscle atrophy or various spastics and deformations accompanying other conditions, such as in the Down syndrome. The task of orthosis is then not to make the limb rigid, but to make it properly oriented, to enable physiologically correct standing and walking, leading to improvement (or at least, slower deterioration) of patient's motor functions and general comfort of life.

The second consideration, further increasing the level of difficulty put before the orthotic devices for legs, comes from the specificity of the human lower limb. By common sense, we would attribute more complexity to the upper limb, which has to perform more delicate, precise operations, thus requiring proper force control and sophisticated limitations of degrees of freedom. The lower limb seems much more simple in that regard. However, there are two main problems – in the author's opinion – hindering the rapid development of modern, automatically designed, 3D printed devices. The first one is related to geometrical constraints. The upper limb can be more or less approximated as a set of links that can be laid along a straight line – the limb, the wrist and the fingers all can follow the same axis. As such, to extract geometry needed for designing a device for the upper limb, it is enough to "travel" along the single axis of a straightened arm – and that is how automation was achieved in AutoMedPrint system considering hand orthosis or upper limb prosthesis – series of planar sections along a single axis are enough to obtain a working data set for a fitting device. In the case of the lower limb, that is much more difficult, as the ankle joint's default position is 90 degree bend – with feet being positioned perpendicularly to calves. To gather the geometry properly, rotation of planes for sections is required and that poses a certain set of problems.

The second problem is related to physical aspects of operation of the lower limb. There are much stronger forces working on the ankle joint and its proper positioning is crucial for maintaining a stable, physiologically correct gait, not leading to further injuries or deformations. This has two implications: the requirements towards the proper shaping of the orthopedic devices are generally higher than in the case of the hand orthoses, and the devices themselves must sustain much higher loads than in the case of the upper limb. This leads to longer, more careful processes of design, with more details to account and concerning the manufacturing, simple extrusion of popular materials is often not enough, save for small children – for the adult patients, the orthoses made of PLA, ABS or even PETG cannot sustain enough load to enable comfortable operation (of experiences by the author and the team), so better materials are needed. Moreover, the orthoses are also quite sizeable, so it is much more difficult to produce them as single pieces on typical machines – specialized 3D printers with large chambers and fast extrusion mechanisms are required to perform properly here.

As such, it is much more difficult – in the author's opinion – to properly design a leg orthosis manually than in the case of hand orthosis – and it is also significantly more difficult, if not impossible, to realize proper automation of such design using the same approach as in the case of upper limb orthoses or prostheses. This automation – if AGI systems are not involved – will always have at least some activities required to be performed by human operators, making certain decisions in the design process, having in mind requirements of specific patients, which differ significantly from case to case. It is also noteworthy, that in the case of leg orthoses, multiple iterations are frequently required, as it is very difficult to create properly functioning design without feedback gathered from the fitting process. This process usually involves three-stage operation: first, fit – then standing and finally walking. At each stage the patient should provide feedback (either directly or by external observation), that would be used in improving the design.

The methodologies, briefly presented in this chapter are based on a number of attempts, applied to cases of various patients. These descriptions are based on previous work and papers published by the author and members of the team (Rybarczyk et al. 2024a, Rybarczyk et al. 2024b), including student diplomas (Dorna 2023). As many of issues and problems are similar to the upper limb orthoses – already covered – the further part of chapter will focus solely on assumed approaches and selected patient cases, bringing forward methodologies and results of these approaches.

6.5.2 Patient case #1 - mesh-based design approach

The mesh-based approach, as previously mentioned, does not assume usage of parametric design, surface or solid modeling and other CAD-specific techniques. Instead, 3D scan of a given patient is a basis here and, using various free modeling techniques, the orthosis is shaped. In the final stages, after the main anatomical shape has been acquired, the CAD systems might play a vital role in increasing a standardized set of features, e.g., circular or rectangular holes for mounting, cutouts, additional standardized parts etc. But most of the work is done in a mesh-processing software, starting from anthropometric measurements realized by 3D scanning techniques.

The course of operation in the case of modeling a leg orthosis will be dependent on a given case. However, a rough set of guidelines, forming a general procedure, can be formulated here. These involve the following:

1. Analyzing the patient's case and making basic assumptions about device type and requirements.
2. Selecting a proper methodology of 3D scanning (device, limb position) and realizing the process.
3. Basic (technical) data processing – cleaning and repairing the scanned mesh, possibly running a reconstruction program to obtain a clean mesh of a limb.
4. Optional – advanced data processing. This is necessary in several cases (that can occur all at once):
   a) patient was not cooperating (e.g., due to young age or mental deficits) and there is a number of partial scans, requiring fusion and smoothing out,

b) the deformation is severe and the limb was not kept in a proper physiological position (straight degree in ankle joint, no twist), so that the scan requires positional correction to "fake" a proper position,

c) in the case of two limbs – they are non-symmetrical (e.g., due to swelling resulting from the ongoing injury) and partial shape of one limb must be mirrored and used in the other.

5. Creating basic shape by using offset operations (twice, to create an initial surface distanced from the scan and then to add thickness to the orthosis shell).
6. Cutting proper orthosis shape out of the created shell.
7. Finalizing the shape by smoothing out certain regions, adding advanced shape features (e.g., flattening out the area of toes or adding space in the ankle area).
8. Adding technological features – holes for straps mounting, openwork cutouts, etc.
9. Verifying the fit using original or reference dataset and iterative improvement of the design.

This process can be realized in two ways. One of them assumes use of standard set of engineering and graphic design tools for working with meshes – the examples involve MeshLab, GOM Inspect, Blender, 3DS Max, Rhinoceros, and various CAD software (CATIA, Fusion, Inventor, etc.). The problem of design of leg orthoses usually requires switching between various software, utilizing different functions to obtain the proper result. The other way involves using dedicated software for design of orthopedic devices. Examples of such software were already mentioned (Mecuris, Spentys, etc.)

As an example of the universal way of conduct, a project of an individual lower limb orthosis aimed at stiffening the ankle joint of an adult, female patient was selected. This particular case example has been described in previous publications, supervised and co-authored by the author of this book, but on a practical level realized mostly by BEng. Piotr Dorna as one of case studies in the EMERALD project (Dorna 2023, Rybarczyk et al. 2024b). The project focused mainly on the mechatronic aspect of the device, i.e., electronic sensor components and the movement data gathering. Here, the focus will be put on the iterative design and manufacturing phase, conducted by the whole AutoMedPrint team. The orthosis model is based on 3D scans of the patient's lower limb performed earlier, and then printed using additive techniques. Digital fitting tests of the orthosis to the patient's scan and actual fitting were conducted.

The individual AFO was made for a 27-year-old volunteer patient who suffered an ankle sprain on the right lower limb and a second-degree tendon tear of the left lower limb's ankle. A comprehensive treatment schema was formulated, rooted in the specific rehabilitation requisites, the geometrical contours of the patient's lower limbs, and anticipated biomechanical forces during movement. The project necessitated the rigidification of both ankle joints, with limb geometries procured through 3D scanning methodologies. The anticipated forces related to the patient's movement were estimated based on her weight, which also was crucial in selecting the material from which the orthosis was made (composite materials were also considered (Rybarczyk et al. 2024a)).

The first stage of design involved capturing the patient's limb geometry, utilizing two distinct 3D scanning technologies: an AutoMedPrint station employing structured light technology (David SLS-3 scanner) and an industrial-grade EinScan Pro scanner. The subsequent data manipulation entailed processing both point clouds and 3D models through a suite of specialized software tools, including CloudCompare, Meshlab, Meshmixer, Blender, and SolidWorks, to refine the raw scan data into actionable design models. Additive manufacturing was executed using two 3D printers, Zortrax M200 Plus and FlSun V400, chosen for their respective cartesian and delta printing configurations. Printer programming was facilitated through Ultimaker Cura and Z-SUITE, tailored for Zortrax printers.

Post-processing of scans, involving techniques described in Chapter 3 (cleaning, hole filling, repairing surface, etc.), led to the creation of highly detailed 3D models of limbs. Ultimately, it was opted to utilize the geometry captured by the EinScan Pro scanner, as the measurements were realized by a skilled, experienced operator, thus containing reliable, trustworthy data points, while the automated scanning system did not cover the whole limb properly (Dorna 2023).

After acquiring the initial limb geometry, Blender and MeshMixer software was utilized to create an initial shell model, very closely fit to the patient's anatomy. This model was created by a simple set of operations:

a) offsetting and adding thickness to the original anatomical surface (akin to operations as in the hand orthosis, described in the previous subchapter),
b) cutting out the unnecessary geometry – top of the foot and front of the calf,
c) refining the obtained geometry, closing holes and obtaining a solid model.

This basic shell model was 3D printed using the Zortrax M200 machine, using PLA material and quick process (low infill, high layer thickness). It was tested by the patient and evaluated – problematic surfaces and areas were marked. The process of development and testing of the initial, shell model is presented in Fig. 6.20.

After the initial evaluation, the orthosis was re-designed. Length of the calf part was set appropriately, the foot part was flattened out, ankle area was appropriately modeled and the edges in the foot and calf part were modified to increase comfort of use. Moreover, the flat part at the bottom was added for more stable standing and walking, rectangular holes were introduced for mounting straps. The shape was also optimized – variable thickness was introduced for better balance and weight control. Digital fitting was also realized, confirming the proper fit. The model is presented in Fig. 6.21.

The final orthosis model was produced using additive manufacturing techniques, specifically utilizing the FLSUN V400 printer for its capability to produce entire orthoses in a single print run (due to large available printing space in the vertical direction). Material selection was informed by preliminary strength analyses, leading to the choice of PET-G for its balance of durability and flexibility.

However, the initial prints were unsuccessful (Fig. 6.22A). During the additive manufacturing process, material flow-related issues were encountered on both

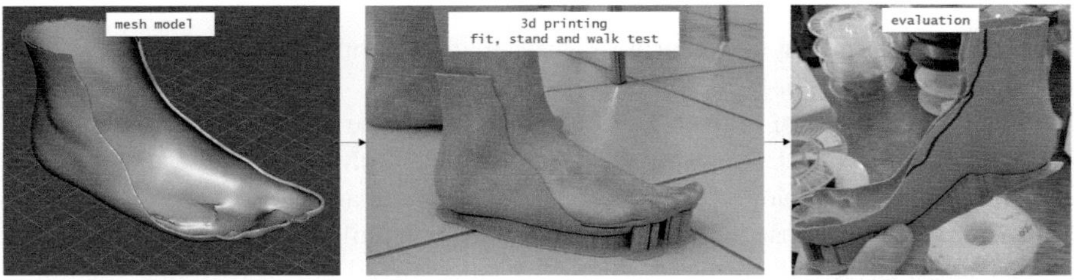

Figure 6.20 Development and testing of initial shell orthosis, on the basis of (Dorna 2023, Rybarczyk et al. 2024b).

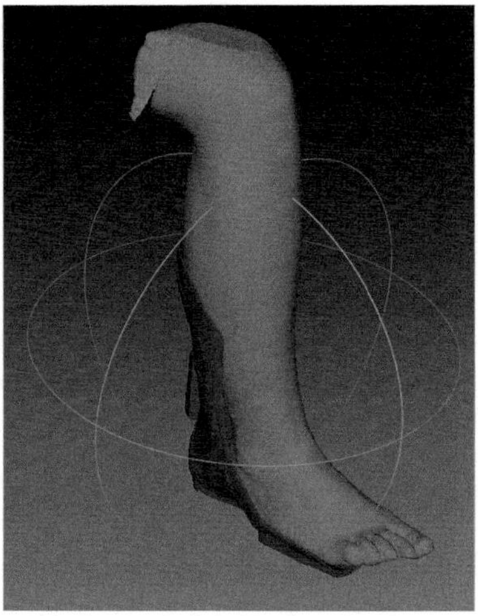

Figure 6.21 Final mesh-based orthosis design (digital fit).

FLSun V400 printers. Initially attributed to incorrect printing parameters, subsequent adjustments in nozzle temperature, printing speed, and a nozzle change did not rectify the issues. However, on the third attempt, successful prints were achieved. The chosen 3D printing parameters were appropriate (consulted and confirmed with expert process engineers), suggesting that the defects were not due to the 3D printer's technological aspects. Thus, the contaminated PET-G material was identified as the source of the defects by the original investigator (Dorna 2023). It was presumed that the material had been exposed to external factors such as moisture and dust, leading to weakened material and "fogging" effects during extrusion. The ultimate version of the 3D printed orthosis, subjected to further tests and usage, is presented in Fig. 6.22B.

The functional efficacy and mechanical properties of the orthosis were assessed through a series of mobility tests conducted by the patient along with the main designer (without assistance by a physiotherapist), encompassing a range of typical movements to evaluate the stabilization provided by the orthosis. The AFO orthosis was fitted successfully, with the patient confirming its comfort and effective stabilization of the ankle joint. The orthosis's durability was validated through full

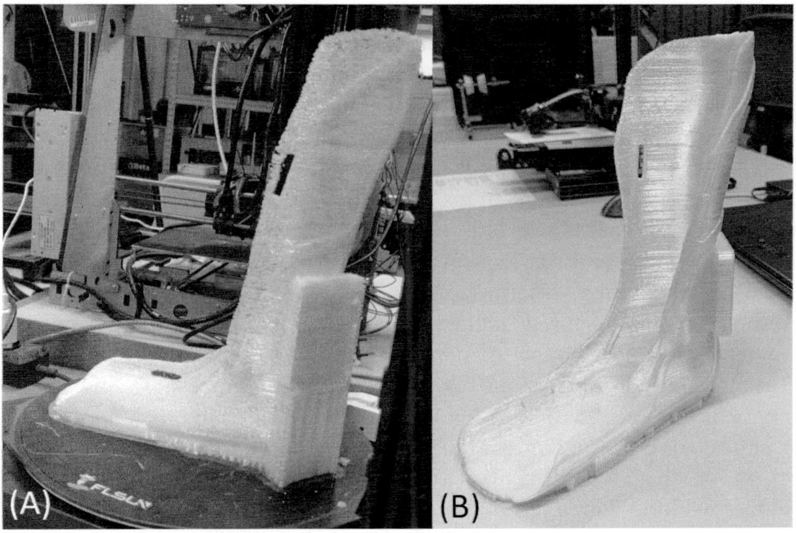

Figure 6.22 The produced orthosis, (A) unsuccessful print (contaminated filament), (B) successful print.

weight bearing and walking tests. However, the results of this validation are mixed – prolonged use showed that the orthosis demonstrated a tendency to break at some point. Current material and manufacturing choices are therefore not fully suitable for adult patients, requiring use of alternative, more durable materials, such as composites or ultra-durable polymers, such as of PEEK and PEAK family (demonstrated to be much more durable in previous studies (Pacurar et al. 2024)).

The economic analysis revealed that the significant expenses were related to the design and technological labor hours. The manual design of the orthosis was the most costly and time-consuming aspect – taking approx. 3 working days to realize, much in accordance with considerations presented in the Chapter 6.2.3. It highlighted the potential cost-effectiveness of automated design systems like AutoMedPrint for orthotic design.

6.5.3 Patient case #2 – parametric CAD design approach

The second case study details the research and development process behind the design and additive manufacturing of individualized Ankle Foot Orthoses (AFOs) for a teenage patient with spina bifida. The primary goal was to create a functional, stable, and usable pair of 3D printed leg orthoses suitable for walking, swimming, and various water activities. This project utilized the prototype AutoMedPrint system for design automation and employed Fused Deposition Modeling (FDM) technology for manufacturing the orthoses in different materials. This research has been described in a previous publication (Górski et al. 2022b) and has been realized by the author alongside the whole AutoMedPrint project team.

The patient in this case study is a 13-year-old male diagnosed with spina bifida, resulting in lower limb paralysis. Despite these challenges, he has been able to walk with the aid of specialized orthoses, which need frequent replacement due to growth and the unique requirement of being suitable for use in water. Commercially available orthoses do not meet these specific needs, highlighting the necessity for a customized solution.

The design process began with 3D scanning the patient's legs in various positions to create accurate digital models of his anatomy. This task required about 45 minutes for scanning and an additional 3.5 hours for joining, cleaning, and reconstructing the scans. The AutoMedPrint system prototype, mentioned multiple earlier times was used in automating data processing and preliminary design. However, the final orthosis design required manual adjustments based on the physiotherapist's recommendations and a couple of iterations, so the automation achieved here was only partial. First and foremost, the scanned geometry required substantial correction, as the patient was unable to keep his legs in a position which would be suitable to reflect in orthosis construction. As such, triangle meshes were adjusted, using GOM Inspect, MeshLab and 3DS Max software. The adjustment was realized manually by an engineer working with a physiotherapist, until satisfying geometry was obtained, which allowed to follow with the CAD design phase.

The design aimed to address several key requirements: water compatibility, anatomical fit, comfort, and asymmetry due to the patient's condition. The orthosis was designed in multiple parts for easy assembly and adjustments, with considerations for water flow, weight reduction, and stabilization of both the calf and foot. Three iterations of the design were tested, with modifications made to improve fit, comfort, and functionality – these are shown in Fig. 6.23. The design was realized in the CATIA software and was based on a hybrid approach – as discussed in Chapter 4. Point coordinates taken from the scan were used to construct curves and a surface model was created between these curves. Then, thickness was added to obtain solids, which were further processed, adding holes, fillets, and other geometrical features. The design took 1–2 working days per single iteration.

The orthoses were 3D printed using FDM technology, initially using PET-G material for testing, followed by nylon for its superior mechanical properties. Manufacturing times and material usage varied across iterations, reflecting changes in design and material choice. The manufacturing process was followed by post-processing, which included support removal, grinding, and lining the orthoses with foam for comfort.

Testing with the patient involved visual inspections, fitting tests, and assessments of standing and walking capabilities with and without support. The first iteration did not have a fully proper fitting, requiring shape correction. Also, two-part design with a substantial load put on the 3D printed nylon connector resulted in gait instability, leading to reconsideration (the patient was able to walk, but needed additional support). Orthoses in the second iteration were perfectly fit and stable, but they quickly failed mechanically when the patient started to walk, with fractures happening at the joining between the carbon fiber connector and the lower part of the orthosis. As such, a single-shell approach was assumed – previously avoided for

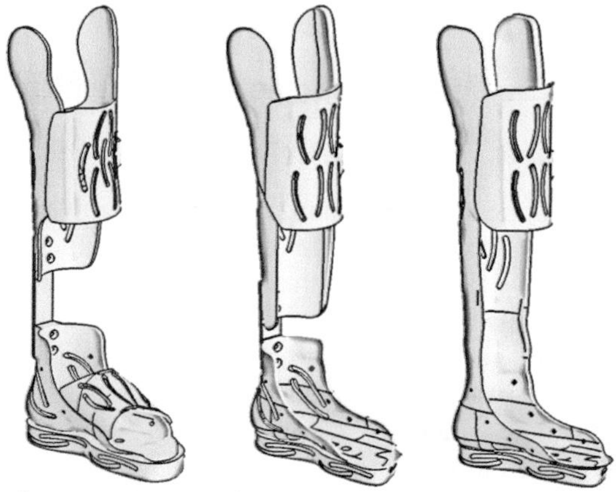

Figure 6.23 Leg orthosis for spina bifida patient – design iterations.

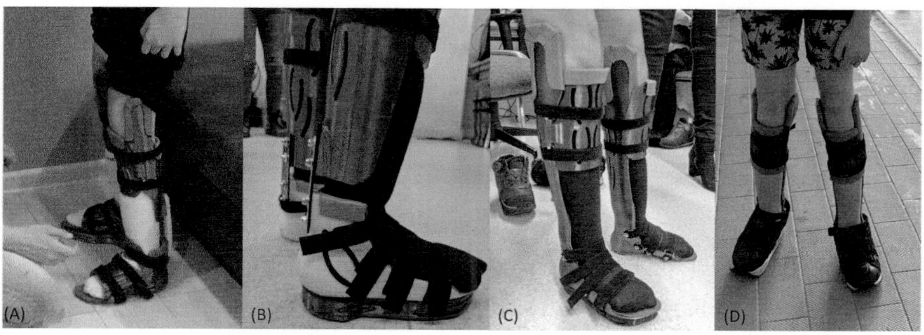

Figure 6.24 Leg orthosis for spina bifida patient – testing: (A) version 1 (problems with fit and gait stability), (B) version 2 (poor mechanical strength), (C) version 3 – testing, (D) version 3 – usage.

the benefit of more healthy walking pattern in two-part orthoses, but as the orthosis was not designed for all-day use, this consideration was not of a big issue. The final iteration of the orthosis – accepted happily both by the patient and a physiotherapist – allowed walk with minimal or no support, marking a significant improvement over previous versions and the commercially available alternatives – as presented in Fig. 6.24.

This case study demonstrates the feasibility and advantages of using 3D printing and automated design systems like AutoMedPrint for creating customized orthoses. The ability to quickly produce and iterate on designs based on patient-specific anatomical data represents a significant advancement over traditional manufacturing methods. However, challenges remain – full automation is very difficult in the cases where anatomical data in form of triangle meshes require manual corrections, also unknowing the requirements of a particular patient – this particular application of leg orthosis was not found in existing literature, so know-how was built from the scratch.

Also, material challenges must be addressed accordingly, to bring forth the products of high durability and stable properties, enabling dynamical movement of patients.

6.6 Case study #3 – modular upper limb prosthesis

6.6.1 Case and problem definition, development scheme

Prosthetic limbs, as we already established, are aimed to replace some functions of the lost or undeveloped biological limb. In recent decades, there has been a notable progress in that matter – we can build prostheses that are smart, precise, and powerful. We can also highly personalize them, obtaining the best results. However, it comes with a cost. The current generation of prosthetic devices are very expensive – a luxurious car or a small house could be bought for the price of some of these devices. The problem is even bigger when considering children – they grow very fast.

The rapid growth of children necessitates frequent replacements of individualized orthopedics or prosthetics, often every few months, to accommodate their changing bodies. This process is hampered by the prohibitive costs of professional prosthetic devices and the extensive time it takes to manufacture them. The absence of a suitable prosthesis or orthosis can lead to a decline in functional abilities and the development of anatomical pathologies, making the need for accessible prosthetic solutions even more critical. There are only a few parents that could afford buying new prosthetic limbs each several months. As such, most disabled children, and unfortunately also most disabled adults, especially in developing countries, cannot afford any prosthesis or for their daily functioning.

Furthermore, there is a limited range of customized devices available, with this problem being particularly acute in niche applications such as transhumeral amputations. This situation underscores the importance of designing and implementing a new class of affordable prosthetic limbs for children, which would not only ease the financial burden on families but also prevent the deterioration of children's health and functionality due to a lack of proper assistive devices.

That was the motivation behind creating a class of solutions described in this case study. The product that was created by the author and his team is a 3D printable mechanical prosthesis, intended mostly for use with personal transportation devices, such as bicycles or scooters, but also for general functions, such as holding items, providing counterbalance and symmetry to prevent postural defects and facilitate cooperation with other people, e.g., in a school environment. The prosthesis is a mechanical device, anatomically adjusted to a specific patient by a set of constraints and dimensions. It is intended first and foremost for patients with transhumeral amputations or defects (above or at the elbow level), although patients with a short forearm stump can also use it, after certain modifications. The current, basic version of the prosthesis is presented in Fig. 6.25.

The prosthesis concept was initialized as part of the AutoMedPrint system development, mainly for children below the age of 10, to help with some sports and daily activities and help maintain proper body balance and posture. It has been patented (in the Polish Patent Office) and implemented in various cases, for more than a dozen users over a course of 3 years preceding this book. In 2023, the

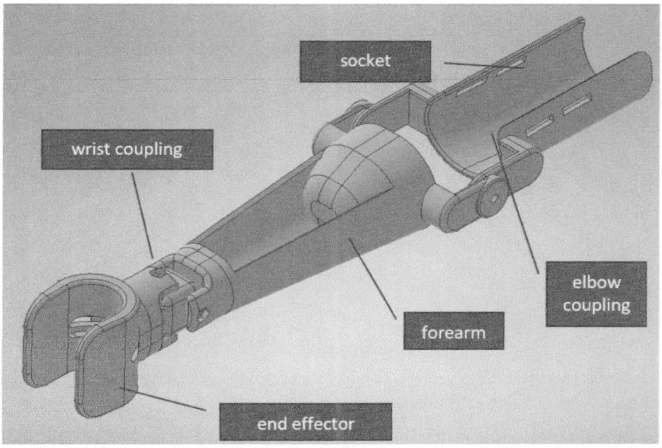

Figure 6.25 The modular hand prosthesis with the basic components indicated.

prosthesis was awarded as a best invention of the year in the Polish contest Eureka! DGP (Polish Science 2023). It has been also used as a basis for multiple studies and publications by the author and the team, also in international projects (Górski et al. 2023a) and by other teams than of the author's (Dudkowiak et al. 2022). The recognition, interest and implementation success shows that there is a need for such cheap and simple solutions, allowing disabled people (especially children) to regain mobility and engage in certain activities. Further work is ongoing, to improve, develop and expand the concept and make it available for more and more recipients.

This chapter, focused on the prosthesis as a case study, describes its development in the scope of life of intelligent CAD models and automated design systems. The process shows how the model was initially created as a non-intelligent one and then upgraded to level 2 by adding design tables. Further stages of work included working with logic, modularity, and scope of possible configuration to create a fully modular, customizable CAD model, operating as a level 3 intelligent model. On top of that, complete design automation was made possible by developing proper algorithms of data extraction and friendly user interfaces to work through them. Many user cases were considered along the way, and two of them are presented in the last two subchapters, to present how the complete development process of a prosthesis for a single patient – a typical and non-typical one – can look like.

6.6.2 Development of parametrized CAD models

The development of the prosthesis started with creating an initial variant for the first patient – a 3-year old girl with a congenital transhumeral defect (lack of forearm, no elbow joint). The initial version was realized with a single function – as a tool for operating the scooter or bicycle with a proper posture, balance, and ergonomics, allowing the initial patient to safely use her vehicles (Fig. 6.26A,B). The prosthesis was 3D printed by extrusion with PLA material, using a low-cost machine for that purpose (FlashForge Creator Pro, which has become a typical, standard machine

Figure 6.26 Initial modular prosthesis, (A) use with scooter; (B) use with bicycle; (C) corrected iterations for first two child patients.

for prosthetic production in the author's team, incorporating the machine into the AutoMedPrint system).

The initial prosthesis design was quickly reiterated to improve ergonomics, safety, and easiness of use. The basic CAD model was created as a parametric model by an engineer (working in a team of the author), but without automation capabilities, with a standard approach (non-intelligent model). The model was prepared for another two patients, with similar defects and of similar age, but with various dimensions (Fig. 6.26C). At that time, knowledge about the design process and the patient's requirements were gathered and written down by a knowledge engineer (following the rules of methodologies of creation of intelligent CAD models), along with the feedback coming from other stages of the lifecycle, including manufacturing, assembly, disassembly, maintenance, usage, disposal. This helped to develop a streamlined process, which was incorporated into the original CAD model. As such, an intelligent model with automation capabilities was created, belonging to the level 2 (see Chapter 4.3 for classification), connected to an external design table. This model is presented in Fig. 6.27.

The first intelligent model has capabilities of rescaling itself, based on 11 basic dimensions – lengths of sections and pairs of diameters of ellipsoid-like shapes forming the open prosthetic socket. It is built as a part in Autodesk Inventor, containing total of 7 solids – three main parts (socket, forearm, end effector) and a wrist coupling (designed a la Cardan coupling) containing 4 parts. However, it is not an assembly and subsequent modules of the prosthesis cannot be exchanged, e.g., it would be very difficult to make it possible to switch the end effector to something different. Still, it is possible to create a basic model in mere minutes – measurement of the 11 dimensions can be realized on the basis of 3D scan of a healthy limb and a stump of a given patient, or directly on the patient's own limbs, using tailor's measure or a similar tool, although in such a case the accuracy will not be maintained and the fitting could require improvements.

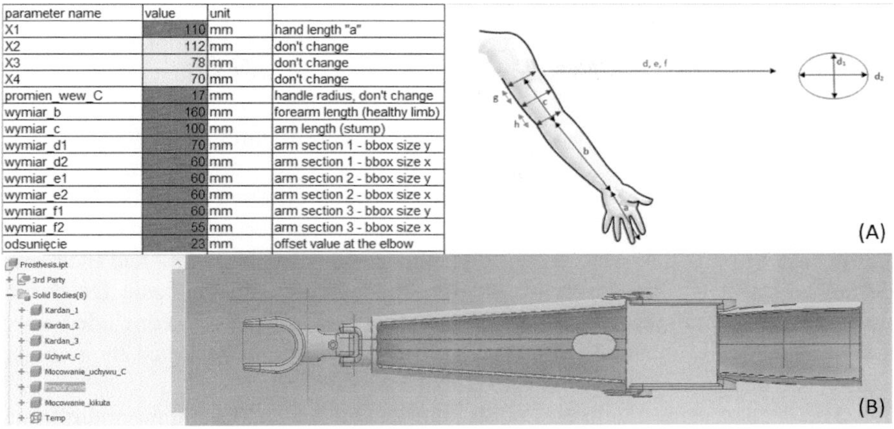

parameter name	value	unit	
X1	110	mm	hand length "a"
X2	112	mm	don't change
X3	78	mm	don't change
X4	70	mm	don't change
promien_wew_C	17	mm	handle radius, don't change
wymiar_b	160	mm	forearm length (healthy limb)
wymiar_c	100	mm	arm length (stump)
wymiar_d1	70	mm	arm section 1 - bbox size y
wymiar_d2	60	mm	arm section 1 - bbox size x
wymiar_e1	60	mm	arm section 2 - bbox size y
wymiar_e2	60	mm	arm section 2 - bbox size x
wymiar_f1	60	mm	arm section 3 - bbox size y
wymiar_f2	55	mm	arm section 3 - bbox size x
odsunięcie	23	mm	offset value at the elbow

Figure 6.27 First intelligent model of modular prosthesis, (A) design table with guidelines; (B) view in Inventor software – prosthesis rescaled for an adult patient.

In the further development stages of the prosthesis, various requirements were formulated by patients belonging to different groups. It resulted in developing an entirely new set of solutions by the engineers working in a team of the author and the author himself, basing on the original version of the prosthesis. The subsequent new parts of the model include:

- *new end effectors* – including permanent mount, spring effector and a mechanical hand with adjustable finger positions,
- *simplified wrist couplings* – full Cardan-esque coupling was not needed in all cases, allowing to reduce number of its degrees of freedom,
- *new types of forearms* – for increased durability, closed variant was introduced; for easier bending by patients with forearm stump, a variant with partial stump socked was created,
- *new sockets* – including a closed and partially closed one, as well as a special type – a compressive-release stabilized socket (CRS) (Zawadzki et al. 2022) in different mounting variants,
- new couplings for end effector for faster assembly;
- adjustable elbow joints.

The above-mentioned solutions have been developed in various times and frequently by different designers, validated for different patients (both child and adult). As such, another effort needed to be made, in order to obtain an intelligent model of the now-modular prosthesis, configurable at the level of the design table in Excel, including rules of variant selection and adjustments based on the selected configuration, elevating the intelligent model into level 3, by writing code for these rules (directly in Inventor, using iLogic capabilities). This work was planned and directed by the author and realized in the Inventor software by another engineer cooperating with the whole AutoMedPrint team. Some results of that work were presented in a thesis entitled "Intelligent model of an individualized prosthesis of an upper limb" (Komorowska 2022). The model assembled by O. Komorowska in that thesis was then verified, improved, and completed by the author and his team and the final result is shortly presented in this chapter, as well as in a paper (Górski et al. 2024c).

First stage of the work was to gather all the construction elements (parts and modules), separate them as CAD parts – distinct files – and parametrize them properly, making certain dimensions shared between different variants of a given module (e.g., applying the same, named, length parameter to all forearm types). In the case of parts that were properly parametrized beforehand and had their own design tables, it was necessary to integrate their tables with the basic Excel file, dividing it into appropriate sections. As such, first part of work was concluded when a set of files were obtained, with their geometrical features governed by a single, aggregated design table file, connected to all of them at once.

The second stage involved building an assembly – adding appropriate joints and constraints between various parts of the model, attempting to make them similar between parts of the same classes (e.g., using the same axis or plane to join between end effector and Cardan coupling). Alternative connections had to be created for various future variants of the prosthesis, e.g., with and without Cardan couplings.

After the basic work in the Inventor Assembly module was concluded, all the prosthesis parts were visible and active at the same time. To enable their changes, iLogic capabilities were used in form of VBA scripts. These scripts mainly consisted of conditional instructions, governing activity of specific components on the basis of values stored in yet another sheet of the design table. This secondary sheet contains a list of components (effectors, forearms, sockets, and their couplings), along with numbers (zeros and ones) deciding about their activity in the general assembly. An example of such a dependency – joining state of an open socket component with Excel spreadsheet value through VBA iLogic script – is shown in Fig. 6.28.

Finishing this work enabled to obtain an assembly customizable by changing numbers in the second spreadsheet and prone to anatomical individualization by altering measurements coming from a 3D scan in the first spreadsheet (both sheets are parts of the same file). Then, after careful empirical validation using datasets of many patients with variable geometry, additional iLogic rules were added along with Excel variables, to tweak the model accordingly when different datasets are uploaded. Various formulas were included in the Excel table and the calculated values were then bound within the iLogic scripts, to decide if some features will be active or not, or what set of parameters will be used – this is true especially when changing size of the prosthesis from a child size to teenager and adult size, where certain strength improvements are required. Another set of logical restraints pertains to type of defect (amputation) – if the patient has a forearm stump, some additional features (such as stump rest in the forearm part, visible in Fig. 6.25, not present in Fig. 6.27) might or might not be turned on. It also works the other way, if no forearm stump is present (defect above or at elbow), some features are possible that would otherwise be not available.

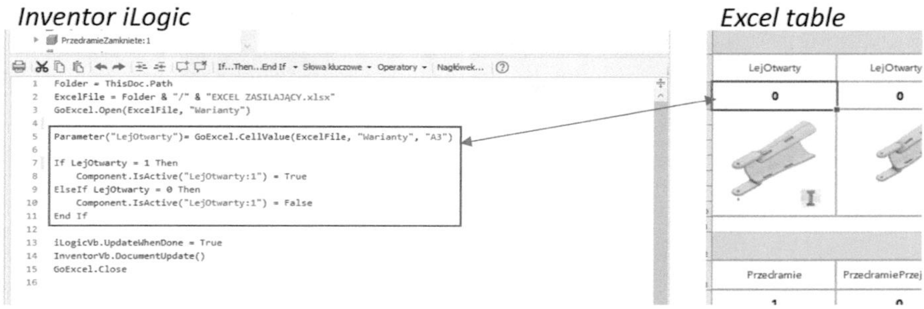

Figure 6.28 Creating iLogic scripts – activity of a component conditionally bound to value taken from the external design table.

The process of improvement of the prosthesis design and its intelligent CAD model is a continuous one. As more and more patient cases are being processed, knowledge gathered from each new unique case will be incorporated within the model, either as new parts (modules) or as new rules or geometrical features in the existing parts – some examples of the currently realized studies might be parametrized cosmetic hand, additional strength-ins in the forearm part, a new type of elbow or CRS socket made for patients with short forearm stumps (to be locked on the arm, above functional elbow). Some solutions, known for their lesser extent of use, may be also removed if necessary. Anyhow, each new feature must be thoroughly validated using an existing dataset (on the basis of historical patients), to not lose existing capabilities and prevent (or eliminate) errors in the design. For intelligent models with such numbers of options combined with high degree of individualization, it is inevitable – in the author's opinion – to continue improving the model throughout the whole lifecycle of the product.

6.6.3 Complete design automation implementation

To make it possible to generate subsequent variants of the modular prosthesis by inexperienced users, or implement it in a production company, it was decided to create a fully working prototype of an automated design system around the intelligent model. As mentioned in Chapter 4, for the automated design system to function, it is necessary to create various interfaces, to enable automatic preparation of data acquired from the individualization and customization processes. In the case of the modular upper limb prosthesis, the following additional system components were created:

1. *Product configurator* – based on previous solutions, a new visual configurator was created, with Virtual Reality (VR) features. Users are able to select their variant and some aesthetic features using a touch screen and additionally, they are able to feel the prosthesis in their own hands by use of VR controller with the stump, as presented in Fig. 6.29. This configurator is described in more detail in the paper (Górski et al. 2024d). Data saved in the configurator are in form of a text file, used by the automated design table.

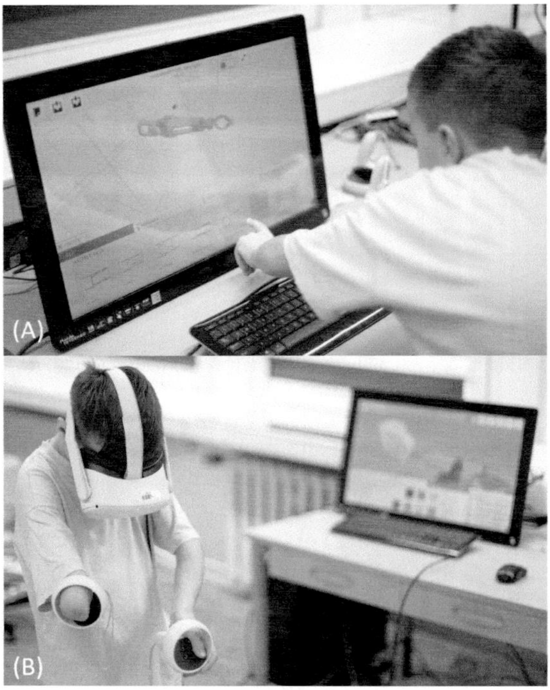

Figure 6.29 Interactive configurator of modular hand prosthesis, (A) touch mode; (B) immersive VR mode.

2. *Algorithms for anatomical data extraction* – MeshLab software and its automation capabilities were used here to create a set of macros for computing planar sections of healthy hand and a residual limb and acquiring point coordinates, distances, and diameters from them. These macros were recorded in MeshLab using its proprietary MLX format (using XML notation) and then stored in external files, with the MeshLab server solution used to launch them in the console mode, from a batch file. Launching them results in obtaining a set of XYZ files with point coordinates and TXT files with various measurements (e.g., diameters of a given section).
3. *Limb calibrator* – a visual application that enables the user to select positions of the most important planes sectioning the scanned limb – wrist, end of hand, limb, etc., facilitating the whole process. This application was created using Unity game engine and it has a simple GUI with sliders and buttons, enabling object navigation around the limb and controlling of the sectioning planes positions. The positions are then saved in an external file and used to adjust values stored in MeshLab macros, individualizing them for a current patient. This application is very similar to the one used in the WHO model, described in one of previous subchapters.

4. *Automated design table* – using a basic design table for the modular prosthesis, an automated table was created in Excel, using Visual Basic for Applications. After launching, this spreadsheet gets data from the MeshLab-generated files and the saved configuration and pastes them inside the original design table, effectively transforming it into a new design table, for another (current) patient.
5. *Batch file* – it was used to gather all the above mentioned algorithms – MeshLab macros, custom visual application (limb calibrator) and Excel spreadsheets with macros – and put them into a single sequence, launched from AutoMedPrint system main interface. Separate batch files were created for residual limbs (stumps) and healthy limbs, as the procedure is slightly different.
6. *Main operator's interface* – the AutoMedPrint system interface for the operator (as discussed in 6.3 – a custom application made using the Unity software) was extended with features enabling performing the batch, automated operation of generating data for the intelligent model of the modular prosthesis. Selecting a proper product of the starting screen enables further selection – the user decides if the data will be extracted of the stump or healthy limb. If both are needed, the process can be repeated twice, starting with the stump.

Implementing all the above-mentioned modules and applications enabled to obtain a fully automated procedure of prosthesis generation. It has been tested with the real patients – in several tested cases, it was possible to obtain a basic model (similar to those visible in Figs. 6.25. or 6.27) individualized for a given patient in a time less than 15 minutes after processing scan data. This process was also realized by inexperienced users (namely, students), also with a certain degree of success – these users were able to go through all the phases of the process using randomly assigned datasets. However, not always everything went properly (e.g., due to improper 3D scan data, or improper selection of planes, etc.) and in the case of any errors, lack of experience prevented these users into making any corrections within the CAD model or the design table – the only way was to iterate the whole process once again with slightly changed conditions.

The current version of the complete automation system, described here, must be therefore marked as a first prototype and it will be certainly developed further and validated by use of more and more datasets, as well as more and more inexperienced users. So far, the solution requires many various software applications to be installed as prerequisites – namely, Microsoft Office (MS Excel), MeshLab and Autodesk Inventor, plus the custom applications belonging to the AutoMedPrint system. Moreover, certain tweaks in the operating system are required, such as enabling MS Office macros and making MeshLab a default opener for STL files. However, in the future it would probably be required to prepare an own, enclosed solution. It can be achieved, for instance, by using XML file instead of MS Excel for the design table and incorporating MeshLab not as separate program, but rather a part of new one (thanks to the open-source approach of MeshLab creators). However, Autodesk Inventor and its iLogic features are there to stay – to avoid necessity of installing this software in each local instance of automated design system, cloud or server-based access could be used.

A separate issue is automation of 3D printing processes, especially program preparation. It is also achievable by means of specially designed software, that reads STL files and makes NC code automatically, which could be then uploaded into an appropriate machine (through cable or wirelessly) and started by an operator. However, in the case of multi-part device such as the considered prosthesis – which can have more than a dozen of single parts – it is, in the author's opinion, not worth the effort, as it has been proven in the past that almost any person can be trained in basic work with the extrusion 3D printing systems – and the medical personnel can efficiently prepare programs for NC machines by themselves, not needing a fully automated solution. That is why, in the development of the AutoMedPrint system branch focusing on modular hand prostheses it was decided to focus entirely on the CAD aspect, omitting the CAM part for practical reasons.

6.6.4 Patient case #1 – the bicycle prosthesis

The first patient to undergo the fully automated design procedure was a 10-year old girl, born without left forearm (transradial congenital defect). The patient had no previous experience with prosthetic devices – she did not use any prosthesis whatsoever throughout her life so far. The aim of the prosthesis was mainly to help her operate a large, "adult" sized bicycle, instead of one made for children, which was easier to operate with the right arm and the stump of the left arm. Moreover, the parents also hoped for better balance and stability, also in everyday activities. It has therefore been decided to perform a 3D scan procedure of both limbs. As the patient was living relatively far from the AutoMedPrint laboratory, it has been decided to perform the virtual try-on and fitting, instead of a physical one, thus not requiring the patient's presence in the laboratory. The parents were asked to bring in also the bicycle and it was 3D scanned as well. The 3D scanning process is summed up in Fig. 6.30. For the both scanned limbs, David SLS-3 scanner mounted in an automated workstation was used, as well as the manual scanner – EinScan Pro. The bicycle was scanned only using the latter, due to lack of space to fit patient on a bicycle inside the scanning rig.

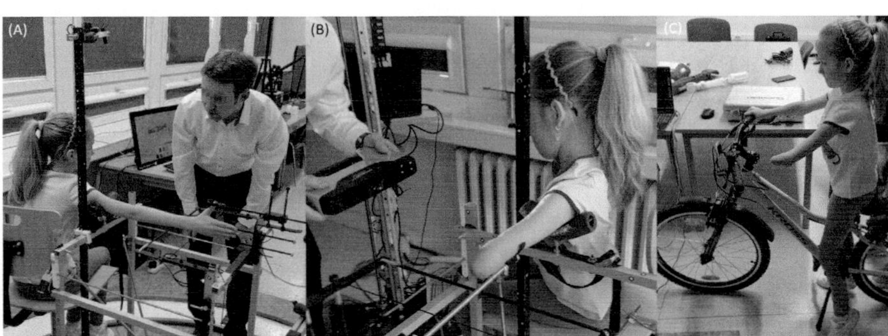

Figure 6.30 Patient during scanning: (A) healthy hand, preparation; (B) stump, manual scanning; (C) bicycle, scanning in target position.

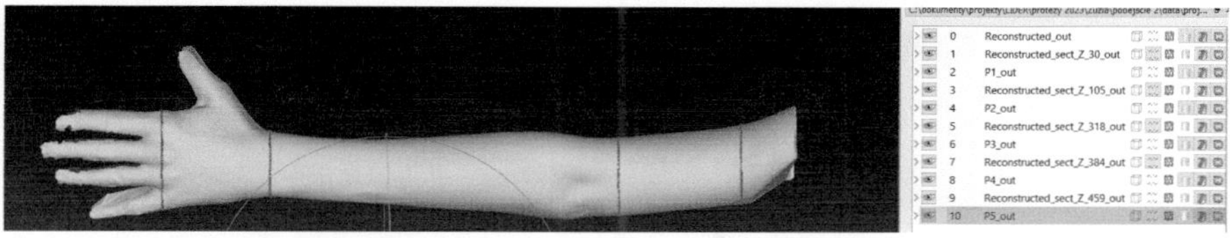

Figure 6.31 Automatically created MeshLab project with computed planar section of patient's right hand (mirrored to serve as "pattern" left limb).

The 3D scanning data were automatically processed. Data obtained by EinScan Pro were selected for further processing due to being of better perceived quality (smoother scan). After careful consideration, it was found that the stump, in the arm part (which is, the one where prosthetic socket would be holding on to the limb) has very similar geometrical features as the corresponding area in the healthy hand. That is why it was decided to use only the right hand features for generation of the prosthesis model, omitting the stump, except for measuring its total length from the elbow joint to the tip. The right hand scan was automatically mirrored (the limb calibrator application has this capability as an in-built one) and the planes were then selected for further computations. The result of the selected basic computed sections previewed as MeshLab project is shown in Fig. 6.31.

Going through the complete automated procedure as described in the previous subchapter, a new design table in Excel was generated. The configuration process in the previously mentioned visual configurator was also realized. However, the configurator was missing the color that was preferred by the patient, which was a color corresponding to the bicycle frame paint (turquoise) – but it was only a minor hindrance (not affecting the CAD model). Moreover, the patient (along with her parents) was not able to decide about the end effector, due to her lack of experience with prosthetic devices. That is why it was decided to try two possible solutions – a spring end effector and a permanent bicycle mount (closed bracket, screwed together around the bicycle handlebar). The second one was proposed considering a specific requirement for the patient – to be able to quickly release her hand from the prosthetic socket in the case of an accident.

The first generated version of the prosthesis normally would be probably manufactured and given for the patient to try-on. However, in this particular case, it was decided to use the virtual try-on. The generated model was converted to an appropriate format and uploaded into the Unity game engine with SteamVR plugin installed. Along with the prosthesis, all the measurements were placed in the virtual reality as well, allowing free manipulation. That made it possible to check the fitting of the stump and length of the prosthesis against the healthy hand and also compared to the bicycle. This virtual try-on is described in a recent paper (Górski et al. 2024e).

After the try-on, it was decided to make some alterations in the prosthesis, The length was cut short a little bit, the space for elbow was slightly widened and the stump rest was added in the forearm part, which was also reinforced a bit. All this was achieved by selecting the planes slightly different in the limb calibrator tool and altering some parameters in the design table. Finally, two variants were

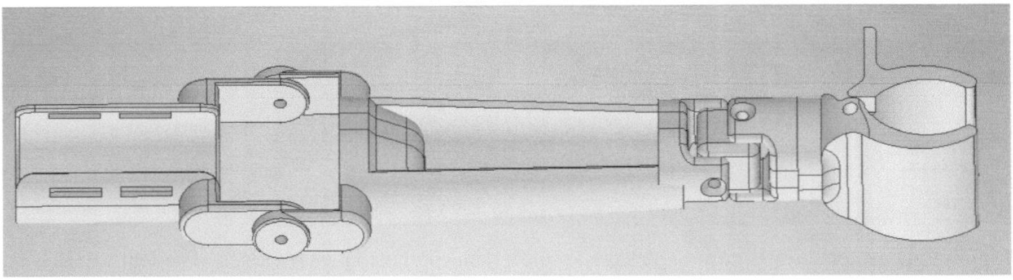

Figure 6.32 The second, virtually verified variant of the CAD model of the prosthesis with spring end effector.

generated, with two different end effectors. The basic variant with spring end effector is presented in Fig. 6.32.

Subsequently, all the parts of the prosthesis were manufactured. For that purpose, FlashForge Creator Pro FDM machine was used, with PLA material of turquoise color, uniform for the whole prosthesis (except for the alternative end effector, which was manufactured as green). The standard settings of infill equal to 35%, layer thickness of 0,25 or 0,2 mm (for smaller parts), 2 contours and 3 bottom and top closing layers were applied. The 3D printing process was realized smoothly, without any disturbances and over a course of 2 working days (using single machine), all the parts were obtained. After removing them from the build chamber, they were subjected to basic, typical post processing – support structures were removed, rough surfaces were manually polished and sharp edges were filleted. The socket and the stump rest were laid out with EVA foam of 1 mm thickness, for better comfort and ergonomics (the foam was glued to the PLA material, for constant connection). The final device weight was approx. 0.3 kg and the approximated cost of manufacturing, involving post processing and assembly elements, was roughly assumed at US$100. The device is shown in Fig. 6.33.

The prosthesis was sent to the patient and was validated and operated without direct contact with any technical personnel – teleconsultations were realized for the patient and her caretakers to be sure they are properly mounting and using the device. The validation was successful – the patient managed to comfortably ride a bike, even in an uneven terrain, and managed also to quickly learn how to dismount from the

Figure 6.33 The prosthesis after printing, post processing and assembling, before sending to the patient.

prosthesis in the case of any accident. On the basis of materials shared by the parents and subsequent discussions, the following observations were made:

1. The spring end effector was not useful for the bicycle, as the spring was too "weak" for the user. It could be more useful if it was exchanged. This end effector was used for everyday activities instead.
2. The permanent mount was very useful for the bicycle, but the wrist ("Cardan") coupling was found to be unnecessary, even after making it rigid by tightly screwing the screws. Finally, it was removed and the patient used the bicycle without it, directly mounting the end effector to the forearm. Such modifications are fully possible to realize in home conditions by the parents or sometimes the patients themselves.
3. The bicycle riding was comfortable and proper balance, as well as posture, was maintained. However, in the author's opinion, even after learning how to dismount quickly, this is a less safe option for patients with transradial defect. As opposed to patients without elbow, dismounting requires the forearm stump to be in straight line with the arm, which might not be intuitive while using the prosthesis and riding a bike.
4. The design of prosthesis itself was very quick, while the virtual try-on and subsequent improvements took some time – however, the patient's time on transport was saved that way. This approach allowed building a functional, fitting prosthesis at first attempt, from the patient's perspective, and is being continued. It has also been proven as a viable option to work entirely remotely, without physical contact with the patient – as long as 3D scans of acceptable quality can be supplied, this is possible. An example of such patient and his case is described here (Górski et al. 2024f).

6.6.5 Patient case #2 – a prosthesis for wheelchair operation

The second case of patient to describe was selected as one of the unique ones, where both full design automation and fully individualized (Engineering-To-Order) approaches were used. This case would not be possible to fully realize without the automated design system, or rather the knowledge in-built into it – working with the patient was not relied upon automated systems for measurement and design, created at the earlier stages of development described in this chapter. However, existence of these systems and a large number of previous, more typical patients, allowed the engineers from the author's team to create a fully personalized and useful solution.

The patient was a 25-year old female, born with several congenital limb defects. The lower limbs are not fully and symmetrically developed, rendering the patient unable to walk and forcing her to use a wheelchair. It is similar with the upper limbs – the left hand does not contain an elbow joint, but there is some mobility – the patient has one single movable extremity at the end of the arm. The right upper limb is dexterous, while in itself also being not developed properly – the patient suffers from a form of ectrodactyly. Figure 6.34 presents both upper limbs of the presented case.

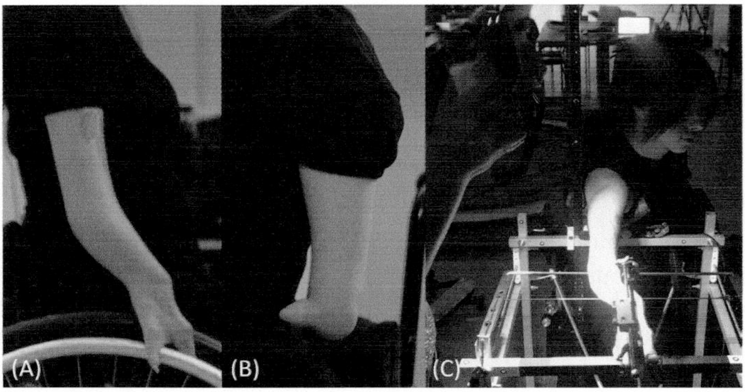

Figure 6.34 The second case study patient – (A) right hand; (B) left hand; (C) process of 3D scanning.

The need formulated by the patient – otherwise being able to cope with daily routines, tasks, and work – was to be able to operate a manual, muscle-powered wheelchair, for better flexibility and more dynamic possibilities of locomoting. Due to significant difference in length of limbs, as well as lack of the bending joint in the left hand, it was impossible for her to ergonomically, comfortably and efficiently operate a manual wheelchair and she needed to use an electrical one, operated single-handedly by a joystick. From the start it was apparent that with such atypical anatomical features, a custom design would be required – so an engineer was tasked with preparing it, basing on available, previous solutions. It is noteworthy that at the time, there were no designs for such a prosthesis in available literature, besides general, universal myoelectric prostheses.

It was also decided to parallelly implement the AutoMedPrint system capabilities and, as a form of experiment, try to design automatically a standard case of modular prosthesis to check if it would properly fit and be operated by the patient.

The first stage was 3D scanning – and it was also the first stage where only partial automation could be used, as the patient was unable to use the automated design station comfortably due to her disability. That is why it was decided to employ only the scanning rig and perform manual scanning, using the EinScan Pro manual scanner (Fig. 6.34C). Apart from 3D scanning of both limbs, in various positions, also the manual wheelchair, brought by the patient, was 3D scanned, with and without the patient. Some data captured during that process is presented in Fig. 6.35.

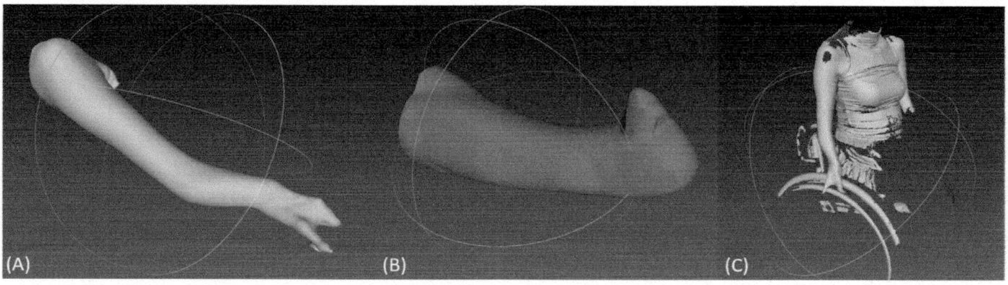

Figure 6.35 Data acquired in the 3D scanning process, (A) right hand, (B) left hand, (C) patient on a wheelchair.

As the 3D scanning was realized manually, it required post processing. Using MeshLab and GOM Inspect software, both limb scans were separated, cleaned, and placed in a proper coordinate system. From that point onward, the design activities were twofold:

1. a complete set of scans was taken by the aforementioned engineer (experienced designed, author of some previous prosthetic designs in the AutoMedPrint system) to create several variants of a new prosthetic device for the wheelchair,
2. the two limb scans were merged in the MeshLab software to form a "model" left limb (of appropriate length) and they were used for automated generation of a universal prosthesis, using capabilities of AutoMedPrint system, similarly as in the case of the first patient (previous subchapter).

The first process – custom engineering – was realized in the CATIA v5 software. The scan was uploaded and used as a basis for a surface model, with a faithful representation of patient's residual limb geometry. Three different proposals of a prosthetic socket were designed – an open one, a semi-closed and a fully closed one, leaving space for the movable extremity at the end of patient's left limb. Apart from this, two different end effectors for wheelchair pushing were proposed, similar in geometry – one with 3D printed surface made of elastic material, the other with the same surface made by gluing a piece of rubber inside, 3D printed out of single material. The interface between a socket and an effector allowed angular adjustment, to help customize the prosthesis to patient's needs.

The second process – automated design – was based on the result of scan merge. The used scan, as well as processing using AutoMedPrint algorithms and selection of section planes is presented in Fig. 6.35A. As a result of this automated workflow (similar to the one presented in the previous case), a set of data for a model of a new prosthesis was generated, without need of operator intervention in the CAD system. It is noteworthy that even such a non-typical geometry of the patient's limb did not pose a problem in generating a dataset that was used for generating a correctly-looking (positively verified) prosthetic device. This device was subjected to virtual try-on. As a result of the virtual try-on, some things were changed – the prosthesis was evaluated in VR as slightly loose, so the dimensions in the socket were slightly decreased and the socket length, as well as some other parameters were slightly modified, obtaining a prosthesis model as presented in Fig. 6.36B.

All designed prosthetic devices were 3D printed using extrusion (FDM) technology. PLA material was used for the automated design, as well as for the open and semi-closed socket and the end effectors of the customized prosthesis. The fully closed socket was printed out of TPU (elastic) material, with the adapter for end effector printed of PLA material, in the same process (by switching the material from PLA to TPU mid-printing). A typical set of properties was applied in the process – 0.25 mm layer thickness, infill of 35% and temperatures and velocities adjusted to specific branches of materials used. The printed parts were post processed by removing the support structures and polishing rough surfaces. In the automatically designed prosthesis, selected surfaces were lined up with EVA foam

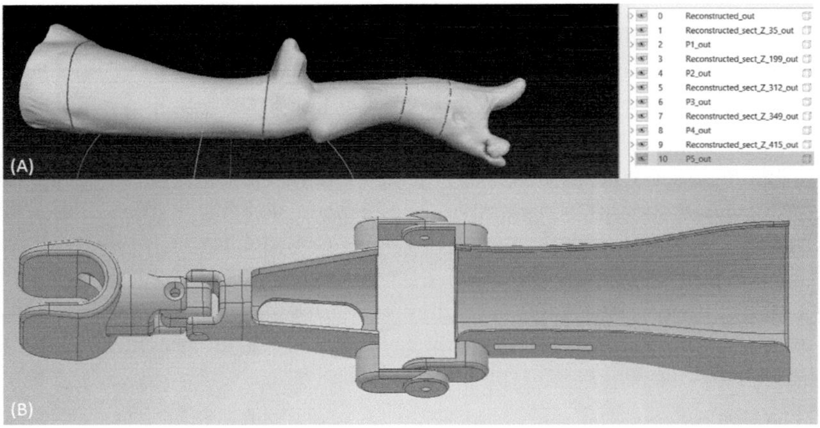

Figure 6.36 Automated prosthesis design: (A) scan merge of two limbs and basic section planes; (B) final variant after modifications from virtual try-on.

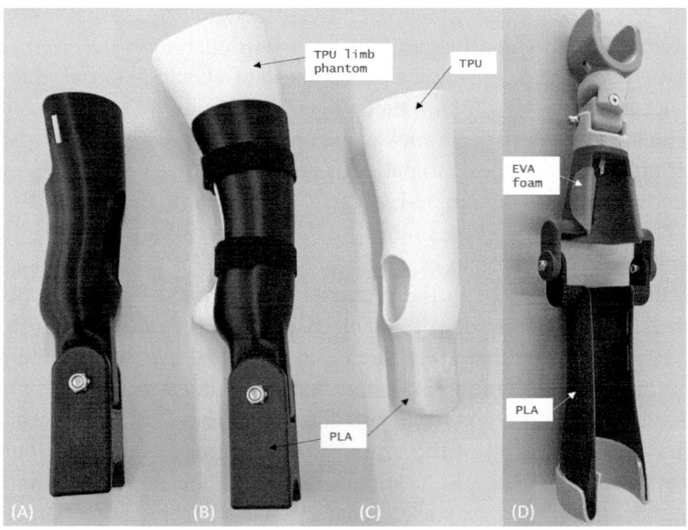

Figure 6.37 3D printed prosthetic devices for the wheelchair patient, (A)–(C) custom designed prostheses; (D) automatically designed prosthesis.

for better comfort. Then, parts were assembled using standard, metric nuts, and bolts. Figure 6.37 presents all the printed and assembled prostheses, with designations of used basic materials.

The patient was invited for try-on of all prepared variants (Fig. 6.38). In general, all the proposed designs had correct fitting and were functionally usable, enabling

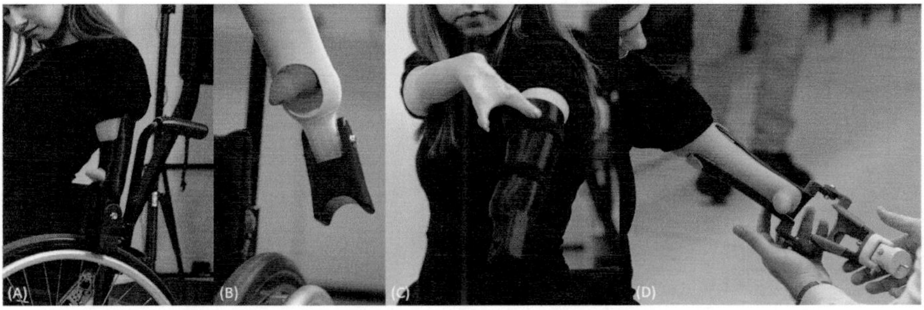

Figure 6.38 Try-on of the prostheses, (A)–(C) custom design; (D) automated design.

the patient to realize certain, basic activities with them. The detailed observations made and conclusions drawn are presented below:

1. The custom engineered prosthesis was well fit, although it is seemingly slightly too short (visible partially in Fig. 6.38B), independently on the used socket variant (they all have the exact same length) – the patient had problems to comfortably reach the pushing hoop of the wheelchair.
2. Of the three custom prosthetic sockets, the elastic, closed one was evaluated as the best in terms of comfort, despite having slightly too much space at the end (again, visible in Fig. 6.38B). The two other sockets were also acceptable in terms of fit and comfort (Fig. 6.38A and C), with the open one being slightly problematic for the patient to put on by herself, due to need of tightening the Velcro straps. However, all three were also slightly too loose.
3. The automatically designed modular prosthesis was also correctly fit, as visible in Fig. 6.38D. It has been evaluated as slightly too long, with possibility of removing the wrist coupling – then it would have an acceptable length for the patient to remain properly balanced.
4. The automatically designed socket was slightly too tight, especially in the bicep area. Here, the perceived "looseness" in the virtual try-on mislead the researchers into making it tighter, which was not necessary. The problem was later identified as resulting not from the wrongly done VR try-on, but rather from the slightly faulty manual 3D scan of the socket, realized by an operator with only intermediate level of skills – it was found out that the actual limb has smaller cross-sectional dimensions than the reconstructed limb model based on the 3D scans.
5. The custom end effectors need improvement for proper pushing of the wheelchair. The straight position (Fig. 6.38A) is not optimal for the pushing movement – the slightly angled position (Fig. 6.38B) was both observed and evaluated as bringing better, more stable movement of the wheelchair. However, shape of the effector must be improved, as to increase the range of dragging the hoop – so that it does not run away when set at an angle. Interestingly, the patient was able to effectively push the wheelchair by using the prosthesis directly on

the tire – which is technically not wrong, except from the certain usage problems (dirt can gather in the end effector and it can wear faster).

6. The automated, universal, modular prosthesis was evaluated by the patient as not really useful – she does not need such a mechanical device in daily functioning. The patient stated a need for a mechatronic prosthesis, which could be controlled by the movements of her "finger" at the end of the left arm. Such a prosthesis is currently under design, based on existing solutions for low-cost mechatronic prosthetic devices (Górski et al. 2024f).
7. Costs of both types of prosthetic devices did not exceed US$100 per complete prosthesis, considering manufacturing time, materials, standard parts, post processing and assembly. However, design time of the custom variant was significantly longer – it took more than a week of work of a skilled engineer to come up with the proposed variants, while the automated prosthesis was designed in less than one working day (including the virtual try-on).

Overall, it has been proven that even for a seemingly "difficult" and unique – from the anatomical viewpoint – case of a patient, it is still a viable option to pursue automated design solutions, which will be able to fit when proper algorithms are applied. However, it must be noted that much attention must be paid to proper realization of anthropometric measurements and processing their results, to get a prosthesis with a proper fit. The manually designed solution was still not perfectly fit (also due to imperfect 3D scans), but interestingly, the character of the slight misfits was different – while the automated design brought prosthesis that was too tight and too long, the manual design had the opposite results – the prosthesis was too short and slightly too loose. Anyhow, the patient expressed interest in using the prostheses so they were given for further use and observations, with improved versions being developed as part of further studies.

6.7 Discussion and recommendations

The presented case studies – as in, a case of an automated design system (AutoMedPrint) and its various use cases and types of products – show that the transition between the old ways of manufacturing of orthopedic products into a new, modern methodology is almost done at the moment of writing of this book. The problem of the "skillset issue" can be omitted, partially by conceiving formal and detailed methodologies of use of available commercial and open-source software, which is continuously developed, and partially by creating automated algorithms, that facilitate the work with anatomical data and make it easy and quick to create basic – and more advanced – shapes of orthopedic products.

Summarizing, in author's opinion, it is fully viable to build a fully automated system, operated only by the patient assisted by medical personnel, where a full anthropometric data acquisition and design process would be realized in time less than 15 minutes. Such system, by the author's estimation, if leaving intelligent models at level 3 (no advanced AI included), would be probably able to properly realize automated design for more than 90% patients with amputations and congenital defects, which already would make an enormous difference in patients' care and would open entirely new technological, logistical and societal possibilities, if these systems were widespread.

It is noteworthy that in the author's team, new ways of use and new functions of 3D printed orthopedic devices are still invented. An example is converting a 3D printed corrective orthosis into a Virtual Reality game controller, giving the patients an entire world of possibilities for improving their hand therapy through the gaming experience (Górski et al. 2023b). In terms of hand prostheses, these are also developed continuously, adding new and new modules to the already expanded database. As such, it is possible to automatically design very complex, modular hand prostheses that can be used for sport activities, as well as in daily life.

During more than 5 years of research in this topic, the AutoMedPrint team was able to work with several dozens of patients and help them by creating hundreds of cheap, 3D printed, personalized devices. The AutoMedPrint system – done as a prototype for proving an idea of fully automated systems for medical devices development – now is used for research and improvement of design automation and 3D printing techniques. It was used to prove that it is possible to design and 3D print a useful, customized prosthesis or orthosis which is many times cheaper than the current commercial solutions and also can be delivered very quickly, in the matter of days, or somehow even hours. Knowledge of medical and engineering specialists can be stored in a computer algorithm, so that these specialists are no longer needed in a daily process of creating repetitive, yet still customized devices and their talents may be efficiently used elsewhere, to create new ways of dealing with health issues.

Future investigative and developmental directions should lean towards the integration of automated orthosis modeling frameworks, aiming to minimize the economic and temporal expenditures associated with manual design methodologies. Furthermore, an in-depth exploration into material sciences and advanced manufacturing technologies could catalyze improvements in subsequent project iterations, enriching the robustness and reliability of produced orthoses.

In terms of the prosthetic devices – it was found out that even the cheapest, simplistic devices are able to transform lives of patients affected by limb loss or born without it. As the prosthetic arms race continues, with the advanced bionic solutions often assessed as impractical (Young 2022), it is noteworthy that simple yet accessible solutions may help thousands of patients, even if not advanced on a mechanical or mechatronic level. Many challenges remain in this domain – the most pressing issues, in the author's opinion, are:

- rapid development of fully customizable prosthetics, especially for children (quickly growing and changing size),
- rapid and cost accessible production of prosthetic parts with sufficient strength and surface quality, as well as aesthetic features (multi-colored),
- designing and implementing precise, strong actuated fingers and main joints, with reliable power sources, built on inexpensive components,
- designing and implementing efficient methods of controlling the actuated prosthetics,

- designing and implementing prosthetics that are comfortable and ergonomic for longer use,
- making prosthetics widely available and cheap by widespread commercial implementation.

Many of these challenges, on a technical level, are far beyond the scope of this book. Some of them are also more related to "soft" topics, such as logistics, economy, and healthcare systems management. This book focuses on technological and technical aspects and does not provide answers on how to introduce these solutions to the market effectively – there is a room for another study in another domain.

To sum up, it should be stated that the synergy of 3D scanning, intelligent CAD and additive manufacturing presents a fertile ground for innovation within orthopedic applications, particularly in the customization of medical devices such as orthoses and simple prostheses. The AutoMedPrint project and the team's successful endeavors at creating partially or fully automated processes of design and manufacturing underscores the potential for automated manufacturing systems to revolutionize the orthopedic sector by enhancing the accessibility and affordability of advanced medical supplies, laying foundational stones for future advancements in patient-centric orthopedic solutions.

CHAPTER 7

Design and 3D Printing of Anatomical Surgery Supplies

7.1 Pre- and interoperative supplies in surgery

7.1.1 Basic definitions and concepts

Modern surgery is often challenged with a necessity of highly personalized approach to any given patient, as anatomy greatly varies and so does pathology which is a reason for surgical interventions (Vendittoli et al. 2023). Apart from a number of mostly routinely performed surgeries (such as appendix or tonsils removal), there is a whole class of surgeries that are very difficult to perform successfully, very frequently applied to patients that are not in good condition, further increasing the risk. These surgeries mostly pertain to removing of cancerous tumors located in various areas of the body (de Ponthaud et al. 2021), be it hard or soft tissues, as well as interventions aimed at saving lives, after serious injuries related to physical damage (Aseni et al. 2023).

Some of the most demanding surgeries are these where part of a body that will be removed (e.g., due to cancer or irreversible damage) should be replaced with an either artificial or biological structure, formed into an anatomically adjusted shape, to fulfill the functions and mimic geometry of the removed tissues (Zenn and Jones 2012). The reconstructive surgeries, as well as precise cancer removal surgeries, require personalized approach, meticulous preparations and very extensive diagnostic procedures, for the doctors to be fully familiarized with all the details of the given patients' case, in as short time as possible, with as informative manner as viable, considering the factor of time (which can be seriously limiting the possibilities of a detailed diagnosis). Often they also require cooperation between surgeons of varied specializations, e.g., neurosurgeons and maxillofacial surgeons for some facial reconstructions involving eye manipulation (Kim et al. 2015), or orthopedists and neurosurgeons with spine implantation surgeries (Lebude et al. 2010).

Ever since the 3D printing and digital design technologies gained significant interest and a number of practical implementations in machine building industry, there were speculations about their possible use in surgery. First medical use of 3D printing in surgery – manufacturing a piece of human skull – took place in the 90s (Su and Al'Aref 2018, Tsoulfas et al. 2019). However, only the popularization of SLA and FDM in the second decade of 21st century, as well as increase in computing power and being able to process very large datasets of mesh models based on human anatomy helped to realize a larger number of surgical cases, where 3D printed anatomical models could be used to help the doctors realize the patient's treatment. The author, together with a team, started manufacturing first anatomically adjusted models as early as 2013 (Banaszewski et al. 2018), as described in an overview – case study no. 1 in this chapter.

In the initial phase of development, these models were generally focused on hard tissues, e.g., bones. This is due to two main reasons – first of all, hard tissues are relatively easy to separate using CT imaging and creating a 3D model does not present too many problems – unless the bone has very complex, thin-walled structures (like in the case of maxillofacial area) or is penetrated with a cancerous tumor – but this still is not a hard limiting factor on creating such models. Effectively, such models – ready for 3D printing – can be generated by engineers, not requiring substantial medical knowledge and assistance from qualified medical doctors. The second reason is that it is relatively easy to roughly mimic properties of hard tissues with the 3D printed models, which – using SLA, FDM or SLS – are also usually "hard" (not elastic, often brittle), behaving similarly as the bone, being susceptible to cutting and fracture during simulated surgery (Ganguli et al. 2018, Mowry et al. 2015).

It is much more difficult to realize usable working models of soft human tissues – first of all, it requires using MRI, or multi-phase CT, or sensor fusion, to obtain proper imaging. Processing the imaging is very hard, as it is often very difficult to differentiate between subsequent tissues (Alshipli et al. 2022), bringing the necessity of real-time cooperation between engineers, radiologists, and surgeons – and such a cooperation is usually challenging logistically, especially considering the time constraints (short time between imaging and planned surgery, usually counted in days). That is why such cases are much rarer and are still relatively new. What is more, it is difficult to represent the behavior of human soft tissues – muscles, arteries etc. – using a standard 3D printing material portfolio. Only the development of multi-material printing, such as PolyJet, facilitates obtaining such models (Birbara et al. 2019), there is also an increasing number of solutions enabling printing from soft materials, such as silicones (Duran et al. 2023). Alternative approaches exist, such as in work by the author's team, using vacuum casting techniques combined with FDM printing of molds, to obtain soft tissue models using very soft silicones (Żukowska et al. 2022). One alternative approach – or a complementary one, depending on used scenarios – is to use the anatomical models only in a digital form, as holograms used before or during the operation, without giving them a physical form. This has also been proven to be an effective approach by available literature (Saito et al. 2020).

Anyhow, the increasing number of cases and excellent results proven by a still small sample allows to consider the 3D printed (and virtual) models used for aiding personalized surgeries a viable and necessary approach. There are more and more cases, where 3D printing is often the only way to minimize risk and increase chances of success of some very complex surgeries. It is also often a method to relieve the doctors, giving them tools enabling a slightly more "carefree" approach, making the surgeries shorter, more effective and more safe for the patient (Banaszewski et al. 2018). This chapter presents some methodologies and cases focused on specific, different types of surgery and human body areas, basing only on the author's firsthand experiences, to give the reader an insight into how modern digital and manufacturing technologies can be used directly in a hospital to help the doctors obtain better clinical results.

7.1.2 Classification and use of anatomically adjusted surgical supplies

Classification, applications, materials, and processes related to anatomically adjusted surgical supplies have been a particular interest for the author and the team, and detailed information can be found in a review publication (Żukowska et al. 2023). This information will be partially and shortly shown in this subchapter.

The anatomically adjusted models used throughout a process of patient treatment involving surgical procedures can be classified using several criteria. The first, basic criterion is related to the type of tissues that will be represented (hard or soft) and whether the model will be a single-tissue model or a multi-material one. As such, we can distinguish the following classes here:

1. *Single material hard (bone tissue) models* – frequently used in maxillofacial and laryngological surgeries (Fig. 7.1A) due to the complexity of human skull and oral cavity, examples include mandibles (Pabiszczak et al. 2016), eye sockets (Żukowska et al. 2024a), skull or the whole maxilla (Meglioli et al. 2020). These usually are not used in oncological surgery, but rather in a reconstructive one, aimed at decreasing negative effects of injuries, deformations, or previous tissue resections.
2. *Bi-material models* – these typically involve two tissues: a healthy tissue of a given body part and a pathological tissue, usually a cancerous tumor, penetrating the otherwise healthy body part. Bi-material models are a foundational, basic form of models aiding oncological surgeries, including only the two tissues, with the one being intended to be removed during the surgery. Examples include tongues (Żukowska et al. 2024b) (Fig. 7.1B) or jaw bone tumors (Fig. 7.1C), so these can be two soft materials or hard/soft combination.
3. *Multi-material models* – these involve many various tissues being represented, usually in multiple colors and materials with varied properties, often being used in oncological surgery of internal organs, such as kidneys (Żukowska et al. 2019a) (Fig. 7.1D), liver or lungs (Żukowska et al. 2023). Here, apart from the healthy and the pathological (cancer) tissue, a number of surrounding tissues are represented, usually aimed at minimizing risk for the patient. These additional tissues are often arteries and veins, as well as nerves or surrounding,

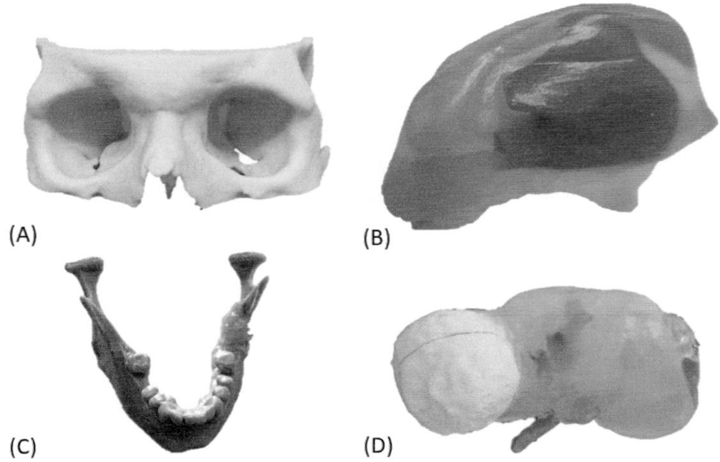

Figure 7.1 Basic types of surgical anatomical models by material: (A) single material model (eye socket), (B) two material model – soft tissues (tongue), (C) two material model – hard tissues (bone tumor); (D) multi material model.

closely neighboring organs (such as the spleen near the liver) that could be affected during the surgery if sufficient precautions are not taken. These models require advanced manufacturing techniques to obtain and requirements are high, although usually hard/soft material combinations are used, with only the basic materials remaining fully soft, with the additional tissues (such as arteries) represented as hard materials.

It is worth noting, that by "hard" material we mean here, generally, a material, with properties not resembling the original body tissue properties, especially regarding hardness and elasticity. In the case of bone tissues, usually surgical models do not need to be fully representative in terms of properties, as the simulated surgery is not a main concern in these cases (as demonstrated in the case studies in the further part of this chapter). In the case of soft tissues, such as muscles and internal body parts (e.g., kidney cortex), mimicking of real body part properties can come in very handy, mostly in the likely scenario of pre-surgical use and simulated surgery, where various activities are trained by the doctors before the actual surgery. However, due to lack of reliable data on mechanical properties of living soft tissues in the human body, creating properly behaving models is a difficult task (Żukowska et al. 2022). In mid-surgery use, however, there is no real need of full representation of the body tissue mechanical properties, so these models, albeit representing soft tissues, are frequently made of hard materials (i.e., polymers typically used in extrusion or UV 3D printing technologies) (Żukowska et al. 2023).

This brings us to the second main criterion of classification of the surgical models, where they are divided by their purpose and time of use – before, during or after the surgery, as mentioned in the previous review work (Żukowska et al. 2023). These models and their characteristics are briefly presented below.

1. *Pre-surgical models* – these are mainly aimed at reviewing and using before the operation. The activities focused on these models may include:
   - reviewing and discussing patient's case and familiarizing with the surgical tasks,
   - simulated surgery – cutting through the tissue models, sewing, drilling etc.
   - preparing of implanted material – titanium meshes or other metallic endoprostheses which require bending and shaping before or during the operation,
   - working with the patient to educate them on their case and treatment procedure, for increase of psychological comfort, as briefly mentioned in Chapter 5.
2. *Mid-surgery models* – these models are taken directly into the operating room and are used during the actual procedure as a part of tooling. Increased requirements, especially regarding sterilization and material selection, are implemented here. The main activities involve:
   - using anatomically shaped models as patterns for reconstruction of missing structures out of the patient's own tissues (bones, muscles),
   - using templates based on the patient's anatomy as tools for realizing precise surgical activities, such as bone incision or endoprosthesis mounting.
3. *Post-surgery models* – this is a least frequently utilized category of anatomical models, which are constructed a specific time after the surgery was conducted, on the basis of post-treatment medical imaging (Zhao et al. 2020b). These models are usually used to assess the patient's condition and recovery level and are mostly purely visual (no simulated cutting or other destructive use).

Some examples of these models used in various stages of treatment, based on the work of the author and team members, are shown in Fig. 7.2A (pre-surgical models) and B (mid-surgery models).

The third main criterion of distinguishing between various categories of anatomical models for surgery is whether they are physical, tangible, usually 3D printed models or just virtual, digital ones, being accessed and utilized by one of the technologies of the Extended Reality (XR) spectrum (Fig. 7.3). Here, all the previous criteria of division also apply – the virtual models can also be used before, during and after a surgical procedure and be composed of one or more materials, where the "materials" are purely a visual term, related to graphical representation. Physics simulation, e.g., simulated cutting, is also possible thanks to use of haptic devices with force feedback (Escobar-Castillejos et al. 2016), but due to their high cost and difficulties in proper reproducing of cutting forces and reactions from human tissues, this usage is limited, especially considering the short time available for preparation to a specific surgery. However, the haptic approach has a wide set of applications in medical education and training (not focused on a specific patient, but rather on specific surgery cases) (Rangarajan et al. 2020).

On that occasion, it is worth noting that all these models can be also used in medical education and training, as noted by many authors beforehand (Smith et al. 2018, Lim et al. 2016), also briefly mentioned in Chapter 5. It can happen during the

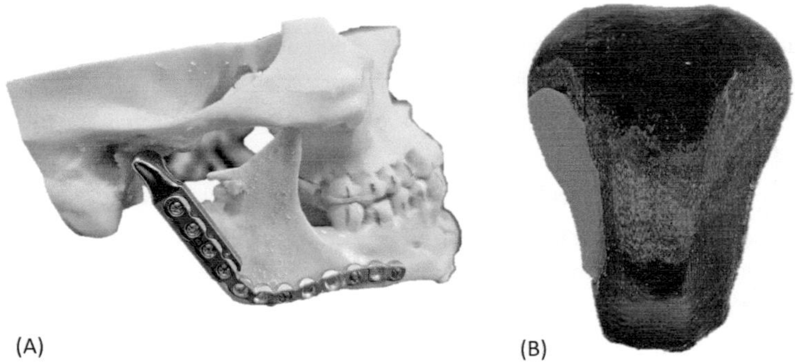

Figure 7.2 Basic types of surgical anatomical models by use stage: (A) pre-surgical model for lower jaw endoprosthesis implantation (photo courtesy of Dr Łukasz Słowik); (B) mid-surgery model for tongue tumor resection and reconstruction.

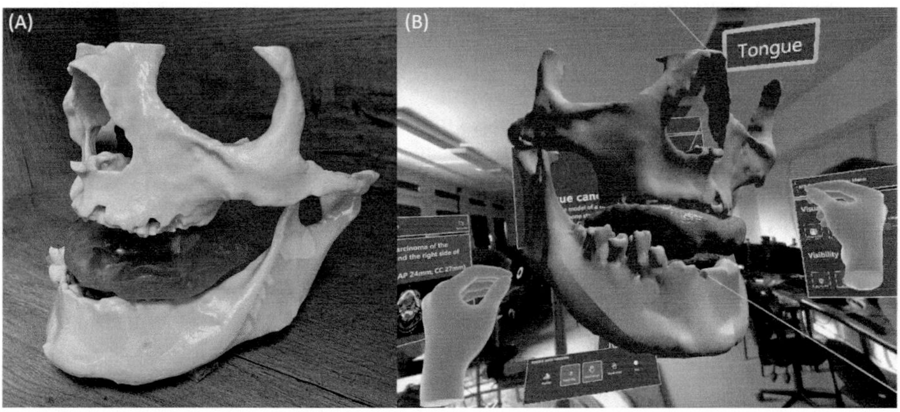

Figure 7.3 Types of anatomical models by form: (A) physical model – tongue cancer, (B) virtual model of the same case, in Mixed Reality.

patient case – where younger doctors and students can observe the expert surgeons doing the procedure and learn. However, it can also happen after the case is actually realized, especially if the models were not damaged or purposely destroyed during the pre- and mid-surgery activities. A library of previous cases can be also re-produced and shown to students, both in virtual and physical form, to enhance the educational effect on the young adepts of medicine.

7.1.3 Current limitations of anatomical surgical models

The current generation of additive manufacturing technologies, as well as XR visualization techniques used for building models helpful in the process of surgical treatment personalization certainly have big possibilities and enable significant improvement in the course of these procedures. By the experience of the author and his team, even the inexpensive 3D printing technologies can be used to produce useful models that could improve surgeon's work.

However, there are certain limitations to today's methods, that disallow to unleash full potential behind the concept of these models – which is, significant reduction of required qualifications for surgery, meaning easier operations by less experienced medical staff, making potentially difficult and costly surgeries more accessible to a larger group of patients.

These limitations, shortly, are as following:

1. *Medical imaging limitations* – currently available techniques, mostly CT and MRI, are not fully suitable to present the human's internal tissues in full details, with possible extraction of the tiniest details in the form of 3D models, with differentiation between various tissues. This limitation reduces the scope, resolution and accuracy of created models, as certain geometrical information (about size and shape of some tissues) are simply not accessible with current imaging technology. A way to overcome this limitation is further development of medical imaging techniques, to increase their resolution and the possibility of differentiating of even very tiny structures inside humans' vital organs.
2. *Modeling and processing limitations* – in the case of soft tissue cancers or deeply penetrating bone tumors, it is very difficult for an engineer to differentiate between various tissues presented in medical imaging and create a proper set of 3D models with clear boundaries. That is not a limitation of the imaging itself (only partially), but rather a logistical constraint – locating and separating (segmenting) particular tissues in the medical imagery requires skill, time and effort that sometimes are inadequate to the scope of the task at hand. Use of artificial intelligence (AI) in processing of medical imagery might be a solution here (Kim et al. 2018, Li et al. 2023), but these algorithms are not yet as popular and accessible and still, the human eye is necessary.
3. *Material limitations* – this is true especially for soft models used in simulated surgery, as it is very difficult to obtain adequate mechanical properties that would help a doctor get a realistic feeling while realizing the cutting, drilling, and sewing the modeled tissues (Muramoto 2020). Extensive material studies, mapping synthetic material properties with certain human tissues would greatly contribute overcoming this limitation, but what makes it really troublesome is that gathering properties of human tissues is not easy – the tissues taken from cadavers behave much differently than the living tissues with blood circulation, and realizing material studies on tissues in a living human being can be considered very unethical. Bioprinting and tissue engineering could help here, to "grow" tissue samples, of which certain properties could be divulged.
4. *Manufacturing limitations* – additive manufacturing technologies have their own limitations, especially regarding surface quality, accuracy, and strength of produced products, as well as time, cost, and labor consumption of post processing activities. These could be overcome by careful selection of methods and materials, as well as proper training – e.g., extrusion technologies (FDM)

are not suitable for certain types of models, as shown in the case of eye socket (where UV printing is the only way).

5. *Visualization limitations* – use of XR technologies can help obtaining greatly in looking at models with interaction, but it is actually very difficult to get models which are free to interact with and also contain proper force feedback. Moreover, mid-surgery uses rely on just a couple of devices (e.g., HoloLens 2), which have limited capabilities, especially regarding field of view, resolution, and clarity of displayed holograms (Palumbo 2022) and can also be bulky and ergonomically unsatisfying for prolonged use. Development of high-resolution, transparent displays built into lightweight, easy to don headsets or glasses can help to introduce virtual and mixed reality into operating rooms on a more regular basis.

This list of limitations is by no means complete, and it is also up-to-date at the moment of writing of this book. However, recent new developments and continuous progress both in XR technologies and additive manufacturing, as well as material engineering, allow to hope that these limitations will be gradually less troublesome and that the anatomical models will be more and more used in the personalized surgery.

7.2 Methodology of design and 3D printing of anatomical surgical supplies

7.2.1 Methodology of surgical supplies development

The process of development of anatomical models used for surgical purposes can be divided into several small subprocesses, which all need to be realized in a relatively short time span, in between initial patient diagnosis and actual surgery. Depending on the class of the model – decided upon analyzing a specific case – the processes will be much different from each other, regarding data sources, processing and design tools, used materials and technological processes, as well as tools for post processing and use scenarios of obtained models. The methodologies and block diagrams presented in this chapter are based on previous knowledge, presented in Chapters 3 (medical imaging and data processing), 4 (design tools) and 5 (additive manufacturing and supporting techniques), so it is assumed that the reader is familiar with them.

A framework methodology, incorporating results of previous work and experiences in different cases and techniques, is presented in Fig, 7.4.

This methodology can be also labeled as a lifecycle of a surgical anatomical model. Its "life" starts when a patient is diagnosed and their medical image is taken. After that, the design phase is commenced, where the model only has a digital version, which is shaped through use of various software – mostly for DICOM segmentation, STL mesh processing and sometimes CAD (if parametric models are, for some reason, desired). Then, after obtaining the 3D model, one of two paths can be taken, or both, simultaneously or at different points in time, depending on the need.

The first path is building the physical model, usually with various rapid manufacturing technologies, most profoundly 3D printing. Then, the model can be

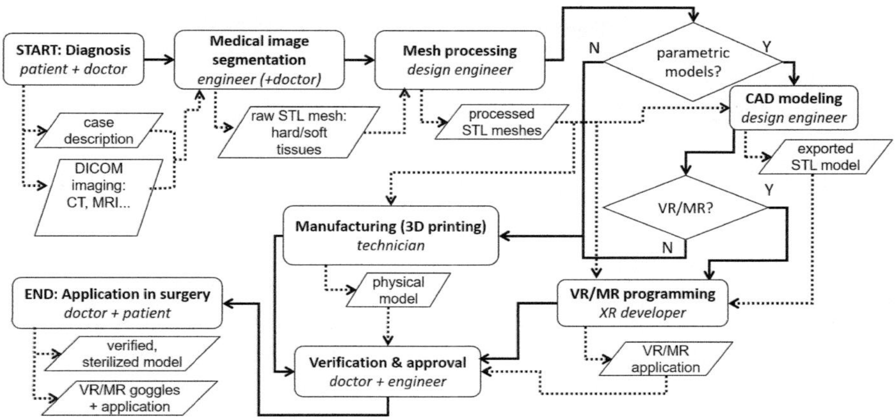

Figure 7.4 Framework methodology of development of anatomically personalized surgical aids.

subjected to various post processing techniques (alongside sterilization for medical use (Kuczko et al. 2018)) and used – directly or indirectly, before or during the surgery. In some cases, an additional, third path can be followed, where the 3D printed model becomes a base for design of another set of surgical tools (for clarity, not shown in Fig. 7.4).

The second path is the virtual one – the obtained model is placed within a game engine or other 3D visualization software with XR capabilities. Visualization is applied (materials, textures and so on), interaction is programmed and the resulting virtual model is deployed for pre-surgical use, sometimes also with capability of using it mid-surgery. After use, the models – both physical and virtual one – could be stored for further use, e.g., in education, or disposed of, depending on the assumed strategy.

The details and technical solutions of these possible paths are based on methodologies already presented in a number of previous chapters. Some hints at possible methods to use are presented in the next subchapters, as well as in the specific case studies.

7.2.2 Design methodologies and tools

Building digital models of surgical aids requires a number of software solutions, depending on the input data and taken path of development (Fig. 7.4). The basic functionalities required to realize the design activities are as following:

1. Extracting mesh models of anatomical structures from medical imaging, obtaining STL or compatible file type out of images stored in the DICOM format or its equivalent. Requires performing segmentation using an appropriate toolset, as described in Chapter 3.
2. Processing obtained mesh models: limiting down the scope by separating only the necessary structures and deleting excessive areas; dividing structures into separate parts and improving their shape by operations realized on surfaces or

volumes (closing holes, removing artifacts, smoothing out unevenness, etc.), as described in Chapter 3.

3. Optionally, creating parametrically designed shapes using CAD software, by techniques of solid and/or surface modeling, using the processed meshes as a point of input – as described in Chapter 4.
4. Exporting the models into a polygon/triangle mesh format for further activities, either in preparation for 3D printing or for visualization in virtual or mixed reality.

It is uncommon for a single software package to contain all these capabilities, however, such software exists. An example is Materialise MIMICS, and there are plenty of cases of its use in model and implant design (Mohammed et al. 2017). Still, the multi-purpose professional software like MIMICS usually is expensive and difficult to implement easily, containing an excessively large number of functions. There are many other software packages enabling work with medical data and CAD design on unusual shapes. These applications are mentioned and characterized in Chapters 3 and 4. Here, it is worth mentioning that the following classes of software, with examples, can be selected and utilized:

1. *DICOM segmentation* – 3D Slicer, InVesalius (free to use), MIMICS, Simpleware ScanIP (commercial).
2. *Mesh processing software* – MeshLab, MeshMixer, GOM Inspect (free to use).
3. *CAD software* – Fusion 360 (free for hobby use, otherwise low-cost), SolidWorks, CATIA.

Use of a specific software tool is almost always dictated by requirements of a given case, as well as quality and amount of input data. In the case of dealing with hard tissues (bones), the segmentation can be usually done semi-automatically or in a relatively simple manner, by filtering out the tissue using the Hounsfield scale (Hohlweg-Majert et al. 2011) and simply exporting it. As such, the simplest tools, such as InVesalius, should be working just fine for that purpose. More work would be required in the mesh processing stage, as there are plenty of operations to realize – exporting the whole set of bone structures out of medical imaging brings volumes of unnecessary data, that must be removed, plus bones must be separated from each other. As such, in the hard tissues case, it is better to select tools with an intuitive interface and plenty of options of working with meshes, such as GOM Inspect. In terms of a CAD software, any software capable of drawing directly on the triangle meshes will do, but if possible it is better to look out for software that has hybrid modeling capabilities (surface and solid modeling with seamless connection between the two), CATIA is one example of such software.

In the case of soft tissues and multi-tissue models, the focus of work is usually put into the segmentation process. It often relies on fusion of imaging series gathered in various measurements (e.g., three-phase computed tomography, or CT scans superimposed with MRI scans, etc.). In the first phase of work, high experience and careful, meticulous approach is needed to properly differentiate between various soft tissue types, e.g., in oral cavity it is very difficult for an inexperienced engineer to differentiate between the tongue, salivary glands and tongue tumors. That is why often it is required for the engineer to work side by side with an experienced radiologist, or to train the radiologist or another experienced doctor to perform the segmentation themselves. In the next stage, if a sensor fusion was required, models obtained of various imaging series must be superimposed and positioned appropriately, to make sure that the models show them in right locations (e.g., arteries going through the liver). As such, in this case it is more worthy to work with an advanced DICOM processing software. Mesh processing is often limited in the case of soft tissues – only some basic operations, like reduction, smoothing etc., as well as positioning and rotating are performed – as such, software such as MeshLab or MeshMixer can be used.

The two general approaches presented above are reflected in the case studies presented in the further part of this chapter. However, the details may vary from case to case and from one body area to another. In the author's opinion, no universal methodology is available and no methodology proposed in literature is definitely better than the other, as the process of designing these models is still a developing one, with plenty of new solutions, approaches and model types unraveling from year to year.

7.2.3 Materials and processes – requirements and methods

The pre- and mid-surgical models used by the doctors at various stages of treatment and work with the patient have various requirements, depending on the body area, tissue type, method of use and purpose. However, usually one requirement is common and it is a short time of production. In the case of oncological patients, the surgeries often save lives and can be ordered and realized in a matter of days between acquiring medical imaging and the actual surgery, leaving a small window for engineering work related to production of appropriately shaped tooling. The same requirement may or may not apply to injuries and reconstructive surgeries – in some situations, the surgeries are aimed at correction of pathological states lasting for years (e.g., lack of bone part, wrongly fused bones preventing movement, etc.), but in the majority of cases, as soon as it is possible, the patient should be treated, leaving limited time for preparation. The least demanding are post-surgery models, which could be really urgent only if the patient's condition was rapidly worsening after the surgery, requiring another intervention.

Taking the above-mentioned into account, regarding the manufacturing processes, it can be assumed that the whole manufacturing process should be conducted over the course of one, maximum two days, considering also that a given team of doctors may be tending to more than one patient at once. This narrows down the possible choices mostly to additive manufacturing technologies, and rather the quicker ones. Casting processes can also be utilized, if it is possible to quickly create a mold that would be filled in with a material with appropriately quick hardening (or cooling, if it is realized by temperature and not chemical reaction). Also, it is noteworthy that the anatomical models used as surgical tooling will not be required to be produced out of metallic, or other very durable and heavy materials. The aim is to represent mainly the proper shapes – for decision-making before and during the operation – and also creating models that resemble human tissues in their mechanical behavior. No tissue in the human body is mechanically comparable with metals used in engineering – except maybe the bones in some cases – so polymer processing additive manufacturing technologies are sufficient.

Simultaneously, the processes should be as simple and straightforward as possible (easy to learn, without lots of possibilities of making errors), and able to be placed inside a given healthcare facility, to avoid logistics and transport. The concept of "hospital bed printing" will, in the author's opinion, be an imminent direction of development. At the same time, the machines should not take too much space and consume too much energy, as the 3D printing of the considered models is usually not a lifesaving activity – at least not directly.

In terms of material requirements, they have already been basically stated in the Chapter 7.1. Many authors in previous publications have considered various materials and their combinations to create anatomically individualized surgical models (Żukowska et al. 2023). The main requirement is the similarity of materials' mechanical behavior to the human tissue, especially regarding hardness, susceptibility to cutting (tear strength) and to dynamical loads, weight, stability, and bio-neutrality enabling sterilization and brief contact with human tissues during a surgical procedure. From a processing point of view, the materials need not be toxic and should be easily processable.

The above mentioned requirements are summarized in Table 7.1 for selected, characteristically demanding types of models. Detailed requirements and results of experiments with various materials can be found in available literature or other works of the author and the team (Żukowska et al. 2024a, Żukowska et al. 2024b).

Meeting these requirements, we can consider the main basic technologies to be used in surgical preparation to be the following:

- UV curing – SLA, DLP,
- jetted photopolymer (PolyJet),
- extrusion (FDM),
- vacuum casting.

Each of these would have different fields of applications and range of materials. This juxtaposition is presented in Table 7.2.

Table 7.1 Requirements for common types of surgical anatomical models.

Type of model	Material requirements	Process requirements
Pre-surgical Oncological Hard tissue	- hardness - susceptibility to cutting - density	- rapid (1 day) - low layer thickness - low-cost
Pre-surgical Oncological Soft tissue/multi	- softness - susceptibility to cutting - transparency	- rapid (1 day) - high surface quality - multi-material
Mid-surgery Reconstructive/implantation	- hardness, durability - transparency - sterilizability, biocompatibility	- rapid (1 day) - high surface quality - realizable in hospital conditions

Table 7.2 Processes and materials used in production of various types of surgical models.

Process	Materials	Application - model types
Extrusion (FDM): up to 2 materials at once	- thermoplastic hard polymers - rubber-like thermoplastics	- molds for soft tissue pre-surgical models - mid-surgery models for reconstructive surgeries with lower accuracy acceptable
UV - SLA/DLP: single material	- photocurable resins (hard)	- pre-surgical models for hard tissues - mid-surgery models for some reconstructive surgeries requiring high accuracy - surgical templates placed inside the patient
PolyJet: up to 6 materials at once	- photocurable resins - hard/soft	- pre-surgical models, all kinds - mid-surgery models for precise and complex reconstructions - surgical templates placed inside the patient
Vacuum Casting: sequential multi-material	- chemically hardened resins - additive silicones - wax and other low-melting materials	- pre-surgical models, soft tissues, and multi-material models

In the further part of this chapter, use cases showing various processes, materials and applications are briefly presented.

7.3 Mixed and virtual reality in surgical visualization

7.3.1 XR technologies – basic concepts

The extended reality (XR) technologies are a group of digital simulation methods, that have been developed over the course of the last several decades, originating from the famous inventions of Morton Heilig over 60 years prior to this book (Gutierrez 2023), and philosophically known since ancient times. The idea is to create an unreal world, populated with imaginary objects, but possible to feel and interact with and, as such, learning new insights and testing concepts and activities that would be too dangerous, cumbersome, or expensive to recreate in real life. In the recent decade, the technologies of virtual reality (VR), Augmented Reality (AR) and Mixed Reality (MR) have gained considerable interest and their popularity skyrocketed, both in terms of mainstream awareness and in a number of professional implementations, especially in medicine (Lie et al. 2023).

The main idea of extended reality continuum has been introduced as soon as the all three technologies were conceived, that is in 90s, by Milgram (Skarbez et al. 2021) in his well-known paper. The continuum spans between the physical, real, non-digital world we live in and the purely virtual, digital world, in ideal conditions completely replacing the real world. The latter is known as VR – this technology is a dominating one in the XR spectrum and has plenty of applications in engineering, education and also medicine. It allows to immerse the user into a responsive simulated environment, where various interactions can be made with objects populating it and, as a result, the user can gain some new knowledge, skills, or experiences. The VR was the first technology of the spectrum to be conceived and named. The most popular ways to enter VR are either through personal projection devices – Head-Mounted Displays (HMDs) (Jensen and Kondradsen 2018) or through large-scale stereoscopic projection screens, often in form of rooms (CAVEs) (Muhanna 2015), while interaction these days is mostly assured by use of motion controllers or alternatively by hand tracking cameras.

The second technology is at the other end of the spectrum, as close to the real world as possible – this one is known as AR. Here, the user is interacting with real-world objects – other humans, machines, tools, books, toys, etc., – and the task of the technology is to supply the users with contextually accurate visualizations that can enhance their perception and similar to VR – enable them to gain some new knowledge, but in connection with experiencing the real world together with some digital objects. The visualizations (augmentations) are displayed when the user sees a specific object or part of it – the triggers for visualizations are known as markers and they can be images, color codes, texts, or 3D shapes, recognized by a camera put inside the AR device. Usually the augmentations are somehow interactive, but not always – often they have a form of data summary, enabling the user to make some decisions (Martins et al. 2022) or visual clues, pointing the user to their next destination or activity in a training scenario (Yount et al. 2022). AR technologies have plenty of uses similar and complementary to VR, mostly in engineering (Pejić et al. 2014) but also in medicine as well (Yeung et al. 2021). Augmented reality is accessible in a number of very diverse ways, starting from simple applications for cellphones or tablets, through stationary systems with large screen displays and cameras, to smart glasses and goggles with transparent screens.

Going through the continuum in between the real world and the virtual reality, various shades of mixed reality – the third XR technology – can be found. Mixed reality is the youngest of the three, mostly because for its full scope of work, very advanced computer technology is required, that has not been possible to squeeze into a goggle-sized contraption until the last couple of years. The main idea of mixed reality is free interaction with digital objects – just as in VR – but without losing contact and perception with the real world – just as in AR. Therefore, the user's location must be tracked in real world and users are allowed to freely manipulate the interactive objects – often called holograms – and leave them wherever they feel suitable, in the environment of real-world objects (Pell 2017). Moreover, often interaction between real and digital objects is possible through the means of "terrain mapping" – real-time 3D scanning of the user's environment, enabling, for instance, to throw a digital ball against a real wall and simulate it bouncing off. Mixed reality has a unique set of advantages and the number of its applications is growing, especially in medicine (Hu et al. 2019), but also in regular engineering, especially engineering education (Minoufekr and Plapper 2020).

To enter the world of MR, head-mounted displays are most frequently used. These can come with two different flavors: either with a semi-transparent screen, where real-world perception is undisturbed, but the holograms are themselves semi-transparent and field of view is smaller, or with a non-transparent screen, with

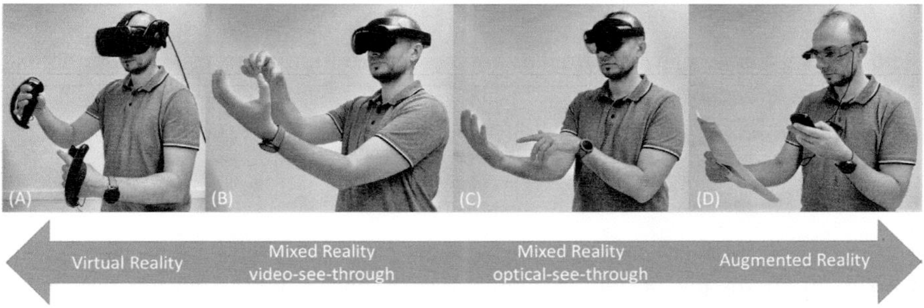

Figure 7.5 XR continuum based on Milgram et al. [1995] concepts, (A) VR – immersion into purely digital world; (B) Mixed Reality – holograms with real world perceived through video stream; (C) Mixed Reality, with real world perceived directly; (D) Augmented Reality – recognition of markers in real world.

real world image displayed through video cameras. In both cases, interaction can be realized through hand tracking and free hand manipulation, without controllers – which often proves itself to be much more intuitive and better adopted by fresh users. The selection of a hardware platform (semi-transparent or video passthrough) would be dependent on specific application and use scenario – if interaction with real world objects is necessary (e.g., in the operating room) then the transparent screen would be recommended, but in the case where digital objects are of prime interest – the passthrough would probably yield better results.

Figure 7.5 presents how the different technologies from the spectrum are used – with examples of specific devices, with division between two main subtypes of mixed reality also distinguished.

In the context of this book and its field of interest it is noteworthy that virtual reality and other XR technologies are often considered as a possible early replacement of physical objects, such as prototypes in the process of product development – this concept is known as Virtual Prototyping (Post and Kendall 2021). Many of earlier studies and implementations by the author involved work with virtual prototypes (Żukowska et al. 2019, Górski 2017), as well as medical uses of VR (Buń et al. 2018, Hamrol et al. 2023). Naturally, XR technologies are utilized in the team also for work with the patients and doctors, for displaying of anatomical models – before the physical ones are made or sometimes simultaneously, for enhanced and more diverse results, depending on the need. That is why it was decided to show an example case involving mixed reality and notify the readers – through the presence of this chapter – that use of XR is a perfectly viable option for perceiving anatomical models as aids before a planned surgery or other medical procedure.

7.3.2 Medical uses of holograms and VR

Virtual reality has been long associated with medicine and medical engineering and to this day it is one of the most prominent fields where it can be implemented, along with other technologies from the continuum (Skarbez et al. 2021). The main stack of medical applications of XR can be divided into three main groups, focused on different recipients (users) of XR systems:

1. *Engineer-centered XR* – virtual prototypes and configurators, product simulators, training applications.
2. *Patient-centered XR* – physical rehabilitation, psychotherapy, product configurators.
3. *Doctor-centered XR* – education, training, surgical simulation.

Engineer-centered XR applications can be used in the whole product lifecycle (Górski 2017). In the design and development phase, virtual prototypes enable engineers to visualize and manipulate medical devices in a virtual space before physical prototypes are created. This can significantly speed up the development process, reduce costs, and improve the final product's design (Verlinden et al. 2003). Product simulators in XR, on the other hand, allow engineers to test the functionality and usability of medical devices in a virtual environment. This can include simulating how a device will perform under various conditions or how it will interact with the human anatomy. Companies such as GE Healthcare, Medtronic and many more are using XR regularly for development of their medical devices (Deakyne and Iaizzo 2021).

Training applications for engineers are another class of applications – essential for understanding complex medical systems and ensuring their proper maintenance and operation. XR provides immersive training experiences where engineers can learn how to assemble, disassemble, and troubleshoot medical equipment. These virtual training sessions can replicate real-world scenarios, allowing engineers to practice without the risk of damaging expensive equipment (Checa et al. 2022). In the production scenario or in the case of maintenance and servicing, MR and AR can also be a valuable asset.

Patient-centered XR applications usually focus on various methods of increasing medical treatment effects by utilizing effects of immersion and interaction. In physical rehabilitation, virtual reality provides engaging and motivating environments for patients to perform exercises (Górski et al. 2023b). Interactive VR or MR not only makes rehabilitation more enjoyable but also encourages consistent practice, which is essential for recovery.

In psychotherapy, XR is used to treat conditions such as anxiety, PTSD, and phobias through exposure therapy, also known as VRET (Bush 2008). Virtual environments can simulate anxiety-provoking situations in a controlled manner, allowing patients to confront and manage their fears safely. One of the first instances of this approach was to help veterans with PTSD by immersing them in virtual scenarios that gradually desensitize their traumatic responses (Rizzo et al. 2009), but other conditions can also be treated, such as arachnophobia or fear of heights.

Product configurators for patients are another innovative use of XR. These systems allow patients to visualize and customize medical devices or prosthetics before they are manufactured. For instance, virtual reality can be used to show how different prostheses or orthoses will look and function, helping patients make informed decisions about their options (Fig. 7.6). This personalization ensures

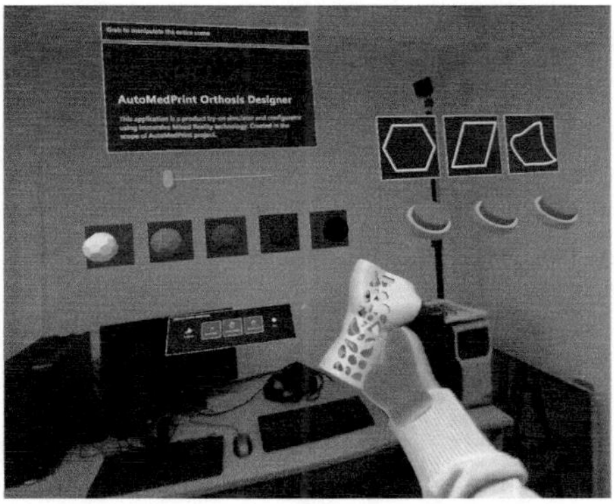

Figure 7.6 Configuring an orthopedic product using Mixed Reality.

a better fit and greater satisfaction with the final product. Such configurators are embedded into the previously discussed AutoMedPrint system, and studies on them were a part of evaluation and development of the system, bringing promising results, as mentioned briefly in Chapter 6.

Doctor-centered XR applications are crucial for education, training, and surgical simulation. In medical education, virtual reality provides immersive environments where medical students can explore the human anatomy in three dimensions, as proven by an earlier work of the author with the team (Hamrol et al. 2013). This method surpasses traditional 2D illustrations, offering detailed and interactive experiences that enhance comprehension and retention. Existing platforms allow students to virtually dissect cadavers, observe organ functions, and understand complex physiological processes in a more intuitive manner (Bartoletti-Stella et al. 2021).

When considering medical professionals, they can use virtual reality to practice surgical procedures without the risk of harming real patients. Surgical simulation in XR enables surgeons to practice complex operations, refine their techniques, and prepare for rare or complicated cases. There are VR surgery platforms that offer high-fidelity simulations that replicate real-life surgical environments, complete with haptic feedback to simulate the sensation of tissues and tools (Arbabtafti et al. 2008, Escobar-Castillejos et al. 2016). This immersive training improves surgical skills, reduces the learning curve for new procedures, and potentially enhances patient outcomes (Javaid and Haleem 2020).

A separate class of applications are the pre-surgical and mid-surgery XR visualizations. Here, medical imaging of a given patient is used to create interactive 3D models. They can be used either in a pre-surgery scenario – using VR or passthrough MR – or mid-surgery, using transparent MR or AR with markers. These applications are counterparts for the physical, 3D printed anatomical models and are the main reason of discussing XR technologies in this book.

7.3.3 Building Mixed Reality surgical aids – basic concepts

In a previous work by the author, lots of attention was paid to methodologies of creation of virtual reality applications, especially for industrial applications (Górski 2017). The general assumptions described in these two papers remain the same since the last decade, however details need clarification in terms of building application based on real medical imaging of a specific patient, as well as utilizing the possibilities of modern Mixed Reality systems.

Building of a VR/MR application for use as a surgical aid can be divided into several basic stages:

1. *Planning* – selecting technology and hardware (VR, MR, subtypes, and specific hardware to be used), as well as basic software tools. Various methodologies can be used here, in the previous work author applied framework of MOKA methodology to construct the own way of medical VR app planning. This stage will not be detailed here. A result should be a list of tools to be used and requirements of a future application.
2. Obtaining and processing medical data of a specific medical case (diagnosis and measurement). This might be dependent on use of various medical imaging techniques, segmenting 3D data out of them (Fig. 7.7A) and polishing them using mesh processing tools (Fig. 7.7B), as widely described in Chapter 3, and in many previous elaborations.

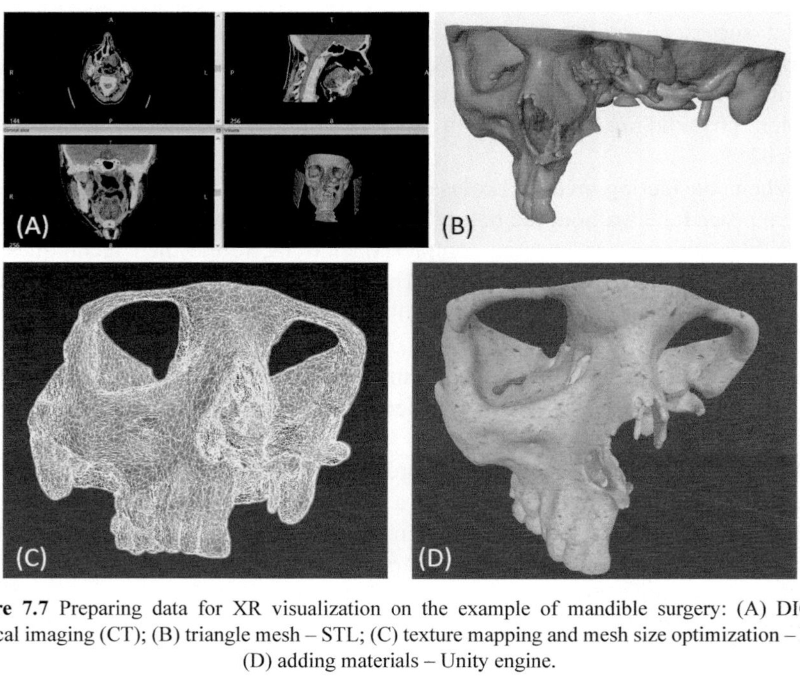

Figure 7.7 Preparing data for XR visualization on the example of mandible surgery: (A) DICOM medical imaging (CT); (B) triangle mesh – STL; (C) texture mapping and mesh size optimization – FBX; (D) adding materials – Unity engine.

3. Preparing the 3D data for visualization in VR (optimization, texture mapping – Fig. 7.7C) and importing the data into a selected game engine (or other software type) capable of building and deploying an interactive visualization into selected piece of hardware (Fig. 7.7D). The most popular solution here is the Unity engine, but other viable alternatives exist (e.g., Unreal engine).
4. *Visualization in the game engine* – creating and applying textures and materials, composing lights, shadows, and special effects, among other things (Fig. 7.8A). In general, at this stage the anatomical models should have the desired look and feel and be possible to display at a MR device. For some surgeries, that is already enough, as visualization is the most important purpose of surgical mixed reality.
5. *Interaction programming* – creating and applying scripts, animations, etc., to create interactive capabilities (Fig. 7.8B), such as free or constrained manipulation (explosive view), adjusting transparency of selected body parts, showing contextual descriptions, measuring selected dimensions in 3D, among many others. This can be done by standard programming (e.g., C# language used in Unity), but also by visual programming.
6. *Interface* – creating and programming Graphical User Interface (GUI) used in the mixed reality for launching appropriate functionalities, consisting of buttons, sliders, panels, labels, icons, etc. (Fig. 7.8C).
7. *Verification, validation, and implementation* – testing of the application's main functions, gathering feedback from professional testers (alfa testing) and potential users (beta testing), finally implementing it in hospital conditions. Conducting of this stage depends on if the application is deployed for the first time in a given environment (thus requiring full testing and verification procedure), or is it just another patient case in the same framework (then only basic testing is needed before clinical use).

Details of implementation of this methodology sketch will be highly dependent on the selected hardware (goggles and their compatibility, as well as operating system), software – mainly the game engine, its programming capabilities and available plugins and software development kits. All the examples shown in this chapter, including case study #2, were visualized using the Unity game engine and

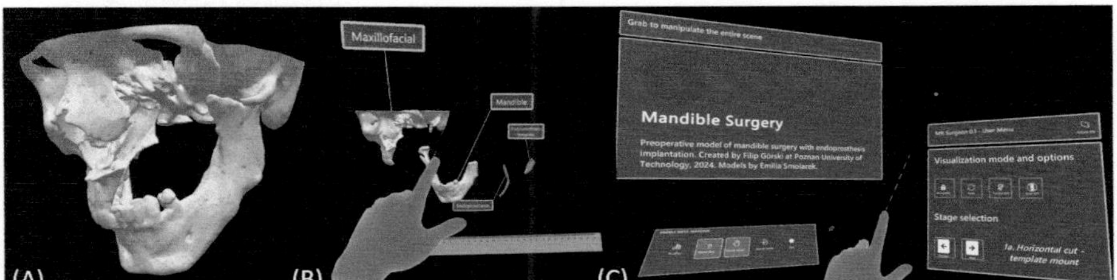

Figure 7.8 Stages of building mixed reality applications for pre-operative support, (A) visualization (materials, lights, shadows); (B) implementing interaction; (C) building user's interface.

Microsoft Reality Toolkit, which is reasonably accessible even for users with low experience. Building Mixed Reality applications with Unity and MRTK can be easily learnt , as shown in the previous work by the author (Górski et al. 2023c). Moreover, Unity software, as well as the toolkit, is generally free to use for small size companies and institutions, as well as personal users, under currently valid licensing policy. As such, these applications can be implemented in an easy and costless manner, greatly helping doctors in preparing for the operations. However, it should be also considered that such applications would need approval before clinical use, for ethical reasons – there are a number of concerns to be raised here and formalities to meet, which go beyond the scope of this book.

7.4 Case study #1 – lower jaw reconstruction

7.4.1 Basic concepts and patient cases

The first case example – or the whole class, to put in a wider perspective – is focused on a series of surgeries realized in a Heliodor Swiecicki Poznan University Clinical Hospital, in the Department of Otolaryngology. The team of the author and the team of doctors (lead by Dr Jacek Banaszewski) from that department are cooperating since 2013, mostly focusing on various 3D printed surgical aids in the surgeries focused on the lower jaw, tongue, throat, and other laryngological areas of interest. Results of this cooperation have been described in numerous papers and theses (Banaszewski et al. 2018, Pabiszczak et al. 2016, Żukowska et al. 2022) and some of the contents of this chapter are based on these publications, as well as direct experiences of the author.

The main distinctive feature of surgical processes, aided by models described in this chapter, is lesser use of artificial materials in the process of reconstructing of missing or removed tissues in the patient's oral cavity. Instead, the doctors of the department are using bony free flaps for reconstruction of vital parts in the oral cavity, namely the lower jaw and tongue (Banaszewski et al. 2018). It can be labeled as autotransplantation and can also be described as creation and implantation of bioprostheses – these prosthetic devices, shaped to fill in the missing part of patient's mandible (or tongue) are made of patient's own tissues, taken from another location in the body, where the bone tissue presence is of lesser importance and impact on the patient's life comfort than in the oral cavity. Most frequently, the fibula and the scapula bones are used to construct the bioprostheses implanted as parts of mandible.

Realizing such surgeries is a demanding process – it requires the doctors to be prepared perfectly, knowing well the anatomy and geometry of the different body parts of a patient, also requiring to make surgical interventions in different locations at the same time (bone removed from the leg to create a part of a jaw, over the course of one surgery). As such, these surgeries last particularly long (possibly over 10 hours) and require precision, focus, and care from the doctors – as well as long times of careful preparation with personalized approach.

The main task of the engineering team, led by the author, was to create 3D printed templates that can be used both as pre-surgical models (to pre-shape the titanium plates used to keep the various parts of bone together until they grow a connection) and also mid-surgery. These templates had to be based on 3D models of the patient's original bone structures. They were used to properly cut and shape the bones (and soft tissues associated with them) taken from the patient's body, facilitating, and speeding up the process, reducing time the patients remained under anesthesia and improving obtained clinical results (Banaszewski et al. 2018).

The main case study described in this chapter investigates the effectiveness of using three-dimensional (3D) printed models in reconstructive surgery for patients with mandibular defects caused by advanced oral cavity Squamous Cell Carcinoma (SCC). The research compares the outcomes of reconstructive surgeries using 3D printed models with those using conventional planning methods (Banaszewski et al. 2018, Pabiszczak et al. 2016). It was hypothesized by the combined teams of engineers (including the author) and doctors that physical 3D models will facilitate precise surgical planning and reconstruction, potentially improving both aesthetic and functional outcomes. The further part of the chapter is focused on technical aspects of proceeding with the varying patient cases, with different types and shapes of templates produced and supplied into the operating room by means of medical imaging processing, CAD and 3D printing.

7.4.2 Design process: segmentation, mesh processing, modeling

As mandibular defects resulting from malignancies such as squamous cellcarcinoma (SCC) pose significant reconstructive challenges due to the complex three-dimensional structure of the mandible, use of additively manufactured anatomical models is a noteworthy approach. First of all, the models need to be prepared, on the basis of medical imaging. High-resolution CT scans enable the segmentation and then creation of physical 3D models.

All the considered patients underwent tumor resection followed by simultaneous mandibular reconstruction. High-resolution CT scans of the mandible and donor sites (fibula and scapula) were used to create digital 3D models, by engineers directly in the author's team (Dr Wiesław Kuczko and Dr Radosław Wichniarek). Two examples are shown in Figs. 7.9A and 7.9B, in one case scapula was used to reconstruct the central part of the mandible and in the other – fibula was a building material for reconstruction of a side part of the jaw.

The 3D models were processed and converted to STL format using software tools (Slicer v4.4.0, GOM Inspect v7, and CATIA). The detailed methodologies of how the segmentation processes were realized are described, in general terms, in Chapters 3 and 4 of this book. The more detailed overview is available in the paper (Banaszewski et al. 2018) and others. A short summary of realized activities is as following:

1. The obtained high-resolution CT scans in DICOM format, obtained from the hospital, were uploaded to the 3D Slicer software.
2. Using the thresholding operation, with a proper values of Hounsfield scale, the bone structures were generated and exported into STL format.

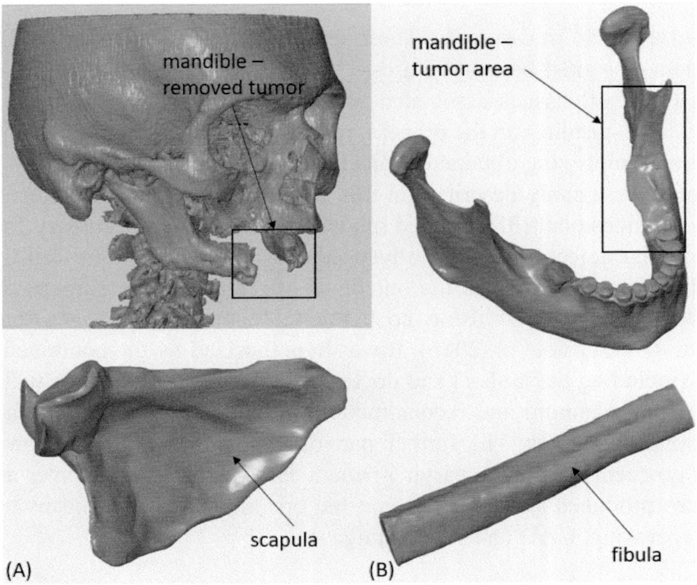

Figure 7.9 Mandible reconstruction – patient cases: (A) scapula-based reconstruction of mid-part of mandible; (B) fibula-based reconstruction of side part of mandible with visible tumor.

3. GOM Inspect software was used for manual work with the mesh, to clean it and reduce it to the area of interest only – which was the mandible itself, or rather its parts (due to tumor resection, patients' mandibles were divided into two parts, separated by the space remaining after tumor resection, as visible in the Figures). Various tools for cutting, cleaning, hole filling and artifact repairing were used by the engineers, to obtain a "clean" mandible mesh for further use.
4. Steps 1–3 were repeated for the donor bones (fibula or scapula).
5. Both the mandible and the donor bones (in each separate case) were uploaded into the CATIA v5 software and converted into a non-parametric ("dumb") surface model, using Digitized Shape Editor module.
6. Using hybrid (surface and solid) modeling techniques, the donor bones were cut into appropriate shapes to fill the missing parts in the mandibles.
7. As an output, a whole mandible model was exported (the desired effect after the surgery), along with the template for cutting the donor bone. The exported files were in STL format, ready for printing.

The obtained models (visible in Fig. 7.10C) were then subjected to processing in the additive manufacturing software and manufactured, as described in the next subchapter.

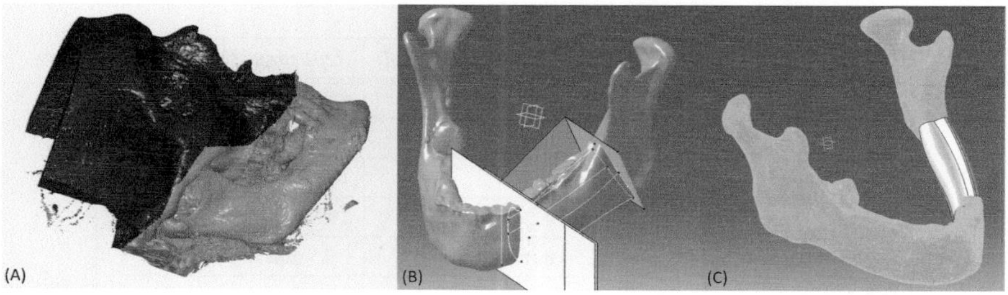

Figure 7.10 Mandible reconstruction design – process course example: (A) model cleaning; (B) reconstructive modelling; (C) final assembly.

7.4.3 FDM 3D printing and properties evaluation

The presented case study used both professional and low-cost FDM machines to manufacture a mandible template, comparing the accuracy and usability of the products from each machine. The details are described in a previous publication (Górski et al. 2019a).

The study employed two FDM machines: the Stratasys Dimension BST 1200, a professional-grade printer, and the MakerBot Replicator 2X, a low-cost alternative. Both machines used similar parameters and materials to ensure a valid comparison. ABS material was used in both cases, coming from the Stratasys company and supported in an enclosed cartridge with an electronic chip. The BST 1200 machine is programmatically locked so that it can use only materials supported by the producer, recognized by reading the chip, which also stores the current material volume available. The MakerBot machines have an open material system, where any filament can be loaded into the machine, provided that the user is able to properly select the temperature and velocity parameters to achieve the extrusion course enabling to obtain proper products in a stable process.

The manufacturing process involved preparing the FDM process, using similar parameters and materials for both machines. More specifically, as the BST 1200 machine is somewhat limited in the scope and variability of the process parameters, the MakerBot print parameters were adjusted so that the prints were comparable. Layer thickness of 0,254 mm was used, with a solid (100%) infill to obtain monolithic parts.

Post-processing consisted solely of mechanical support removal. No additional operations were realized, before handing the 3D printed models to the doctors, who followed with their own operations, mostly sterilization (experimental description on studies of this process are presented in the publication (Kuczko et al. 2018)). The results of the printing processes themselves for the second patient case, comparing the low-cost to professional FDM approach, are presented in Table 7.3.

The printed models, after post processing and before the doctors' evaluation were checked for dimensional accuracy by use of 3D scanning. ATOS GOM I 3D scanner was used to digitize the 3D printed models and the resulting meshes were superimposed over the nominal CAD geometry to create deviation maps. Eleven

Table 7.3 Additive manufacturing of mandible models using low-cost and professional FDM machines – results comparison.

Parameters	***BST 1200 (professional)***	***Replicator 2X (low-cost)***
Mass of finished product	62,98 g	65,01 g
Layer deposition time	5 h 18 min	6 h 5 min
Model material usage	71,63 g	100,83 g
Support material usage	14,24 g	29,81 g
Layer thickness	0,254 mm	0,25 mm
Infill	100%	100%
Model material	ABS	ABS
Support material	ABS	HIPS
Temperatures	Materials: 300°C Chamber: 75°C	Model material: 230°C Support material: 250°C Model plate: 110°C
Total production cost	~ US$ **139**	~ US$ **63**

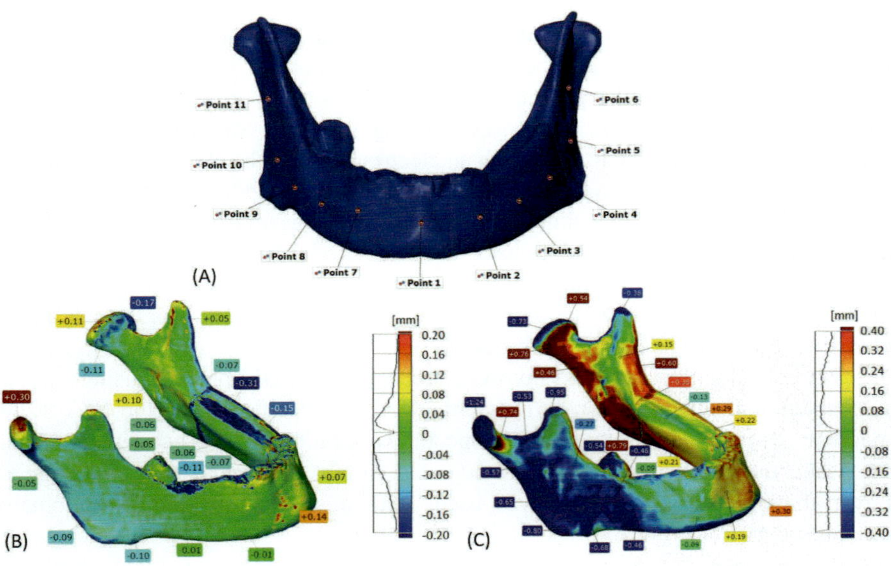

Figure 7.11 Accuracy comparison of 3D printed mandibles: (A) selected points; (B) professional machine; (C) low-cost machine.

characteristic points were selected for detailed comparison. The results are presented in Fig. 7.11.

It was found that average scan fitting error was 0.05 mm for the BST 1200 and 0.27 mm for the Replicator 2X. For the BST 1200, all the points stayed within tolerance of ±0.05 mm, while for Replicator 2X – all the points were outside the tolerance, with maximal deviation of –0.56 mm, in point no. 5. However, such worsened accuracy was still acceptable for the doctors and medical use, as changeability of human tissue dimensions and tolerance of reconstructive surgeries is a lot higher (although there are no precise data in literature about given tolerance ranges). For the doctors, the models were practically indistinguishable, and both models coming from both machines were accepted for sterilization and surgical use. As such, it was concluded that the low-cost technology can be also applied in building pre-surgical and mid-surgery models for the mandible area, with almost twofold price reduction per single model.

7.4.4 Clinical use and results

For the surgical procedures, mandibular defects were reconstructed using free bony flaps from either the fibula or scapula, with attached skin islands for soft tissue reconstruction. The sterilized 3D printed models were used to guide the precise placement of bone grafts, ensuring better alignment and stability (Fig. 7.12).

The study, taken as a complete sample of 12 patients (8 treated conventionally versus 4 with 3D printed models (Banaszewski et al. 2018)) demonstrated that all reconstructive surgeries were successful, achieving complete wound healing. Notably, patients who underwent surgeries planned with 3D printed models exhibited superior outcomes across several metrics compared to those who had conventional planning. In the 3D model group, junction stability was 100%, whereas the conventional group had an 84% stability rate. The average mouth opening was greater in the 3D model group at 3.0 cm, compared to 2.5 cm in the conventional group. Chewing function was also higher in the 3D model group at 90, versus 72% in the conventional group (Banaszewski et al. 2018).

Additionally, the cosmetic (aesthetic) results were more satisfactory, with all of the patients in the 3D model group finding the outcomes acceptable, compared

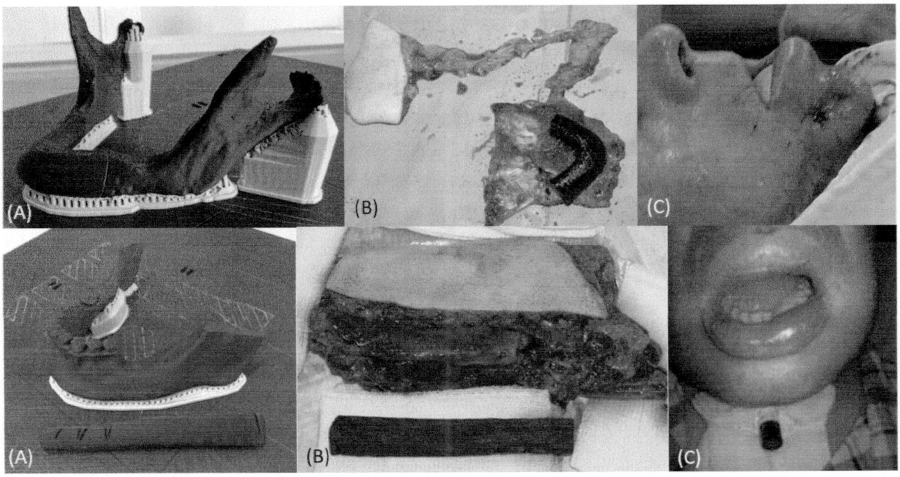

Figure 7.12 Mandible reconstruction in clinical practice: (A) 3D printed models; (B) use in surgery as templates; (C) effect directly after autograft implantation; photographs B and C courtesy of Dr Jacek Banaszewski.

to 60% in the conventional group. The average operation time was significantly reduced in the 3D model group, taking 6.5 hours compared to 8.5 hours in the conventional group. Moreover, the differential angles, which indicate mandibular contour symmetry, were smaller in the 3D model group (7.3 ± 9.1 degrees) than in the conventional surgery group (10.5 ± 12.5 degrees), suggesting better symmetry (Banaszewski et al. 2018).

These findings underscore the significant advantages of using 3D printed models in mandibular reconstruction surgeries. The enhanced stability, greater mouth opening, improved chewing function, and superior cosmetic results highlight the functional and aesthetic benefits of this technology. Additionally, the reduced surgical time, attributable to precise preoperative planning and execution, offers practical benefits in terms of cost savings and reduced anesthesia duration (Banaszewski et al. 2018, Pabiszczak et al. 2016).

Despite the longer pre-surgical preparation time (averagely an additional 3 days) required for creating 3D models, the overall benefits far outweigh this initial time investment. The collaborative effort between radiologists, engineers, and surgeons was crucial for the success of this approach, ensuring that the models are accurately designed and effectively utilized during surgery. It is noteworthy, that the original studies were performed as early as 2015 – and their results still ring true and innovative. Nowadays, the costs of 3D printing technologies significantly decreased and their availability increased, what further supports their integration into routine clinical practice, making advanced reconstructive techniques more feasible – allowing implementation of 3D printers in hospitals.

In conclusion, reconstructing the mandible using the auto transplantation approach (bioprosthetic bone filling made of the patient's own bone) poses significant challenges due to the complex, three-dimensional structure of the bone and necessity of greater anatomical adjustment to maintain proper functions. The application of 3D printed models in mandibular reconstruction provides advantages over traditional methods, resulting in better functional and aesthetic outcomes (Banaszewski et al. 2018, Górski et al. 2019a). This technology not only improves the quality of the surgical results but also enhances operational efficiency by significantly reducing the time required for surgery – and keeping patients under anesthesia, which is quite important considering that these oncological patients are usually not in a good general condition, so risk should be minimized whenever it is possible. However, the scope of work is vast and an interdisciplinary team is required, implementing many technical procedures done in various software – namely, DICOM segmentation, thorough mesh processing, skillful 3D modeling in CAD systems and proper preparation and realization of 3D printing processes.

The study's findings confirm usability of 3D printing technologies in clinical settings, given their visible, objectively confirmed benefits in improving patient outcomes and making surgical procedures smoother and easier for the doctors. It was also a starting point for a long-term collaboration, which is nowadays focusing also on soft tissue models, such as tongues – as described in other papers and theses (Żukowska et al. 2022).

7.5 Case study #2 – liver tumor surgery

7.5.1 Patient case and the model design process

The second case study required design, 3D printing and virtual model of an assembly of soft tissues in the abdominal area, used as tools before a surgical intervention was performed. The case was realized in 2024 by the University Children's Hospital of Cracow (UCH), which is the largest pediatric hospital in the south of Poland. Thanks to the courtesy of the personnel at the hospital, the case was made available to the author to describe here with necessary details related to the diagnosis, treatment, and use of anatomical 3D models throughout the procedure.

The patient was a two-year-old child, with an accidentally diagnosed abdominal tumor. After performing the necessary medical imaging and laboratory tests, it was revealed that it was a liver tumor of the hepatoblastoma type. Chemotherapy was implemented to reduce the tumor mass before the planned partial surgery resection of the liver with tumor. After achieving tumor reduction, reconstruction (segmentation) of the tumor, liver, arterial and venous vessels, and surrounding tissues was performed based on a known three-phase CT examination (Ryu et al. 2009) – arterial, venous, and late phase CT angiography.

The time from the imaging examination to the planned surgery was one week, needed for surgical assessment, preparing the patient for surgery and familiarizing himself with the problem using new technologies (segmentation, VR, 3D printing). It was decided to use a pre-surgical preparation in the form of detailed digital segmentation (creating of a 3D model based on CT and MRI scans), a physical 3D model printed on site, as well as a multi-material physical model printed in the 3D printing laboratory – place of work of the author. Additionally, a Mixed Reality application was created containing the model case, for possible use in pre-surgery.

The segmentation was realized on the basis of supplied medical imaging of a current (post-chemotherapy, pre-surgery) state of the tumor. It was decided to separate the following tissues to be visible as 3D models:

- liver,
- tumor (hepatoblastoma),
- spleen,
- arteries,
- veins,
- bones (spine and ribs – partially).

Segmentation was performed using the 3D Slicer software, using available and previously described tools and methods. The process was realized by Dr Krzysztof Grandys – member of a team of surgeons dealing with the case in the hospital, and a skille. The segmented 3D model is presented in Fig. 7.13. Segmentation process was realized in the course of a single working day.

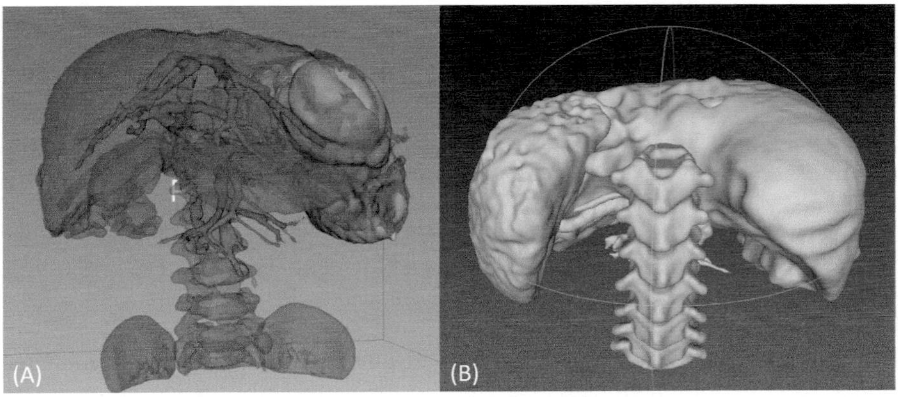

Figure 7.13 The segmented anatomical model of hepatoblastoma case, (A) after segmentation in 3D Slicer, (B) cropped, imported, and visualized in the MeshLab software.

7.5.2 UV 3D printing and properties evaluation

The prepared model of hepatoblastoma and surrounding tissues was decided to be 3D printed as a form of pre-surgical preparation for a team of doctors realizing the medical case. Here, two 3D printing techniques were used:

- stereolithography (SLA), with mono-color printout of the main structures, without the liver tissue,
- jetted photopolymer (PolyJet), with a full representation of the structures, using many materials with different visual and tactile properties, with the liver tissue printed as well.

The SLA printout was realized directly in the hospital, by use of Prusa SL1S SPEED 3D printer, operated by Dr Krzysztof Grandys (no engineering personnel was engaged). It was decided not to print the liver itself, as transparent material was not available and the specifics of the SLA material do not allow to use many different materials. It also was decided not to print the hard tissues (namely, spine bones), as they were not necessary in the pre-surgical preparation. As such, only the tumor, along with arteries, veins and accompanied with the spleen model were printed, of the same material (default, hard photocurable resin), using default printing parameters. The printout result is shown in Fig. 7.14.

The PolyJet print was realized in the laboratory, by the author, using the available MediJet J5 machine with multi-material printing capabilities. The machine in itself is capable of printing hard resins with fully customizable colors and transparency of various model elements printed together. The machine is able to work with a total of 5 different build materials, having 6 slots for material cartridges (the sixth one contains the support material, removable later by water stream). The default machine configuration for hard printouts is one white material cartridge, three color cartridges and one clear (transparent) material cartridge. The latter can be switched into an elastic clear material, enabling to obtain softer printouts, more resembling biological tissues by touch.

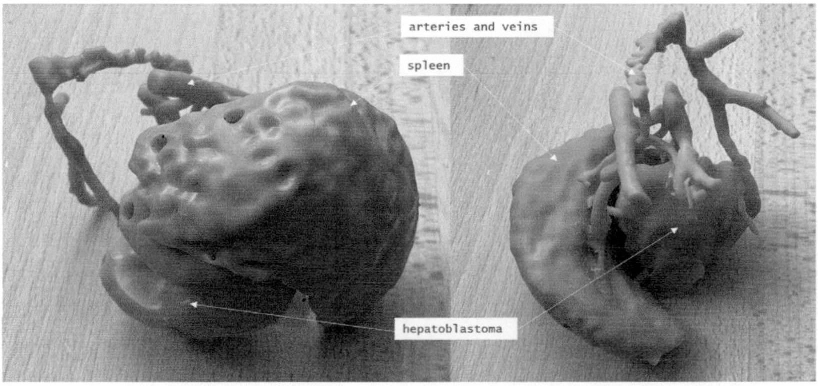

Figure 7.14 The SLA printout (photos courtesy of Dr Krzysztof Grandys).

In the particular case, the configuration with the elastic material was used. A total of 5 different tissues were printed, using different material and infill settings. These were:

- *liver tissue* – clear, elastic material with no color, filled internally with support structure (to improve elastic properties and reduce elastic material consumption), with the shell thickness of 3 mm,
- *hepatoblastoma* – a mixture of clear elastic and a standard resin, with a predefined reddish hue and a nominal hardness of 80 Shore A,
- *spleen* – hard, colored material with a predefined, brownish hue,
- *arteries and veins* – similarly as the spleen, but with different colors (red and blue).

The printing process was planned using the GrabCAD print software connected to the machine directly. The additive manufacturing of the model took approximately 18 hours. The process finished without any disturbances and the model was safely removed from the printer without any damage. The total weight of consumed material was 830 grams. Considering the time and material consumption, a rough cost estimation of production of this size and type of model using the PolyJet technique is US$1000 – a significant amount, requiring careful considerations.

After the product was removed from the machine, it was subjected to waterjet cleaning in a dedicated station, which took approximately 20 minutes. During the process, careful handling was required, as the arteries and veins are delicate structures, prone to damage when printed out of rigid, hard resin. It successfully managed to remove all the support structures and not damage the model. The subsequent stages of printing and cleaning, as well as the final model are shown in Fig. 7.15.

Both models were assessed by doctors and biomedical engineers as properly representing the anatomical geometry (no significant errors were detected), thus possible to be used in the process of treatment of the patient, at the pre-surgery phase.

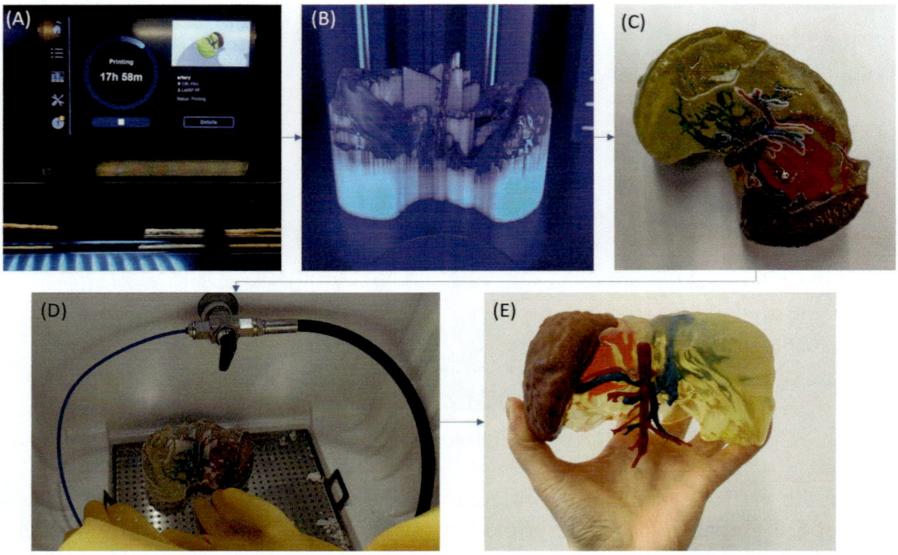

Figure 7.15 The PolyJet printing phases: (A) machine panel view, mid-printing; (B) model in the working chamber after layer deposition; (C) model before post processing; (D) support cleaning with water jet; (E) final model.

7.5.3 Mixed Reality presurgical visualization

The project was initiated with the objective of developing a digital pre-operative visualization tool for the hepatoblastoma case. It was decided to utilize the following software and hardware solutions:

1. Unity game engine (version 2021.3).
2. Microsoft's proprietary Mixed Reality Toolkit (MRTK).
3. Meta Quest Pro goggles – a versatile XR device – in hand tracking mode with passthrough mode activated.
4. Oculus Integration toolkit for Unity.

The first step was preparing Unity with Android build support tools. The Oculus headset was correctly configured with the developer mode activated, capability for hand tracking enabled and mixed reality passthrough view made possible.

The first step involved preparing a basic Unity project setup with MRTK and Oculus configuration. An empty 3D project was created in an appropriate version of the game engine and all the plugins were installed and configured, particularly with the hand interaction examples from the MRTK package.

Then, at a further stage, a basic test scene named was saved under a new name. The scene initially contained test interaction objects, which were thoroughly tested in both the Unity Editor and on the Quest device to ensure the foundational MR capabilities were functional and ensure that the build process was successful. After testing, these objects were hidden rather than deleted, serving as a resource for building custom interactions later.

The core of the project involved the three basic stages of developing XR applications, as established in previous studies (Górski 2017):

1. *Visualization* – at this stage, the model files were properly processed, imported into Unity and given appropriate visual capabilities by creating materials and adding textures.
2. *Interaction* – at this stage, using MRTK's proprietary scripts, basic manipulations of the models were made possible, as well as other interactions – exploding view, transparency adjustment, displaying 3D labels and a simple 3D measurement tool (a ruler).
3. *User's interface* – at this stage, a mixed 2D-3D interface was created, containing text labels, buttons, and sliders, freely interactable using hand movements and gestures.

As a preparation for the visualization activities realized in Unity, the models first had to be processed. Before utilizing human tissue models, specifically STL files representing triangle meshes based on medical imaging (with full fidelity), in a Mixed Reality environment, it is crucial to process these models properly to ensure optimal performance and visual quality. This process involves several key steps, which can be executed in graphic design software such as Autodesk 3DS Max or the free alternative, Blender (here the 3DS Max software was utilized).

The process was commenced by importing the STL files into the chosen graphic design program. This is the initial step where the raw models are brought into the software environment where they will be modified and optimized for Mixed Reality use. Once the models were imported, the next step was to center the pivot point, or the rotation center, of each object. This is important for ensuring that the models rotate correctly around their intended axis during interactions within the Mixed Reality environment.

The next stage was texture mapping. The UVW mapping was applied to the models, opting for a Box mapping method. UVW mapping is essential for correctly applying textures to the models, enhancing their visual appearance and realism in the Mixed Reality space. No advanced mapping techniques were applied (such as unwrapping), as there was no need for photorealism in this case.

The most important stage, though, was reduction of the number of triangles in each model to 10% of the original count. This step is vital for maintaining a smooth experience in Mixed Reality environments, especially considering the hardware limitations of VR and AR devices. The goal is to keep the maximum total number of triangles displayed at any one time to around 1 million. Initially, the hepatoblastoma case models altogether had approximately 3 million triangle count. For test purposes, the model was initially deployed to the Quest Pro device – without any reductions – and the obtained framerate was around 1–2 Frames Per Second (FPS), effectively making it impossible to interact with the models. That is why the reduction was applied. The model did not lose much in terms of the shape fidelity (Fig. 7.16), with the largest changes apparent in the arteries and veins models – that is why it was

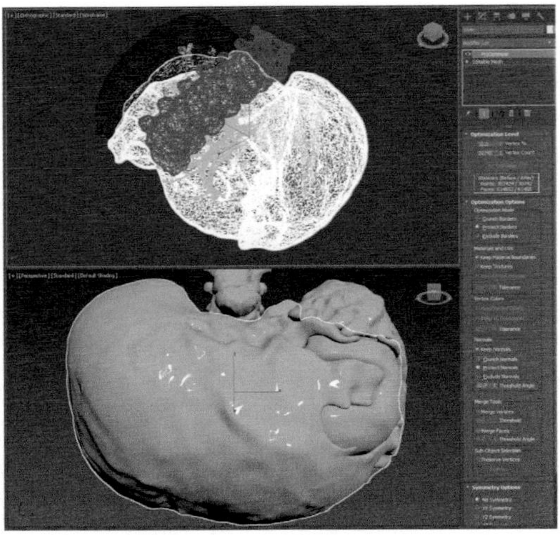

Figure 7.16 Mesh optimization – reduction to 10% triangle size, using proprietary Autodesk 3DS Max algorithms.

decided to leave them in two versions, reduced and not reduced, with the latter made available for interaction in the "unlocked" mode of the application.

Finally, the optimized models were exported in the FBX format. Close attention to the unit scale to ensure that the models are exported with metric units, facilitating accurate scale representation in the Mixed Reality application.

The visualization stage was continued in the Unity software. The materials corresponding to specific tissues were prepared. Freely available textures were utilized (taken from (PolyHaven 2024)) along with textures obtained from a commercially available anatomical model package (used in a previous work (Hamrol et al. 2013)). For the tumor, a leather texture with a custom color mix was used. In the case of the liver, photographs of the surgery (courtesy of the hospital, source material not disclosed in this book) were taken and appropriately processed in graphical software (also mixing with other grayscale structural textures) to obtain a seamless texture file. For the spleen, asphalt-like textures were used with a proper color mix. Arteries and veins were made uniformly red and blue, with increased smoothness. Figure 7.17 shows all the objects with the materials applied, along with the list of these materials. In the case of liver, two versions were prepared – with and without transparency, for the two modes of interaction within the application itself (the semi-transparent material was used as the default one).

After the visualization process was completed, the interactions were programmed. The application was planned to have the following functionalities:

1. Free manipulation of the complete 3D model – rotation, translation and scaling by manipulation using one or two hands.
2. Locked (all together) and unlocked (separate) modes – the locked mode is the default one, while in the alternative mode all the partial models are available separately for manipulation, independent of the others.

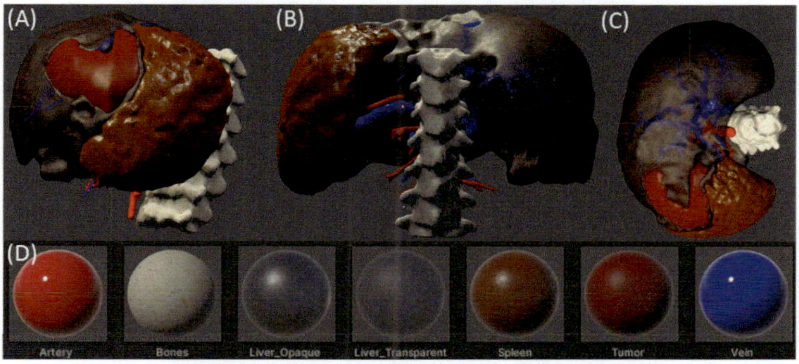

Figure 7.17 Model case visualized in Unity: (A) side view; (B) back view; (C) top view; (D) list and icons of used materials.

3. Resetting of transform (translation, rotation, scale) to default values.
4. Exploding view (only in the locked mode) – a 3D slider controlling the distance between the parts (0 being default value).
5. Transparency – a 3D slider controlling the liver material transparency, with the possibility to turning it completely opaque.
6. Tooltips – 3D labels with names of particular tissues (parts).
7. 3D measurement tool – a ruler (scale in centimeters) to make it possible to measure the models.

All the interactions were achieved either by using MRTK's proprietary scripts (especially related to manipulation with hands) or by own scripts, written by the author (in particular, exploding view, transparency adjustment and resetting). In the first stage, crucial to the project was enabling basic manipulation of the anatomical model (functionality 1), allowing users to interact with the model using hand movements and gestures. This was achieved by adding colliders to the model for physical interaction and copying interaction mechanisms from a template object to enable grabbing, rotating, and scaling of the model. Furthermore, audio feedback was integrated to enhance the user experience during manipulation. This process was made separately for the complete model (available in the locked mode) and the subsequent parts (available in the unlocked mode).

For all the other functionalities, an appropriate user interface had to be created, so the two stages (interaction and interface) were realized in a parallel manner. An interactive 3D interface panel was first created, containing buttons and sliders for activating of certain functions – as presented in Fig. 7.18, divided into two groups, visualization, and visibility. The panel itself can be manipulated – the user can grab it and move, rotate, or scale, as well as make it follow them by use of "follow me" toggle.

The switching between the unlocked and locked mode is realized using a two-state button. Its state is programmatically locked with the state of two models – one with manipulation at the level of the whole group and the other with manipulation capabilities (MRTK's scripts) applied separately to each tissue (part) of the model.

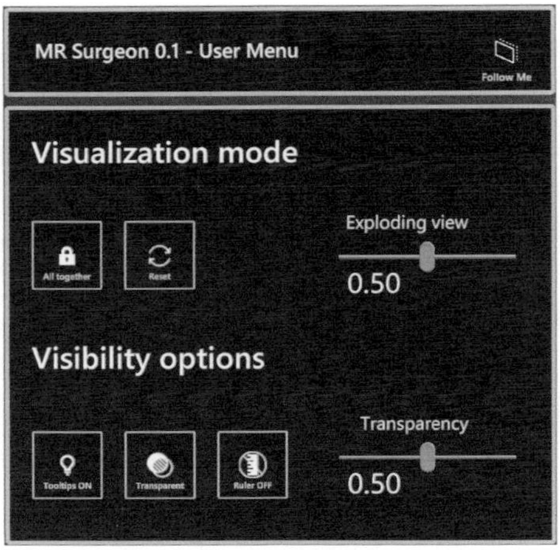

Figure 7.18 Interactive panel with user options in the mixed reality hepatoblastoma model.

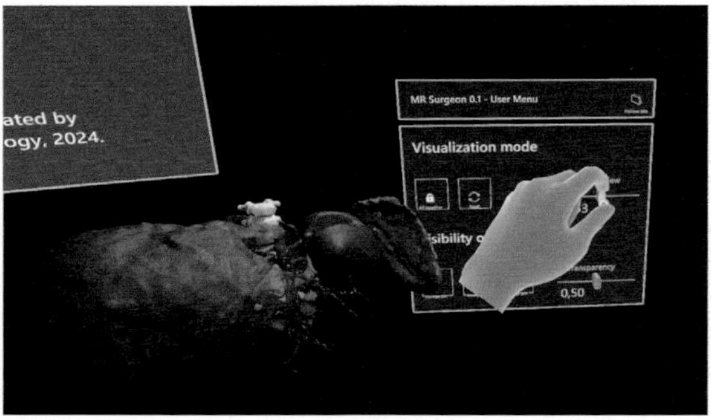

Figure 7.19 Manipulation of the exploding view of the anatomical model.

Both models can be reset to their original position, rotation, and scale by using the reset button.

If the "locked" (all together) mode is active, exploding view is possible to introduce. In this mode, the position of an appropriate slider (manipulated by pinching a handle and moving fingers) corresponds with distance of a given part in a specific axis from the center of the complete model (Fig. 7.19). The exploded model can still be manipulated as a whole (unlike in the unlocked mode).

Among the visibility options, there is a possibility of turning on and off the tooltips with names of parts (Fig. 7.20A), switching the transparency mode on and off (and adjusting transparency by use of 3D slider – Fig. 7.20B), as well as turning on and off a simple measurement tool – ruler scaled in real world units (Fig. 7.20C), making it possible to get to know the size of the objects (of course assuming they have not been rescaled during manipulation). These additional features are shown in Fig. 7.20.

In the final stage of UI preparation, a general panel with description of the application was created, containing title, university's logo and other information.

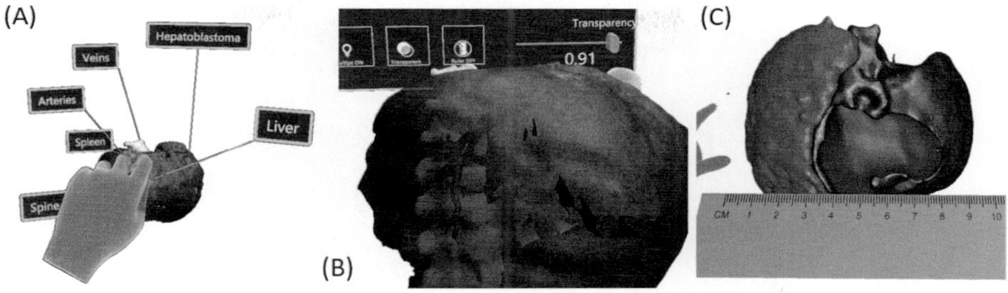

Figure 7.20 Additional features: (A) tooltips; (B) transparency adjustment; (C) simple measuring tool.

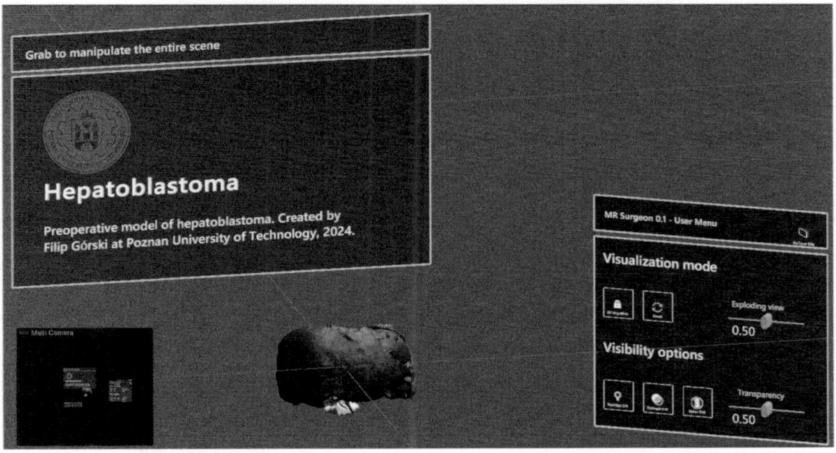

Figure 7.21 Basic scene with UI components and the 3D anatomical model.

This panel can be grabbed by application's user to manipulate the complete scene (e.g., when moving from room to room in real world). The complete scene with all the UI components is presented in Fig. 7.21. The preview from the starting position is also visible – when the application is launched, the model and the title panel are generated in the front of user, while the interactive options panel is generated slightly to the right, a bit down and rotated so that it remains in the reach of user's arms.

The culmination of the project involved compiling the application and conducting thorough testing on the Oculus goggles. This step required adjusting Unity's player settings to reflect the project's identity and ensuring compatibility with the Oculus device. After compiling the application, extensive testing was conducted to verify the functionality of all interactive elements, both with and without controllers, to ensure a seamless user experience. The testing – viewed from the inside of the goggles – is presented in Fig. 7.22.

The testing was performed initially by the author and colleagues from a team. After some initial remarks, some aspects were improved (such as sizes of colliders, materials appearance, and clarity of user interface, among others). Then the application was re-compiled and in the latter phase, it was tested by a group of

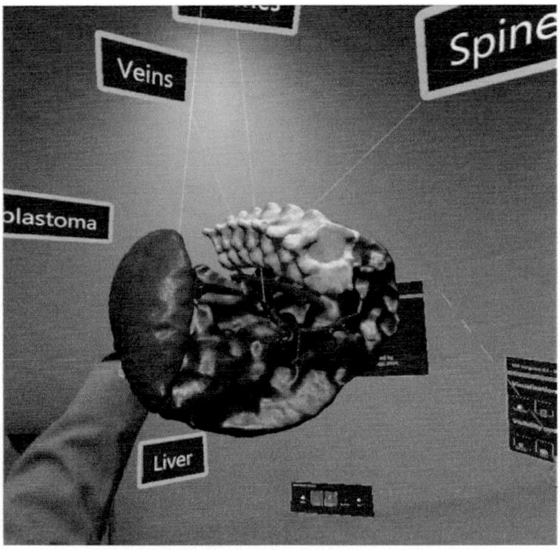

Figure 7.22 Testing the application.

approximately 20 students of biomedical engineering program. Post-surgery it was also tested by the doctors directly involved in the patient's treatment, as described in the next subchapter.

Unequivocally positive evaluation was obtained from all the test participants, marking readiness of application to use it as a support in surgery. Some technical aspects related to the goggles themselves were marked as possible to improve, including the mediocre quality of passthrough using the device's inbuilt cameras, practically eliminating the possibility of using the application mid-surgery. It was also suggested to deploy it on an optical-see-through device, such as HoloLens 2, for possible mid-surgery use. The application was also reviewed post-surgery by the doctors from the team realizing the actual patient treatment, more of which is described in the next section.

The process of creating a Mixed Reality visualization containing interactive anatomical models is a complex one, but – in the author's opinion – not particularly demanding in terms of programming or other highly technical skills. It involves preparation, detailed modification of imported models, and the integration of interactive and animation features to create an immersive tool with realistic capabilities of pre-surgical preparation. The final product not only serves as a testament to the capabilities of current MR technology but also marks a significant step forward in the use of immersive technologies in surgery.

7.5.4 Results and observations

Considering the printed models, the following observations were made:

1. The SLA model, due to its mono-color, makes it a bit difficult to differentiate between the anatomical structures at the first glance and is not a good tool in the hands of the inexperienced.

2. The PolyJet model represents all the structures and the mixed colors help greatly to differentiate between them. However, transparency of the liver is not ideal and the light refractions make it difficult to exactly determine the shape of the hepatoblastoma (it is not clearly visible through the transparent liver). It was also mentioned by the testers that assembly and disassembly capability would be a valuable asset.
3. The elastic material used in PolyJet is capable of deformation, which is more realistic than the use of a hard resin, however it is still not anatomically accurate.
4. The staircase effect and the overall visual quality is better in the case of the PolyJet model. However, it is not homogeneous, as surfaces that were supported with additional material (cleaned in post processing) are matte (as visible in Fig. 7.15E). That could be improved by manual polishing, but it would mean additional workload. It is important to notice, that in the particular case, the PolyJet model was more than several times more costly than the SLA model.
5. Both models, thanks to low value of layer thickness and relatively high 3D printing resolution and accuracy, properly and faithfully represent the segmented shape of the patient's tissues and both were preliminarily evaluated as potentially useful in the surgical process.

As mentioned before, both models were found suitable to use as an additional aid in the process of surgery planning. The hard, SLA model, after sterilization, was possible to be carried into the operating room. Its comparison with the real tumor (cut with a proper margin oh healthy tissue to prevent cancer recurrence) is shown in Fig. 7.23.

The PolyJet model, due to complex logistics (being printed in another city and a separate institution), could not be delivered before the surgery, so it was assessed only after the surgery was completed. The logistics of supplying the 3D printed models to surgeons is a very important factor, not to be neglected. In the author's opinion, in modern times, it is justifiable for the hospitals to be equipped with advanced additive manufacturing devices, so that all the models can be 3D printed on site, not requiring outsourcing the process into the hands of an external institution – this makes it longer and in the case of life-saving surgery (as in the presented case of a small child) time is of the essence.

The Mixed Reality application was tested by the doctors realizing the surgery, but due to logistical reasons, it happened only post-surgery. However, the feedback was unequivocally positive. It was stated by some doctors that in the case of an appropriately created virtual model, with properly working manipulation tools and transparency, 3D printed model could be skipped entirely in the pre-surgical phase. This is, however, not a unified voice and some doctors, interviewed in the process of testing both solutions (3D print and Mixed Reality) were more leaning towards "why not both?" approach, with possible realization of two forms of modeling – physical and virtual – both having their specific and invaluable roles in preparing the surgeons for the operation. Anyhow, further tests and studies are required on a higher number of cases, to decide about the full scope of scenarios of usage.

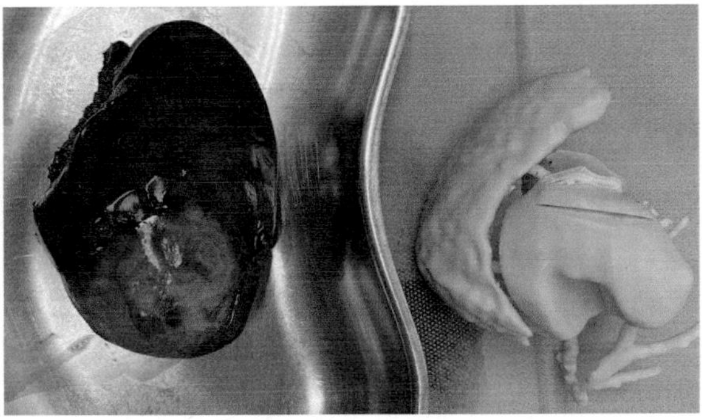

Figure 7.23 The hepatoblastoma (left) and the SLA 3D printed model (right), photo courtesy of Dr Krzysztof Grandys.

In the author's opinion, building MR applications is not difficult and can be pursued even by persons with low experience with VR programming – as proven in previous studies (Górski et al. 2024g). As such, it is fully imaginable for a hospital to employ a biomedical engineer, whose responsibility would be to design, 3D print and create Mixed Reality applications for subsequent surgeries that would benefit from this approach – and that is a most likely scenario of development of hospital applications of these modern technologies.

7.6 Case study #3 – supplies for eye socket surgery

7.6.1 Basic concepts and scheme of conduct

This chapter presents a generalized overview of a case study of an eye socket surgery supported with 3D printed models. The descriptions and materials presented in this chapter are based on previous cases, described in publications (Żukowska et al. 2024a), as well as in Master's Theses realized under the author's supervision (Andrzejewska 2020, Pasierbek 2021, Boberska 2023). This research was realized in close cooperation with Clinic of Maxillofacial Surgery of Hospital of Poznan University of Medical Sciences, represented by Doctor Maciej Okła. The main aim of all these studies was to evaluate usefulness of 3D printed supplies for pre- and mid-surgery uses in the eye socket surgery. The secondary goal was to evaluate and compare various 3D printing technologies, to select processes which are acceptable for clinical use by the doctors themselves. The tertiary goal was to assess potential of automation of design of such models.

The author was responsible for conceptualization of the whole sequence of the studies, as well as their supervision and coordination of cooperation between engineers (who were mostly students of the Master's course on Biomedical Engineering at Poznan University of Technology). One of the overarching aims of the study was also to evaluate if the current software and printing technologies are advanced enough to be helpful even for less skilled and experienced users (namely, students realizing their first clinical cases of surgery supplies).

As the general clinical case, a damaged eye socket was selected. Over the course of the studies, several patient cases were considered. In all considered cases, the eye sockets were somehow injured in the past, usually resulting from a mechanical damage (e.g., a blow to the face). The surgeries, for which the 3D printed anatomical supplies were prepared, were reconstructive surgeries, mostly aimed at restoring facial looks and proper eyesight (reduced due to the lack of eye symmetry), often months or years after the original injury. Considering the surgeries, a necessity of having two basic models was assumed:

1. *"pre" state* – a current state of the patient's eye socket area, with a healed injury, but with some functional or aesthetic defects,
2. *"post" state* – a desired state of the same area – how the eye socket looked like before the surgery and how its geometry would be supposed to be shaped afterwards.

All the patients were examined using a medical grade CT scanner. After obtaining the anatomical data, the engineers were tasked with obtaining either the "pre" or both the "pre" and "post" models, first using segmentation techniques, then – possibly – parametric CAD design. Various approaches were tested, on purpose, by the authors of the subsequent cases. After the design processes were finished, all the designs were subjected to 3D printing, using at least 2 different technologies. In total, 5 different processes were used and evaluated: FDM (extrusion), SLA, DLP, SLS and PolyJet. After obtaining the models, the doctors were asked for their usefulness, filling a survey and discussing their observations. In one of the three realized cases, clinical validation was realized, as the model was managed to be prepared before the operation and it was used before and during the process, having significant impact on the treatment and healing, as described in (Żukowska et al. 2024a).

The three main stages of the study, sequentially assigned to three biomedical engineers realizing the practical part (design, 3D printing program preparation, surveying the doctors) were as following:

1. *Manual design and low-cost 3D printing evaluation* – here, an already treated case of an eye socket surgery was considered, to create a methodology of working with such models. Medical imaging was subjected to segmentation and the models were designed using mesh modeling. 3D printing utilized low-cost FDM and UV photopolymerization processes. This was realized as a Master's Thesis by Paula Andrzejewska, under the author's supervision (Andrzejewska 2020).
2. *Automated design (attempt) and powder 3D printing evaluation* – at this stage, on the basis of previously developed methodology, an attempt was made to create an automated solution for creating the "post" eye socket model. Designed models were subjected to printing using different technologies, including the SLS technology for comparison and evaluation, without clinical use. This was realized as a Master's Thesis by Anna Pasierbek, again under author's full supervision (Pasierbek 2021).

3. *Clinical validation* – manual rapid design of a case of a patient requiring surgery, preparing models and programs and realizing 3D printing directly in a hospital, with later evaluation and comparison between low-cost SLA/DLP and expensive PolyJet device. This was realized as an Engineering Thesis by Daria Boberska, under the author's full supervision, resulting in a subsequent publication (Boberska 2023, Żukowska et al. 2024a).

In the next subchapters, each stage will be presented in greater detail, based on the notions already presented in Chapters 3–5 regarding medical imaging, design and additive manufacturing technologies, drawing on cases described in the theses and referenced previous papers.

7.6.2 Design of anatomical structures

The first stage of the general study, realized by Eng. Paula Andrzejewska under supervision of the author (Andrzejewska 2020) involved two cases: a 44-year-old male who suffered a fracture of the left orbital floor and maxillary sinus after an assault, and a 38-year-old male who required a secondary surgical procedure due to a displaced titanium mesh after right orbital floor reconstruction. Both cases required symmetrical mirror images of the healthy orbital floor.

The CT scans were obtained from the University Medical Hospital in Poznań. The data was recorded in DICOM format. The process involved transforming DICOM files into STL models, editing the resulting meshes, and finalizing the geometry in CAD software. A short software analysis was realized to select the best software tools for each stage. The following programs were selected:

1. *DICOM processing* – comparing MIMICS, InVesalius and 3D Slicer, the last one was selected, due to large set of tools, free license, and automation capabilities, despite unintuitive interface.
2. *Triangle mesh processing* – comparing GOM Inspect, MeshMixer and MeshLab, the MeshMixer was selected as the only software with both intuitive interface and automation capabilities.
3. *Computer Aided Design* – comparing SolidWorks, Inventor and CATIA, the last one was selected, despite unintuitive interface, due to a powerful set of tools for modeling and automation capabilities.

In the first part of the work, the full methodology of working with the 3D Slicer was laid out. The applied segmentation method was crucial for accuracy and efficiency, influenced by DICOM data quality and anatomical complexity. Manual segmentation was assessed to be time-consuming and error-prone, especially for complex structures like the orbital walls, so automatic segmentation was preferred using thresholding. The pixel brightness range is manually defined or based on the Hounsfield scale, verified across different planes.

Segmentation produces 2D sections in three planes, serving as the basis for a 3D voxel model created. The voxel model, with pixel-sized dimensions in X and Y planes and slice thickness in the Z direction, was saved and exported to STL. In both the patient cases, this was done successfully, although in case 1, the automated method did not work properly and manual adjustments were required.

The raw STL model generated in 3D Slicer required finishing work. This process was conducted using MeshMixer. The models were cleaned – unconnected objects were eliminated with an appropriate tool, then the model was smoothed out to improve surface quality, in a way not to lose its crucial basic shape. Afterwards, the mesh errors were detected and cleaned, using another tool available in the software. To minimize processing time, only the skull side with the intact orbital wall for mirror imaging was cleaned. The finished model was exported in the STL format.

After the model was prepared, the next step was to set it up as a preoperative aid using CATIA V5 software. The software allows importing STL files as either a part (CATPart) or a component of an assembly (CATProduct). Importing as a part keeps the triangular mesh intact, enabling surface editing while preserving quality, which is ideal for complex geometry.

The work in the CAD software involved mainly creating a mirror image of the healthy bone structure. The first step was to remove the symmetric half of the damaged orbital floor. Then, the mirror image of the remaining half was created and merged with the other half, filling all the resulting discontinuities automatically using tools available in CATIA modules of Generative Shape Design and Digitized Shape Editor.

Figure 7.24 shows the course of the process in the first stage. The processed models – the "post" state of the eye socket area of patients' skulls – were saved as an STL file, ready for further processing in 3D printing software. Beforehand, the models were trimmed in order to leave only the eye socket area, for shorter print, suitable for lower size of working chamber of the DLP technology selected as one of the manufacturing methods.

The second stage of the overall study, realized by Eng. Anna Pasierbek under supervision of the author (Pasierbek 2021), was focused mostly on checking on design automation capabilities. The model case was based upon DICOM data of a 47-year old patient, already after the eye socket surgery, selected out of four test cases as the data was of best quality. As in the initial stage, 3D Slicer software was used for image segmentation. However, the mesh processing stage was realized purely in the CATIA software and being assessed as less intuitive and more time consuming than in MeshMixer, but still acceptable from the point of view of obtained results.

A simple intelligent model was then built and tested in the CATIA software, enabling automated realization of the symmetry operation, leading to obtaining of a "post" model in a more automated fashion than the previous manual attempts. However, creating the model was troublesome and labor consuming and there are many requirements that the imported mesh must fulfill (mostly related to proper positioning in the coordinate system), so it was assessed as not worth an effort, unless the full automation would be included, relieving the engineer from all the work that leads to the moment of symmetry operation in CAD. A proper model was obtained and prepared for 3D printing, as the second main goal of this second part of the study was also to compare additive manufacturing processes and gain the doctor's opinions on them, as described in the next chapter.

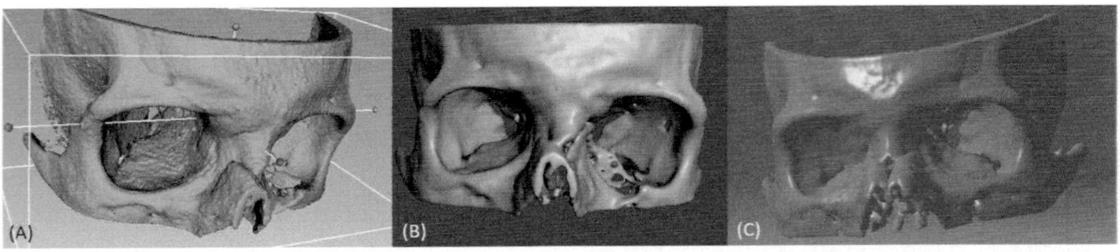

Figure 7.24 Eye socket study, part 1, design: (A) segmentation result; (B) mesh processing result; (C) CAD design result, screens courtesy of Paula Andrzejewska.

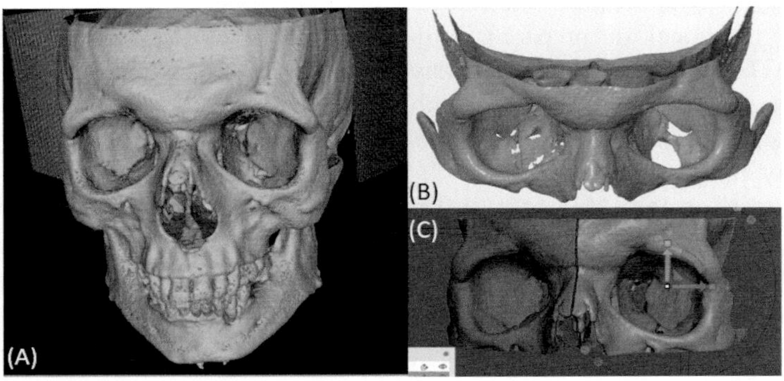

Figure 7.25 Eye socket study, part 3, design: (A) segmentation result; (B) mesh processing; (C) symmetry operation, screens courtesy of Daria Boberska.

The third part of the study, realized mostly by Daria Boberska under supervision of the author (Boberska 2023) focused on clinical implementation of a real patient case and mid-surgery use of the 3D printed "pre" and "post" models. Here, a case of 46-year-old patient was considered. The patient was assaulted and the eye socket was damaged, leading to lowering of one eye and double sight, some time after the event. A time window of 3 weeks was given to the engineering team (the student and the author, along with 3D printing technicians and engineers) to prepare the models. As the intuitiveness and availability of the tools were the key points here (not focusing on automation capabilities), InVesalius and GOM Inspect software were selected as the main tools for the job, with MeshMixer being used as a support in the symmetry operation, leading to obtaining the "post" model. The course of work is roughly presented in Fig. 7.25.

7.6.3 3D printing and technology evaluation

In the first part of the studies, FDM technology was compared to the DLP process. Low-cost FDM machine – FlashForge Creator Pro – along with a popular PLA material were used. In terms of the DLP, XYZ Nobel SuperFine machine was used, likewise a low-cost device, with a standard resin. In the second part of the study, the model was built using SLS printer Sintratec Kit, with PA12 medical grade (certified) material. In the third part of the study, knowing already that FDM and SLS are not recommended, low-cost SLA was used for clinical implementation. The hospital was

Figure 7.26 Eye socket study, manufacturing: (A) FDM; (B) SLS; (C) DLP; (D) PolyJet.

equipped with its own 3D printer at that time (as a result of cooperation and know-how transfer during the first two stages of cooperation), it was Phrozen Sonic Mighty 4K printer. Finally, for wider evaluation, PolyJet printer was used after acquiring it by the laboratory where the author works, and a final comparison was made.

Figure 7.26 presents various models from various stages of study, realized using different manufacturing technologies (photos were taken either by the author or by the authors of the earlier mentioned theses (Andrzejewska 2020, Pasierbek 2021)).

The models produced using FDM, DLP and SLS printing technologies were compared by their economic aspects. They differ in quality, time, and cost. DLP printing with a layer thickness of 0.05 mm took approximately 7 hours, FDM printing with a 0.2 mm layer thickness took a bit above 4 hours and the SLS printing of a similar model took approximately 2,5 hours. The costs were differing greatly, with the FDM being the cheapest - ~ US$15 per single print (covering all the costs related to material, machine, energy and labor), with DLP print costing ~ US$50 and SLS ~ US$25, the latter mostly due to a short building time. The PolyJet technology was not assessed in this comparison, but it is definitely the most expensive of the compared processes.

Each model required cleaning of leftover material or removal of support structures. For DLP prints, the process involved soaking the model in alcohol to remove resin, followed by mechanical support removal, which took approximately 30 minutes. FDM prints had fewer support structures, allowing mechanical removal in approximately 10 minutes. PolyJet requires cleaning of excess material by a waterjet stream, which takes approximately 15–20 minutes. Of the compared processes, FDM is the least problematic and requires only manual tools, with the SLA/DLP being the most troublesome and "dirty", requiring manual dealing with potentially toxic substances (fumes of unbound resin, mixed with alcohol used for cleaning).

The validation process of the obtained models was based on comparing the quality and completion time of models produced using different additive manufacturing technologies, as well as the costs. The models were evaluated sequentially, with comparable results about five different manufacturing technologies obtained at the end of the presented studies.

Clinical evaluation by a maxillofacial surgeon, Dr. Maciej Okła (surveyed by all three students separately in the course of their respective theses), assessed the models' aesthetics, quality, and functionality. A survey was realized, with several

Table 7.4 Survey study – evaluation of 3D printed eye socket models (summary of all three stages).

Question		Score/answer
Is the anatomical structure of the lower orbital wall properly represented?		5
Surface quality	FDM	2
	DLP	5
	SLS	4
	PolyJet	5
General quality of obtained models (on average)		4.5
Potential usefulness in surgery (average)		4.5
Technology ranking from best to worst		PolyJet > DLP > SLS > FDM

questions scored from 1 to 5, covered aspects like anatomical accuracy and surface quality. The questions and results of the survey are presented in Table 7.4.

The conclusions of this stage of study highlighted that modern medical imaging and modeling technologies alongside available additive manufacturing technologies can accurately recreate anatomical structures. However, quality is somehow dependent on imaging data, software, and designer experience and not all additive manufacturing technologies are suitable for delicate bone structures in the eye socket area – FDM was generally rejected as not particularly useful due to relatively poor quality and lack of proper representation of thin bones due to high layer thickness.

7.6.4 Clinical use, results, and conclusions

The clinical validation of the manufactured eye socket models was positive. The models were successfully printed, processed, and sterilized in the hospital and brought into the operating room. The "pre" model was used for shaping an autografted filling, while the "post" model was used for shaping of a titanium mesh. Both parts were then implanted into the patient's eye socket, saving time of surgeons and maximizing the chances of successful recovery, as well as improved functional and aesthetic effects. Figure 7.27 presents these models and their use in the operating room, as shown in a previous publication (Żukowska et al. 2024a).

The clinical validation, after a long buildup of preliminary studies, allowed to obtain a successful result in surgical practice and have become a new methodology of work. This methodology aims to accommodate the main clinical requirements, even for unqualified personnel. CT segmentation and mesh modeling seem relatively easy and straightforward with the available freeware tools, achievable to realize properly and obtain clinical usability even by the students and young doctors, what has been further confirmed in ongoing work – even engineering students, without full knowledge and experience, are able to greatly help doctors in preparation for the surgery with their skills in design, mesh modeling and 3D printers operation. However, the process can be relatively time-consuming, requiring up to 3 working days of design, which can be a drawback when time is seriously limited. To this time, manufacturing time must be added, taking additional several hours, depending on the

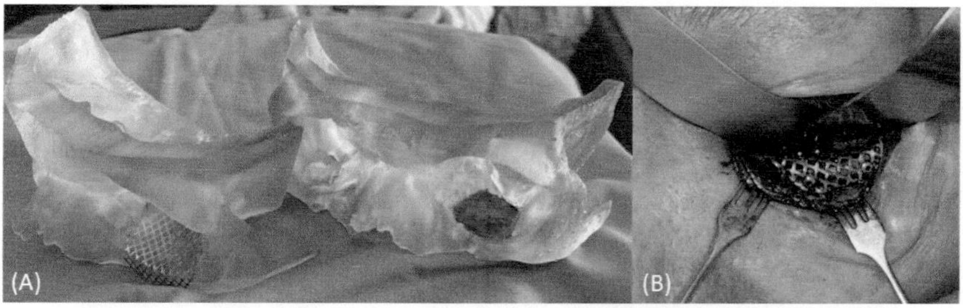

Figure 7.27 Eye socket study, part 3, clinical evaluation: (A) the models in the operating room, (B) patient before and after the surgery, photos courtesy of Dr Maciej Okła.

technology. Of course, quality input data is a must, as wrongly acquired DICOMs, without clear view of the tissues and with low resolution, will not only greatly increase working time, but simply disallow to produce a useful model.

Automation of design (and use of advanced CAD tools, in general) was ditched as a concept at this stage, as even relatively unskilled students are being able to proceed with the mesh-based manual work methodology and bring in models that are positively assessed by the doctors. Automating this case, as opposed to previously described orthopedic devices, is difficult to achieve and not really desirable in this class of problem, would require sophisticated mesh analysis and data extraction techniques to work properly on intricate structures of human maxillofacial area around the eye sockets.

Consequently, a manual line of work was recommended, alongside of using purely UV resin technology for the eye sockets (DLP or PolyJet, if available). It was observed that logistics of preparation of pre- and mid-surgery supplies work much better (in terms of shorter time of obtaining a ready model) in the case where a 3D printer is located directly in the hospital, in vicinity to operating rooms.

As a result of this particular study, the UV printing technology of eye sockets and the presented surgical methods were adopted by the doctors working in the hospital. A combined effort of the team of the author, alongside the medical team, as well as the younger biomedical engineering adepts (Eng. Andrzejewska, Pasierbek and Boberska, realizing three separate theses on the matter, supervised by the author) helped build a know-how that is now, at this point, efficiently used in surgical practice. A 3D printing laboratory was built in the hospital (Fig. 7.28) and the methodology was learned by the doctors, now realizing the eye socket prints by themselves on a daily basis. This is a proof that 3D printing can be successfully used in daily surgical practice, not only as a "wow" factor, for showcasing or educating, but for actual work, helping to gain better results in most patients, tailoring the procedures to their needs.

The eye socket studies have become a successful example of low-cost hospital bed printing in maxillofacial surgery – the cooperation between engineers (represented by the author and his team) and doctors from the clinic are continuing, also focusing on many other cases, such as mandible implantation surgeries – the observations made during the three-stage study helped to develop a generalized methodology

Figure 7.28 The 3D printing laboratory built in a hospital, for daily printing of anatomical eye socket models.

and know-how, enabling more frequent and robust production of supplies useful in operating room conditions.

7.7 Discussion, recommendations, and conclusions

This chapter has thoroughly discussed the evolution, methodologies, and applications of these technologies, particularly focusing on their use in creating anatomically adjusted models for surgical interventions. The case studies presented in this chapter demonstrate the practical benefits of using 3D printed models, from pre-surgical planning to mid-surgery applications. These models not only aid in surgical precision but also improve patient outcomes by reducing surgery time and enhancing the accuracy of procedures. Here, we will sum it up with key points and insights.

First of all, the necessity for a personalized approach in modern surgery must be emphasized. The concept of "personalized medicine" has been mostly used in terms of pharmaceutical treatments suited to a given person (Chan et al. 2011), but it should be also applied for anatomically adjusted surgical treatments. This personalization is driven by the variability in human anatomy and pathology, which demands tailored solutions for effective surgical outcomes.

In accordance with the author's experience, some surgeries would be very difficult to perform without 3D printed models. Complex procedures such as removals of tumors that are penetrating multiple body parts in sensitive areas (such as the oral cavity) or reconstructive surgeries in these areas pose significant challenges, often requiring interdisciplinary teams of doctors, days spent planning and multiple hours of work in operating rooms. These surgeries benefit immensely from pre-operative planning and the use of anatomically accurate models.

So far, these models have primarily been used for hard tissues due to the relative ease of imaging and mimicking their properties. However, the development of multi-material printing (PolyJet) and techniques like vacuum casting has expanded the possibilities to soft tissue modeling as well. Utilizing possibilities and diversity of modern polymeric materials and their properties, it is possible to construct very complex models. Methodologies start to be developed and standardized and 3D printers in hospitals are no longer uncommon, as results of the author's own observations. As such, new possibilities and paths in personalized surgeries open, allowing comfortable realization of surgical tasks that would be very difficult or near impossible to perform in a given way without the 3D printed templates.

However, much remains to be done in terms of technological progress. Further development in medical imaging technologies, particularly in improving the resolution and differentiation of tissues, is essential. Techniques like multi-phase CT and advanced MRI should be explored and refined to provide more detailed anatomical data. Another important scope of potential research and progress should focus on the application of artificial intelligence for processing medical images. When done properly, it could significantly reduce the time and effort required for segmentation and modeling. AI algorithms can assist in accurately distinguishing between various tissue types and creating precise 3D models. Nowadays they are used mostly for bone structures or healthy tissues (Li et al. 2023), and the research should be focused on teaching AI to recognize pathological structures, such as cancerous tumors or injured tissues, to be able to generate their 3D models with low effort and in short time.

Considering the manufacturing of the anatomical models, continued research into materials that closely mimic the mechanical properties of human tissues is crucial. Advances in bioprinting and tissue engineering could provide new materials that offer better performance in simulated surgeries and training scenarios. Ideally, it should be possible to print both implantable and temporarily used anatomical models in very similar processes, directly in the hospital, personalizing treatment for each surgical patient.

In the terms of more widespread use and ensuring of fulfillment of all ethical requirements, it is important to establish clear guidelines and create procedures for obtaining necessary approvals for the clinical use of 3D printed models and XR applications. This includes ensuring that these tools meet safety and ethical standards for medical use. It may prove to be difficult, as each patient case may require use of different materials, manufacturing processes and visualization techniques, as well as various software tools. That is why work should be commenced as soon as possible, to make sure that the doctors can completely legally and without any concerns use the 3D printed models in their operation rooms, for the benefits of patients and themselves.

What is also important to notice, collaboration between engineers, radiologists, and surgeons is vital for the successful implementation of these technologies. Joint training sessions and workshops can facilitate better communication and understanding between these professionals. Also, the use of anatomical models and XR technologies in medical education should be expanded. Creating a repository of physical and digital models from previous surgeries can serve as a valuable resource for training medical students and young professionals.

To sum up, the integration of 3D printing and methods of digital design in surgery marks a paradigm shift in the way complex medical procedures are planned and executed. By providing personalized anatomical models, these technologies enhance the precision, safety, and effectiveness of surgeries. Despite the current limitations, ongoing advancements in imaging, materials, and manufacturing processes hold the promise of overcoming these challenges, making personalized surgical care more accessible and reliable.

The future of personalized surgery lies in the continuous evolution of medical imaging, AI-driven design, 3D printing and XR technologies. By addressing the current limitations and embracing interdisciplinary collaboration, the medical field can significantly advance towards more effective and personalized patient care, potentially saving and prolonging lives on a level unimaginable as of today.

CHAPTER 8

Conclusions

The main aim of this book was to draw readers' attention and increase their level of knowledge in the breaking field of science and engineering, which is use of computer aided design and 3D printing for creation of anatomically shaped medical devices for use by patients and doctors in various stages of treatment. Due to recent emerging and rapid development in digital design techniques, reverse engineering, automation of design, additive manufacturing, and extended reality, the advanced, yet accessible technology can serve people in healthcare, helping in treatment of certain conditions, improving vital processes and generally easily achieving previously unattainable solutions to long-known problems.

Chapter 1 of the book was an introduction to the topic, written primarily to lay out the common, ongoing themes of the book and prepare the readers to what they might expect in the further chapters. Chapter 2 provided a comprehensive overview of basic concepts of computer aided design, biomedical engineering, additive manufacturing, and individualization in production of medical devices. From there, the three subsequent chapters with detailed technology overviews followed. Chapter 3 focused on gathering and processing anatomical data by various sensing and imaging technologies, including medical imaging and 3D scanning. In Chapter 4, basics of CAD were explained, starting with simple concepts, and going through more advanced problems, such as automation of design of medical devices and implementing automated design systems. Then, Chapter 5 provided a comprehensive overview of additive manufacturing technologies used in the creation of anatomically adjusted medical devices, on which this book is focused.

Following the initial 5 Chapters, two subsequent ones focused on the particular applications of techniques presented in the first part of the book, based solely on the author's scientific experiences in the presented matters. Chapter 6 presents one side of the metaphorical coin – devices based on measurements of patients' external body parts, i.e., limbs. Methodology of design of orthoses, prostheses and their additive manufacturing is presented in greater detail, also showing the author's experiences with automating data processing and design of some of these devices (AutoMedPrint system), as well as their implementation in practice with certain patients. A number of real-life case studies are shown for various product types, proving that it is perfectly possible to design and produce inexpensive, accessible orthopedic and prosthetic devices for juvenile and adult patients by a fraction of price of commercial, professional solutions.

Chapter 7 presents models that are based on measurements of internal geometry of a given patient, used as aids in surgical processes, helping mostly doctors to understand a given case, prepare it and realize it in shorter time with better outcomes on patients' health. These models are created on the basis of medical imaging and can be used before a surgery or in the middle of it, to maximize benefits for both the doctor and patient. General concepts of these devices are presented in the first part of the chapter, along with classification and schemes of methodologies based on previous overviews. The second part of the chapter shows specific examples of usage in various scenarios, highlighting cooperation of the author and his team with various healthcare facilities in Poland, realizing various types of surgeries with very different demands. This chapter also introduces a concept of virtual and mixed reality in surgical preparation and presents one case where such applications took place, with a beneficial effect.

The core concepts and basic ideas presented in this book have been known for the scientific and engineering community. However, new, and totally unique insights are presented on how to utilize and mix them effectively in medical applications, to bring a real change and benefit for doctors and patients. Simple concepts of 3D scanning, mesh processing and material extrusion or hardening, combined properly, may evolve into powerful tools enabling reaching new peaks in personalized healthcare – and that is the point that the book is trying to make. A lot of attention is also paid to potential of automation of medical devices based on individual measurements of patients – not an easy task, but demonstrated feasible and perfectly practical by the author and his team on a number of cases, some of which are shown directly in the book.

It is very important to notice, that all the great clinical results and potentially life-changing solutions presented in this book would not be possible to implement if the work was realized only at a university, among engineers themselves. Medical inventions, new devices and therapy methods and breakthrough processes do not come solely from the engineer's perspective – they must be a result of a direct need from doctors and patients. The author and the team assumed a strategy of working with the medical environment from day 0 – notifying possible bottlenecks in personalized treatment methods and proposing potential advancements, using new technologies. It is very important also to listen to the patients and their feedback and try to give them what they want, even if it seems very difficult from the doctor's perspective. This interdisciplinary cooperation would seem obvious, but from the author's experience – it is not, while keeping an open mind on both (or all three, counting patients in) sides of the discourse can bring in truly remarkable results.

Of course, not all is perfect and there are still many challenges to overcome. First of all, technical validation of certain concepts and their testing on a small batch of patients and their devices does not yet make the technology ready for mass adoption in medicine. Clinical studies should be realized, lasting appropriately long, to evaluate long-term consequences of use of 3D printed anatomically shaped devices in general healthcare. Logistical, formal, and economic constraints are still preventing hospitals, doctors, and patients from widespread use of digitally designed and 3D printed prostheses, orthoses, implants, and surgical aids, individualized for needs of a single person. Especially formal constraints, ethical and safety concerns related to risks of potential harmful influence of wrongly designed and manufactured devices must be taken into account. These risks should be minimized by rigorous quality enforcement and it can be done using existing tools, but it requires building an entirely new set of rules and standards, governing Quality Management Systems (QMSs) focused on manufacturing processes realized not in production lines, but in healthcare facilities. In other words, despite initial success and very promising results of implementations of inexpensive, accessible, and easy to work on technology, the road is still long until the patients will be able to fully use its potential.

However, the author's hopes lie in general human ingenuity and observable rapid development towards progress, especially in personalized healthcare. It is hoped that this book will be useful to many specialists – engineers wanting to see how some solutions might be developed and applied, doctors interested in how modern digital technologies can add up to the scope of their portfolio of treatment methods, as well as students learning new ways of conducting work with medical devices.

From the author's personal perspective, medical engineering, science and working with the patients form a marvelous, unique experience, as consequences of scientific and engineering efforts are directly translated into improved health effect of specific patients, as well as satisfaction of the doctors realizing the treatment. Smiles of the child patients obtaining and using their first bicycle prosthetics will remain forever as a magical memory, seeing several years of hard work and intense scientific experiments pay off in a way that makes another person's world more bright and happy. Proper engineering and science can transform lives and help to save the world – and we should use this fact as a motivation for even better and harder work.

Literature

Abdalsadah, F. H., Hasan, F., Murtaza, Q. and Khan, A. A. (2021). Design and manufacture of a custom ankle–foot orthoses using traditional manufacturing and fused deposition modeling. *Progress in Additive Manufacturing*, 6(3): 555–570.

Abeykoon, C., Sri-Amphorn, P. and Fernando, A. (2020). Optimization of fused deposition modeling parameters for improved PLA and ABS 3D printed structures. *International Journal of Lightweight Materials and Manufacture*, 3(3): 284–297.

Aggenbacht, F. (2004). *Development and Evaluation of a Vacuum Casting Process* (Doctoral dissertation, Bloemfontein: Central University of Technology, Free State).

Ahmad, A. N. A., Ahmad, M. F., Hamid, N. A., Chuan, L. T., Rahim, M. K. I. A., Nawanir, G. et al. (2022, November). Innovative in 3D printing: Design and development of a user-friendly wheelchair. *In*: *AIP Conference Proceedings* (Vol. 2644, No. 1). AIP Publishing.

Ahmadi, C. (2023). Software in a Medical Device vs. Software as a Medical Device: Understanding the Differences. Intertek. Online: https://www.intertek.com/blog/2023/2023-12-12-med-devices/

Aiello, M., Esposito, G., Pagliari, G., Borrelli, P., Brancato, V. and Salvatore, M. (2021). How does DICOM support big data management? Investigating its use in medical imaging community. *Insights into Imaging*, 12(1): 164.

Akhnoukh, A. K. (2021). Advantages of contour crafting in construction applications. *Recent Patents on Engineering*, 15(3): 294–300.

Al Hasan, S. and Hoque, M. Z. (2008). Lower limb orthoses: A review. *Journal of Chittagong Medical College Teachers' Association*, 19(1): 33–36.

Alanazi, K. K., Wood, D., Shepherd, J., Stokes, C. W. and Asencio, I. O. (2024). Assessing the suitability of fused deposition modeling to produce acrylic removable denture bases. *Clinical and Experimental Dental Research*, 10(3): e880.

Alcaide-Marzal, J., Diego-Más, J. A., Asensio-Cuesta, S. and Piqueras-Fiszman, B. (2013). An exploratory study on the use of digital sculpting in conceptual product design. *Design Studies*, 34(2): 264–284.

Ali, M. A., Rajabi, M. and Sali, S. S. (2020). Additive manufacturing potential for medical devices and technology. *Current Opinion in Chemical Engineering*, 28: 127–133.

Ali, S., Ravi, P., Williams, R., DiPaola, D. and Breazeal, C. (2024, March). Constructing dreams using generative AI. pp. 23268–23275. *In*: *Proceedings of the AAAI Conference on Artificial Intelligence* (Vol. 38, No. 21).

Alshipli, M., Marashdeh, M. and Oglat, A. A. (2022). Manual image segmentation using dicom format to examine and analysis brain magnetic resonance imaging (MRI) tumor. *Journal of Positive School Psychology*, 6156–6162.

Alsleem, H. and Davidson, R. (2013). Factors affecting contrast-detail performance in computed tomography: A review. *Journal of Medical Imaging and Radiation Sciences*, 44(2): 62–70.

Altwal, J., Wilson, C. H. and Griffon, D. J. (2022). Applications of 3-dimensional printing in small-animal surgery: a review of current practices. *Veterinary Surgery*, 51(1): 34–51.

Alvares, A. J. and Gudwin, R. (2019). Integrated system of predictive maintenance and operation of eletronorte based on expert system. *IEEE Latin America Transactions*, 17(01): 155–166.

Alvarez, C. K. L., Lagos, C. R. F. and Aizpun, M. (2016). Investigating the influence of infill percentage on the mechanical properties of fused deposition modelled ABS parts. *Ingeniería e Investigación*, 36(3): 110–116.

Andrzejewska, P. (2020). Methodology of development of additively manufactured pre-surgery supplies in maxillofacial surgery. Master's thesis (supervision: Górski, F.), Poznan University of Technology.

Animated Physics. (2011). Positron emission tomography (PET scans) Nuclear medicine imaging. Online: http://www.animatedphysics.com/Live/pet_scans.htm.

Aphirakmethawong, J., Yang, E. and Mehnen, J. (2022, September). An overview of artificial intelligence in product design for smart manufacturing. pp. 1–6. *In*: *2022 27th International Conference on Automation and Computing (ICAC)*. IEEE.

Arbabtafti, M., Moghaddam, M., Nahvi, A., Mahvash, M. and Rahimi, A. (2008, October). Haptic and visual rendering of virtual bone surgery: A physically realistic voxel-based approach. pp. 30–35. *In*: *2008 IEEE International Workshop on Haptic Audio visual Environments and Games*. IEEE.

Aronson, J. K., Heneghan, C. and Ferner, R. E. (2020). Medical devices: definition, classification, and regulatory implications. *Drug Safety*, 43(2): 83–93.

Arrighi, P. A. and Mougenot, C. (2019). Towards user empowerment in product design: a mixed reality tool for interactive virtual prototyping. *Journal of Intelligent Manufacturing*, 30(2): 743–754.

Aseni, P., Henry, S., Grande, A. M., Fiore, A. and Scalea, T. M. (2023). Lifesaving and emergency surgical procedures in trauma patients. pp. 901–945. *In*: *The High-risk Surgical Patient*. Cham: Springer International Publishing.

Attaran, M. (2017). Additive manufacturing: the most promising technology to alter the supply chain and logistics. *Journal of Service Science and Management*, 10(03): 189.

AutoMedPrint. (2021). AutoMedPrint: Automatycznie projektujemy i drukujemy w 3D ortezy i protezy. Online: https://automedprint.put.poznan.pl/.

Bae, K. J., Jang, H. S., Gong, H. S., Kang, Y., Kim, J. and Baek, G. H. (2020). Prevalence and distribution of MRI abnormalities in the articular cartilage and supporting ligaments in patients with early clinical stage first carpometacarpal joint osteoarthritis. *Skeletal Radiology*, 49: 1089–1097.

Baechle-Clayton, M., Loos, E., Taheri, M. and Taheri, H. (2022). Failures and flaws in fused deposition modeling (FDM) additively manufactured polymers and composites. *Journal of Composites Science*, 6(7): 202.

Bakker, C. J. G., Bos, C. and Weinmann, H. J. (2001). Passive tracking of catheters and guidewires by contrast-enhanced MR fluoroscopy. *Magnetic Resonance in Medicine: An Official Journal of the International Society for Magnetic Resonance in Medicine*, 45(1): 17–23.

Ban, T. K., Lee, J. L. J., Volha, R., Zamri, M. S. A. B., Azian, N. H. B., Ng, P. L. et al. (2024). Investigating the Strategies employed by Mcdonald's to establish sustainable customer relationships. *International Journal of Tourism and Hospitality in Asia Pasific (IJTHAP)*, 7(2): 76–89.

Banaszewski, J., Pabiszczak, M., Pastusiak, T., Buczkowska, A., Kuczko, W., Wichniarek, R. et al. (2018). 3D printed models in mandibular reconstruction with bony free flaps. *Journal of Materials Science: Materials in Medicine*, 29: 1–6.

Bandutunga, C. P. (2020). *Digitally Enhanced Interferometry for Precision Metrology*. PhD Thesis, Australian National University.

Banga, H. K., Kumar, P. and Kumar, H. (2021). Utilization of additive manufacturing in orthotics and prosthetic devices development. *In*: *IOP Conference Series: Materials Science and Engineering* (Vol. 1033, No. 1, p. 012083). IOP Publishing.

Bar-Cohen, Y. (2006). Biomimetics—using nature to inspire human innovation. *Bioinspiration & Biomimetics*, 1(1): P1.

Barragan De Los Rios, G. A., Ferreira, R., Mariani, F. E., da Silva, E. J. and Coelho, R. T. (2023). Study of the surface roughness of a remanufactured bimetallic AISI 1045 and 316L SS part obtained by hybrid manufacturing (DED/HSM). *The International Journal of Advanced Manufacturing Technology*, 124(9): 3185–3199.

Barreto, Marco, A., Jorge Perez-Gonzalez, Hugh M. Herr and Joel C. Huegel. 2022. ARACAM: A RGB-D multi-view photogrammetry system for lower limb 3D reconstruction applications. *Sensors,* 22(7): 2443.

Barrios-Muriel, J., Romero-Sánchez, F., Alonso-Sánchez, F. J. and Salgado, D. R. (2020). Advances in orthotic and prosthetic manufacturing: a technology review. *Materials*, 13(2): 295.

Bartoletti-Stella, A., Gatta, V., Mariani, G. A., Gobbi, P., Falconi, M., Manzoli, L. et al. (2021). Three-dimensional virtual anatomy as a new approach for medical student's learning. *International Journal of Environmental Research and Public Health*, 18(24): 13247.

Bauld, T. J. (1991). The definition of a clinical engineer. *Journal of Clinical Engineering*, 16(5): 403–406.

Beattie, N., Bock, N., Anderson, T., Edgeworth, T., Kloss, T. and Swanson, J. (2021). Effects of build orientation on mechanical properties of fused deposition modeling parts. *Journal of Materials Engineering and Performance*, 30: 5059–5065.

Bens, A., Seitz, H., Bermes, G., Emons, M., Pansky, A., Roitzheim, B. et al. (2007). Non-toxic flexible photopolymers for medical stereolithography technology. *Rapid Prototyping Journal*, 13(1): 38–47.

Bern, M. and Eppstein, D. (1995). Mesh generation and optimal triangulation. pp. 47–123. *In*: *Computing in Euclidean Geometry*.

Bernhard, J. C., Isotani, S., Matsugasumi, T., Duddalwar, V., Hung, A. J., Suer, E. et al. (2016). Personalized 3D printed model of kidney and tumor anatomy: a useful tool for patient education. *World Journal of Urology*, 34: 337–345.

Berthiaume, F., Maguire, T. J. and Yarmush, M. L. (2011). Tissue engineering and regenerative medicine: history, progress, and challenges. *Annual Review of Chemical and Biomolecular Engineering*, 2(1): 403–430.

Bevanda, M. (2017). Medical Devices Safety (Regulation (EU) 2017/745) and Liability for Damages Caused by Defective Medical Device (Directive 85/374/EEC). *Collection Papers Actualities Civ. & Com. L. & Legal Prac.*, 275.

Bhargav, A., Sanjairaj, V., Rosa, V., Feng, L. W. and Fuh YH, J. (2018). Applications of additive manufacturing in dentistry: A review. *Journal of Biomedical Materials Research Part B: Applied Biomaterials*, 106(5): 2058–2064.

Bibb, R., Eggbeer, D. and Paterson, A. (2024). *Medical Modeling: The Application of Advanced Design and Additive Manufacturing Techniques in Medicine. Chapter 4 – Physical Reproduction*. Elsevier.

Bimbraw, K., Nycz, C. J., Schueler, M. J., Zhang, Z. and Zhang, H. K. (2022, May). Prediction of metacarpophalangeal joint angles and classification of hand configurations based on ultrasound imaging of the forearm. pp. 91–97. *In*: *2022 International Conference on Robotics and Automation (ICRA)*. IEEE.

Birbara, N. S., Otton, J. M. and Pather, N. (2019). 3D modelling and printing technology to produce patient-specific 3D models. *Heart, Lung and Circulation*, 28(2): 302–313.

Black, J. (2005). *Biological Performance of Materials: Fundamentals of Biocompatibility*. CRC Press.

Blaj, M. and Oancea, G. (2021). Fused deposition modelling process: a literature review. *In*: *IOP Conference Series: Materials Science and Engineering* (Vol. 1009, No. 1, p. 012006). IOP Publishing.

Boberska, D. (2023). *Design and Additive Manufacturing of Mid-surgery Supplies in Maxillofacial Surgery*. Diploma thesis (supervision: Górski, F.), Poznan University of Technology.

Bolas, N. (2010). Basic MRI principles. *Equine MRI*, 1–37.

Boopathi, S. and Kumar, P. (2024). Advanced bioprinting processes using additive manufacturing technologies: Revolutionizing tissue engineering. 3D Printing Technologies: Digital Manufacturing. *Artificial Intelligence, Industry*, 4: 95.

Booysen, G. J. (2019). *Developing a Certification Framework to Manufacture Patient-Specific Implants Using Selective Laser Melting* (Doctoral dissertation).

Bormann, T., Schumacher, R., Müller, B., Mertmann, M. and De Wild, M. (2012). Tailoring selective laser melting process parameters for NiTi implants. *Journal of Materials Engineering and Performance*, 21: 2519–2524.

Botsch, M., Kobbelt, L., Pauly, M., Alliez, P. and Lévy, B. (2010). *Polygon Mesh Processing*. CRC Press.

Brooke, R. (2013). Stratasys vs 3D Systems: A heavyweight fight?. TCT Magazine. Online: https://www.tctmagazine.com/additive-manufacturing-3d-printing-news/stratasys-vs-3d-systems-a-heavyweight-fight/.

Bruce, M., Hannah, A., Hammond, R., Khaing, Z. Z., Tremblay-Darveau, C., Burns, P. N. et al. (2020). High-frequency nonlinear doppler contrast-enhanced ultrasound imaging of blood flow. *IEEE Transactions on Ultrasonics, Ferroelectrics, and Frequency Control*, 67(9): 1776–1784.

Buj-Corral, I., Domínguez-Fernández, A. and Durán-Llucià, R. (2019). Influence of print orientation on surface roughness in fused deposition modeling (FDM) processes. *Materials*, 12(23): 3834.

Buń, P., Górski, F. and Turkowska, K. (2018). Virtual reality training application of medical procedure. pp. 49–58. *In*: *Smart Technology: First International Conference, MTYMEX 2017, Monterrey, Mexico, May 24–26, 2017, Proceedings*. Springer International Publishing.

Buonamici, F., Furferi, R., Governi, L., Lazzeri, S., McGreevy, K. S., Servi, M. et al. (2020). A practical methodology for computer-aided design of custom 3D printable casts for wrist fractures. *The Visual Computer*, 36: 375–390.

Burdzińska, M. (2015). *Prototyping of Customized Wrist Orthesis using Additive Manufacturing Technologies*. Diploma thesis, Poznan University of Technology.

Burdzińska, M., Górski, F., Wichniarek, R., Kuczko, W. and Buń., P. (2015). Prototyping a customized wrist orthosis using additive manufacturing techniques (in Polish: Prototypowanie zindywidualizowanej ortezy nadgarstka z zastosowaniem technik wytwarzania przyrostowego). pp. 7–24. *In*: Nowoczesne technologie w projektowaniu, inżynierii i wytwarzaniu : problemy badawcze (monograph), eds.: Jachowicz, T., Klonica, M., Perfekta info, ISBN 978-83-63657-56-7.

Bush, J. (2008). Viability of virtual reality exposure therapy as a treatment alternative. *Computers in Human Behavior*, 24(3): 1032–1040.

Callieri, M., Ranzuglia, G., Dellepiane, M., Cignoni, P. and Scopigno, R. (2012). Meshlab as a complete open tool for the integration of photos and colour with high-resolution 3D geometry data. *Comput Appl Quant Methods Archaeol*, 406–16.

Camba, J. D., Contero, M. and Company, P. (2016). Parametric CAD modeling: An analysis of strategies for design reusability. *Computer-Aided Design*, 74: 18–31.

Campbell, I., Bourell, D. and Gibson, I. (2012). Additive manufacturing: rapid prototyping comes of age. *Rapid Prototyping Journal*, 18(4): 255–258.

Canciglieri, O., Rudek, M., Francesconi, T. and De Souza, T. M. (2010, July). Geometric modeling from the medical images in the CAD system to support prosthesis design. pp. 1–6. *In*: *The 40th International Conference on Computers & Indutrial Engineering*. IEEE.

Carl, A. K. and Hochmann, D. (2022). Comparison of the regulatory requirements for custom-made medical devices using 3D printing in Europe, the United States, and Australia. *Biomedical Engineering/Biomedizinische Technik*, 67(2): 61–69.

Carol, M. (2022). Why It Takes So Long to Develop a Medical Technology. Focused Ultrasound Foundation, online: https://www.fusfoundation.org/posts/the-complex-ecosystem-of-a-medical-device-startup/.

Carse, B., Bowers, R., Meadows, B. C. and Rowe, P. (2015). The immediate effects of fitting and tuning solid ankle–foot orthoses in early stroke rehabilitation. *Prosthetics and Orthotics International*, 39(6): 454–462.

Carvalho, L. R. R. (2024). 3D printed orthopedic prostheses for domestic and wild birds. *Scientific Reports*, 14(1): 7989.

Cazon, A., Aizpurua, J., Paterson, A., Bibb, R. and Campbell, R. I. (2014). Customised design and manufacture of protective face masks combining a practitioner-friendly modelling approach and low-cost devices for digitising and additive manufacturing: This paper analyses the viability of replacing conventional practice with AM method to make customized protective face masks. *Virtual and Physical Prototyping*, 9(4): 251–261.

Chan, I. S. and Ginsburg, G. S. (2011). Personalized medicine: progress and promise. *Annual Review of Genomics and Human Genetics*, 12(1): 217–244.

Chapman, C. B. and Pinfold, M. (2001). The application of a knowledge based engineering approach to the rapid design and analysis of an automotive structure. *Advances in Engineering Software*, 32(12): 903–912.

Chaturvedi, J. and Srinivas, G. (2021). MedTech Innovation using a structured Biodesign process: Barriers and Opportunities. *Karnataka Paediatric Journal*, 36(2): 106–112.

Checa, D., Saucedo-Dorantes, J. J., Osornio-Rios, R. A., Antonino-Daviu, J. A. and Bustillo, A. (2022). Virtual reality training application for the condition-based maintenance of induction motors. *Applied Sciences*, 12(1): 414.

Chen, F., Wu, J. M., Wu, H. Q., Chen, Y., Li, C. H. and Shi, Y. S. (2018a). Microstructure and mechanical properties of 3Y-TZP dental ceramics fabricated by selective laser sintering combined with cold isostatic pressing. *International Journal of Lightweight Materials and Manufacture*, 1(4): 239–245.

Chen, J., Heyse, J. and Lai, T. L. (2018b). *Medical Product Safety Evaluation: Biological Models and Statistical Methods*. Chapman and Hall/CRC.

Chen, R. K., Jin, Y. A., Wensman, J. and Shih, A. (2016). Additive manufacturing of custom orthoses and prostheses—A review. *Additive Manufacturing*, 12: 77–89.

Choo, Y. J. (2022). Orthoses for the management of upper extremity pain: a narrative review. *International Journal of Pain*, 13(1): 25–35.

Choonara, Y. E., du Toit, L. C., Kumar, P., Kondiah, P. P. and Pillay, V. (2016). 3D-printing and the effect on medical costs: a new era?. *Expert Review of Pharmacoeconomics & Outcomes Research*, 16(1): 23–32.

Christensen, J., Matzen, L. H., Væth, M., Schou, S. and Wenzel, A. (2012). Thermography as a quantitative imaging method for assessing postoperative inflammation. *Dentomaxillofacial Radiology*, 41(6): 494–499.

Cignoni, P., Callieri, M., Corsini, M., Dellepiane, M., Ganovelli, F. and Ranzuglia, G. (2008, July). Meshlab: an open-source mesh processing tool. pp. 129–136. *In*: *Eurographics Italian Chapter Conference* (Vol. 2008).

Ciobanu, O., Xu, W. and Ciobanu, G. (2013). The use of 3D scanning and rapid prototyping in medical engineering. *Fiabil. Durab*, 1: 241–247.

Cirello, A., Ingrassia, T., Marannano, G., Mirulla, A. I., Nigrelli, V., Petrucci, G. et al. (2024). A new automatic process based on generative design for CAD modeling and manufacturing of customized orthosis. *Applied Sciences*, 14(14): 6231.

Ciurana, J. (2014). Designing, prototyping and manufacturing medical devices: an overview. *International Journal of Computer Integrated Manufacturing*, 27(10): 901–918.

Clement, R. G. E., Bugler, K. E. and Oliver, C. W. (2011). Bionic prosthetic hands: A review of present technology and future aspirations. *The Surgeon*, 9(6): 336–340.

Cleveland Clinic. (2022). MRI (Magnetic Resonance Imaging). Online: https://my.clevelandclinic.org/health/diagnostics/4876-magnetic-resonance-imaging-mri.

Coletti, P., Aichner, T., Coletti, P. and Aichner, T. (2011). *Mass Customization* (pp. 23–40). Springer Berlin Heidelberg.

Collins, C. M. and Safiuddin, M. D. (2022). Lotus-leaf-inspired biomimetic coatings: different types, key properties, and applications in infrastructures. *Infrastructures*, 7(4): 46.

Comaneanu, R. M., Tarcolea, M., Vlasceanu, D. and Cotrut, M. C. (2012). Virtual 3D reconstruction, diagnosis and surgical planning with Mimics software. *International Journal of Nano and Biomaterials*, 4(1): 69–77.

Conejo, J., Dayo, A. F., Syed, A. Z. and Mupparapu, M. (2021). The digital clone: intraoral scanning, face scans and cone beam computed tomography integration for diagnosis and treatment planning. *Dental Clinics*, 65(3): 529–553.

Corallo, A., Latino, M. E., Lazoi, M., Lettera, S., Marra, M. and Verardi, S. (2013). Defining product lifecycle management: A journey across features, definitions, and concepts. *International Scholarly Research Notices*, 2013(1): 170812.

Coté, J. J., Coté, B. P. and Badura-Brack, A. S. (2022). 3D printed models in pregnancy and its utility in improving psychological constructs: a case series. *3D Printing in Medicine*, 8(1): 16.

Crump, S. S. (1991, June). Fused deposition modeling (FDM): putting rapid back into prototyping. pp. 354–357. *In*: *Proceedings of the 2nd International Conference on Rapid Prototyping*.

Cucos, M. M., Pista, I. M. and Ripanu, M. I. (2018). Product engineering design enhancing by parameterizing the 3D solid model. *In*: *MATEC Web of Conferences* (Vol. 178, p. 05011). EDP Sciences.

Culmone, C., Smit, G. and Breedveld, P. (2019). Additive manufacturing of medical instruments: A state-of-the-art review. *Additive Manufacturing*, 27: 461–473.

Culot, G., Nassimbeni, G., Orzes, G. and Sartor, M. (2020). Behind the definition of Industry 4.0: Analysis and open questions. *International Journal of Production Economics*, 226: 107617.

Cykowska, A., Danalache, M., Bonnaire, F. C., Feierabend, M. and Hofmann, U. K. (2022). Detecting early osteoarthritis through changes in biomechanical properties–a review of recent advances in indentation technologies in a clinical arthroscopic setup. *Journal of Biomechanics*, 132: 110955.

da Silva Vieira, S., Badke-Schaub, P. and Fernandes, A. A. (2010). Replacing routine: reframing design opportunities for creativity and innovation. *In*: *17th International Product Development Management Conference*.

da Silva, R. and Moreira, R. (2021). PET and SPECT in the evaluation of cardiac implantable electronic devices. *Nuclear Cardiology: Basic and Advanced Concepts in Clinical Practice*, 619–674.

Damiao, J., Calianese, N., Cartwright, D., Cherian, C., Lee, E. and Mucek, D. (2023). Comparison of 3D printing and traditional hand orthosis fabrication. *Global Journal of Health Science*, 15(2): 42.

Davies, S. (2020). Invention, optimism & realism: The back story of SLS 3D printing with Dr Joe Beaman. TCT Magazine. Online: https://www.tctmagazine.com/additive-manufacturing-3d-printing-news/back-story-sls-3d-printing-dr-joe-beaman/

de Koning, S. B., Schaeffers, A. W. M. A., Schats, W., van den Brekel, M. W. M., Ruers, T. J. M. and Karakullukcu, M. B. (2021). Assessment of the deep resection margin during oral cancer surgery: A systematic review. *European Journal of Surgical Oncology*, 47(9): 2220–2232.

de Ponthaud, C., Menegaux, F. and Gaujoux, S. (2021). Updated principles of surgical management of pancreatic neuroendocrine tumours (pNETs): What every surgeon needs to know. *Cancers*, 13(23): 5969.

Deakyne, A. J. and Iaizzo, P. A. (2021, April). Interactive computational medical device deployments within virtual reality. *In*: *Frontiers in Biomedical Devices* (Vol. 84812, p. V001T03A002). American Society of Mechanical Engineers.

Design. (2024). In *Cambridge Dictionary*. Cambridge University Press. Online: https://dictionary.cambridge.org/dictionary/english/design?q=Design.

Di Stefano, F., Chiappini, S., Gorreja, A., Balestra, M. and Pierdicca, R. (2021). Mobile 3D scan LiDAR: A literature review. *Geomatics, Natural Hazards and Risk*, 12(1): 2387–2429.

Diesendruck, C. E., Sottos, N. R., Moore, J. S. and White, S. R. (2015). Biomimetic self-healing. *Angewandte Chemie International Edition*, 54(36): 10428–10447.

Dimov, S. S., Pham, D. T., Lacan, F. and Dotchev, K. D. (2001). Rapid tooling applications of the selective laser sintering process. *Assembly Automation*, 21(4): 296–302.

Dong, Y. P., Tang, J. C., Wang, D. W., Wang, N., He, Z. D., Li, J. et al. (2020). Additive manufacturing of pure Ti with superior mechanical performance, low cost, and biocompatibility for potential replacement of Ti-6Al-4V. *Materials & Design*, 196: 109142.

Dorna, P. (2023). Design of Mechatronic Lower Limb Orthoses. Master's Thesis (supervision: Górski, F.), Poznan University of Technology.

Dotremont, K. (2023). From medical images to 3D model: processing and segmentation. pp. 65–91. *In*: *Handbook of Surgical Planning and 3D Printing*. Academic Press.

Dudkiewicz, I., Gabrielov, R., Seiv-Ner, I., Zelig, G. and Heim, M. (2004). Evaluation of prosthetic usage in upper limb amputees. *Disability and Rehabilitation*, 26(1): 60–63.

Dudkowiak, A., Dostatni, E., Rojek, I. and Mikołajewski, D. (2022, March). The environmental analysis of a product manufactured with the use of an additive technology. pp. 76–89. *In*: *International Scientific-Technical Conference MANUFACTURING*. Cham: Springer International Publishing.

Duran, M. M., Moro, G., Zhang, Y. and Islam, A. (2023). 3D printing of silicone and polyurethane elastomers for medical device application: A review. *Advances in Industrial and Manufacturing Engineering*, 100125.

Ekanayake, C., Gamage, J. C. P. H., Mendis, P. and Weerasinghe, P. (2023). Revolution in orthopedic immobilization materials: A comprehensive review. *Heliyon*, 9(3).

Elanchezhian, C. and Sundar, G. S. (2007). *Computer Aided Manufacturing*. Firewall Media.

Encarnacao, J. L., Lindner, R. and Schlechtendahl, E. G. (2012). *Computer Aided Design: Fundamentals and System Architectures*. Springer Science & Business Media.

Enderle, J. and Bronzino, J. (2011). *Introduction to Biomedical Engineering*. Academic Press.

Escobar-Castillejos, D., Noguez, J., Neri, L., Magana, A. and Benes, B. (2016). A review of simulators with haptic devices for medical training. *Journal of Medical Systems*, 40: 1–22.

Fandáková, M., Bučko, B. and Zábovská, K. (2019, May). Parametric versus direct modeling in manufacturing process management. pp. 1543–1547. *In*: *2019 42nd International Convention on Information and Communication Technology, Electronics and Microelectronics (MIPRO)*. IEEE.

Fang, J. F., Wong, Y. C., Lin, B. C., Hsu, Y. P. and Chen, M. F. (2006). Usefulness of multidetector computed tomography for the initial assessment of blunt abdominal trauma patients. *World Journal of Surgery*, 30: 176–182.

Farin, G., Hoschek, J. and Kim, M. S. (eds.). (2002). *Handbook of Computer Aided Geometric Design*. Elsevier.

Faustini, M. C., Neptune, R. R., Crawford, R. H. and Stanhope, S. J. (2008). Manufacture of passive dynamic ankle–foot orthoses using selective laser sintering. *IEEE Transactions on Biomedical Engineering*, 55(2): 784–790.

Fogliatto, F. S., Da Silveira, G. J. and Borenstein, D. (2012). The mass customization decade: An updated review of the literature. *International Journal of Production Economics*, 138(1): 14–25.

Franssen, B. B., Schuurman, A. H., Van der Molen, A. M. and Kon, M. (2010). One century of Kirschner wires and Kirschner wire insertion techniques: a historical review. *Acta Orthop Belg*, 76(1): 1–6.

Fu, P. S., Wang, J. C., Lai, P. L., Liu, S. M., Chen, Y. S., Chen, W. C. et al. (2021). Effects of gamma radiation on the sterility assurance, antibacterial ability, and biocompatibility of impregnated hydrogel macrosphere protein and drug release. *Polymers*, 13(6): 938.

Furdella, K. J., Higuchi, S., Behrangzade, A., Kim, K., Wagner, W. R. and Geest, J. P. V. (2021). *In-vivo* assessment of a tissue engineered vascular graft computationally optimized for target vessel compliance. *Acta Biomaterialia*, 123: 298–311.

Galetti, P. M., Aebischer, P. and Lysaght, M. J. (1995). The dawn of biotechnology in artificial organs. ASAIO J. 1995 Jan-Mar; 41(1): 49–57.

Ganguli, A., Pagan-Diaz, G. J., Grant, L., Cvetkovic, C., Bramlet, M., Vozenilek, J. et al. (2018). 3D printing for preoperative planning and surgical training: a review. *Biomedical Microdevices*, 20: 1–24.

Gao, W., Wen, C. P., Wu, A. and Welch, H. G. (2022). Association of computed tomographic screening promotion with lung cancer overdiagnosis among Asian women. *JAMA Internal Medicine*, 182(3): 283–290.

Garg, P. K. (2021). Overview of artificial intelligence. pp. 3–18. *In*: *Artificial Intelligence*. Chapman and Hall/CRC.

Garlinghouse, A., Rud, S., Johnson, K., Plocher, T., Klassen, D., Havey, T. et al. (2018). Creating objects with 3D printers to stimulate reminiscence in memory loss: A mixed-method feasibility study. *Informatics for Health and Social Care*, 43(4): 362–378.

Gautreaux, S. T. (2021). *3D Printing Path Reallocation for Concurrent IDEX Systems* (Doctoral dissertation).

George, M., Aroom, K. R., Hawes, H. G., Gill, B. S. and Love, J. (2017). 3D printed surgical instruments: the design and fabrication process. *World Journal of Surgery*, 41: 314–319.

Gholamrezanezhad, A., Basques, K., Batouli, A., Matcuk, G., Alavi, A. and Jadvar, H. (2018). Clinical nononcologic applications of PET/CT and PET/MRI in musculoskeletal, orthopedic, and rheumatologic imaging. *American Journal of Roentgenology*, 210(6): W245–W263.

Gilbert, S. H., McConnell, F. J., Holden, A. V., Sivananthan, M. U. and Dukes-McEwan, J. (2010). The potential role of MRI in veterinary clinical cardiology. *The Veterinary Journal*, 183(2): 124–134.

Godec, D., Pilipović, A. and Breški, T. (2022). General process workflow in additive manufacturing. pp. 45–57. *In*: *A Guide to Additive Manufacturing*. Cham: Springer International Publishing.

Goel, R. and Gupta, P. (2020). Robotics and industry 4.0. *A Roadmap to Industry 4.0: Smart Production, Sharp Business and Sustainable Development*, 157–169.

Gold, N. B., Westgate, M. N. and Holmes, L. B. (2011). Anatomic and etiological classification of congenital limb deficiencies. *American Journal of Medical Genetics Part A*, 155(6): 1225–1235.

Gomes, T. E., Cadete, M. S., Dias-de-Oliveira, J. and Neto, V. (2022). Controlling the properties of parts 3D printed from recycled thermoplastics: A review of current practices. *Polymer Degradation and Stability*, 196: 109850.

Gonzalez, F. J. (2016). Theoretical and clinical aspects of the use of thermography in non-invasive medical diagnosis. *Biomedical Spectroscopy and Imaging*, 5(4): 347–358.

Górski, F., Kuczko, W., Wichniarek, R. and Zawadzki, P. (2010a). Use of marker projection method in photogrammetric applications. Scientific Journals of the Poznań University of Technology. *Machine Construction and Production Management*, 14: 31–40.

Górski, F., Kuczko, W., Wichniarek, R., Dudziak, A., Kowalski, M. and Zawadzki, P. (2010b). Choosing optimal rapid manufacturing process for thin-walled products using expert algorithm. *Journal of Industrial Engineering and Management (JIEM)*, 3(2): 408–420.

Górski, F., Hamrol, A., Kowalski, M., Paszkiewicz, R. and Zawadzki, P. (2011). An automatic system for 3D models and technology process design. *Transactions of FAMENA*, 35(2).

Górski, F., Kuczko, W. and Wichniarek, R. (2013). Influence of process parameters on dimensional accuracy of parts manufactured using fused deposition modelling technology. *Advances in Science and Technology. Research Journal*, 7(19): 27–35.

Górski, F., Kuczko, W. and Wichniarek, R. (2014). Impact strength of ABS parts manufactured using Fused Deposition Modeling technology. *Archives of Mechanical Technology and Automation*, 34(1): 3–12.

Górski, F., Wichniarek, R., Kuczko, W., Zawadzki, P. and Buń, P. (2015). Strength of ABS parts produced by Fused Deposition Modelling technology—a critical orientation problem. *Advances in Science and Technology. Research Journal*, 9(26): 12–19.

Górski, F., Zawadzki, P. and Hamrol, A. (2016). Knowledge based engineering as a condition of effective mass production of configurable products by design automation. *Journal of Machine Engineering*, 16.

Górski, F., Buń, P., Wichniarek, R., Zawadzki, P. and Hamrol, A. (2016b). Design and implementation of a complex virtual reality system for product design with active participation of end user. pp. 31–43. *In*: *Advances in Human Factors, Software, and Systems Engineering: Proceedings of the AHFE 2016 International Conference on Human Factors, Software, and Systems Engineering, July 27–31, 2016, Walt Disney World®, Florida, USA*. Springer International Publishing.

Górski, F. (2017). Building virtual reality applications for engineering with knowledge-based approach. *Management and Production Engineering Review*, 8(4): 64–73.

Górski, F., Kuczko, W., Wichniarek, R. and Hamrol, A. (2018). Mechanical properties of composite parts manufactured in FDM technology. *Rapid Prototyping Journal*, 24(8): 1281–1287.

Górski, F., Wichniarek, R., Kuczko, W., Banaszewski, J. and Pabiszczak, M. (2019a). Application of low-cost 3D printing for production of CT-based individual surgery supplies. pp. 249–253. *In*: *World Congress on Medical Physics and Biomedical Engineering 2018: June 3–8, 2018, Prague, Czech Republic (Vol. 1)*. Springer Singapore.

Górski, F., Kuczko, W., Weiss, W., Wichniarek, R. and Żukowska, M. (2019b). Prototyping of an individualized multi-material wrist orthosis using fused deposition modelling. *Advances in Science and Technology. Research Journal*, 13(4).

Górski, F., Zawadzki, P., Wichniarek, R., Kuczko, W., Żukowska, M., Wesołowska, I. et al. (2020a). Automated design of customized 3D-printed wrist orthoses on the basis of 3D scanning. pp. 1133–1143. *In*: *Computational and Experimental Simulations in Engineering: Proceedings of ICCES2019 24*. Springer International Publishing.

Górski, F., Wichniarek, R., Kuczko, W., Żukowska, M., Lulkiewicz, M. and Zawadzki, P. (2020b). Experimental studies on 3D printing of automatically designed customized wrist-hand orthoses. *Materials*, 13(18): 4091.

Górski, F., Osiński, F., Wierzbicka, N. and Żukowska, M. (2020c). Environmental impact of additive manufacturing for individual supplies. pp. 384–393. *In*: *Grabchenko's International Conference on Advanced Manufacturing Processes*. Cham: Springer International Publishing.

Górski, F., Wichniarek, R., Kuczko, W. and Żukowska, M. (2021). Study on properties of automatically designed 3d-printed customized prosthetic sockets. *Materials*, 14(18): 5240.

Górski, F., Wichniarek, R., Kuczko, W., Żukowska, M., Rybarczyk, J. and Lulkiewicz, M. (2022a, April). Evaluation of a prototype system of automated design and rapid manufacturing of orthopaedic supplies. pp. 1–15. *In*: *International Scientific-Technical Conference MANUFACTURING*. Cham: Springer International Publishing.

Górski, F., Rybarczyk, J., Zawadzki, P., Kuczko, W., Wierzbicka, N., Żukowska, M. et al. (2022b, April). Design and additive manufacturing of an individualized specialized leg orthosis. pp. 31–44. *In*: *International Scientific-Technical Conference MANUFACTURING*. Cham: Springer International Publishing.

Górski, F., Sahaj, N., Kuczko, W., Żukowska, M. and Hamrol, A. (2022c). Risk assessment of individualized 3D printed prostheses using failure mode and effect analysis. *Advances in Science and Technology. Research Journal*, 16(4).

Górski, F., Rybarczyk, D., Wichniarek, R., Wierzbicka, N., Kuczko, W., Żukowska, M. et al. (2023a). Development and testing of an individualized sensorised 3D printed upper limb bicycle prosthesis for adult patients. *Applied Sciences*, 13(23): 12918.

Górski, F., Grohs, A., Kuczko, W., Żukowska, M., Wichniarek, R., Siwiec, S. et al. (2023b). Development and studies of VR-assisted hand therapy using a customized biomechatronic 3D printed orthosis. *Electronics*, 13(1): 79.

Górski, F., Łabudzki, R., Żukowska, M., Sanfilippo, F., Ottestad, M., Zelenay, M. et al. (2023c). Experimental evaluation of extended reality technologies in the development of individualized three-dimensionally printed upper limb prostheses. *Applied Sciences*, 13(14): 8035.

Górski, F., Żukowska, M., Kuczko, W., Wichniarek, R. and Siwiec, S. (2024a). Automated design and 3D printing of therapeutic wrist hand orthosis. pp. 24–32. *In*: *Innovations in Biomedical Engineering 2023*. Cham: Springer Nature Switzerland.

Górski, F., Smolarek, E., Żukowska, M. and Okła, M. (2024b). Mixed Reality Versus 3D Printing In Presurgical Visualization of Soft and Hard Tissues: Selected Cases. Conference Proceedings of XR Salento 2024 conference, Lecce, Italy (in press).

Górski, F., Denysenko, Y., Kuczko, W., Żukowska, M., Wichniarek, R., Zawadzki, P. et al. (2024c). Individualized 3D printed orthopaedic and prosthetic devices using AutoMedPrint technology—methodologies and examples. Advances in Science and Technology. *Research Journal*, 18(6).

Górski, F., Gapsa, J., Kupaj, A., Kuczko, W., Żukowska, M. and Zawadzki, P. (2024d, March). Virtual design process of customized 3D printed modular upper limb prostheses. pp. 206–218. *In*: *International Scientific-Technical Conference MANUFACTURING*. Cham: Springer Nature Switzerland.

Górski, F., Denysenko, Y., Kuczko, W. and Żukowska, M. (2024e, June). Automated design and virtual fitting of 3D printed bicycle prostheses for children. pp. 242–253. *In*: *International Conference Innovation in Engineering*. Cham: Springer Nature Switzerland.

Górski, F., Marciniak, A., Wichniarek, R., Kuczko, W., Żukowska, M. and Rybarczyk, J. (2024f, March). Development of 3D printed low-cost individualized actuated upper limb prostheses. pp. 179–192. *In*: *International Scientific-Technical Conference MANUFACTURING*. Cham: Springer Nature Switzerland.

Górski, F., Łabudzki, R., Păcurar, R. and Păcurar, A. (2024g, June). Learning to build extended reality applications for customized devices: a survey study. pp. 357–369. *In*: *International Conference Innovation in Engineering*. Cham: Springer Nature Switzerland.

Goto, L., Lee, W., Molenbroek, J. F., Cabo, A. J. and Goossens, R. H. (2019). Traditional and 3D scan extracted measurements of the heads and faces of Dutch children. *International Journal of Industrial Ergonomics*, 73: 102828.

Grieves, M. (2023). Digital Model, Digital Shadow, Digital Twin. *Preprint*.

Growiec, J. (2022). Mechanization, automation and the labor market. pp. 63–76. *In*: *Accelerating Economic Growth: Lessons From 200,000 Years of Technological Progress and Human Development*. Cham: Springer International Publishing.

Gubin, A. V., Kuznetsov, V. P., Borzunov, D. Y., Koryukov, A. A., Reznik, A. V. and Chevardin, A. Y. (2016). Challenges and perspectives in the use of additive technologies for making customized implants for traumatology and orthopedics. *Biomedical Engineering*, 50(4): 285–289.

Guo, L., Naghavi, S. A., Wang, Z., Varma, S. N., Han, Z., Yao, Z. et al. (2022). On the design evolution of hip implants: A review. *Materials & Design*, 216: 110552.

Gutierrez, N. (2023). The ballad of morton heilig: on VR's mythic past. *JCMS: Journal of Cinema and Media Studies*, 62(3): 86–106.

Guttridge, C., Shannon, A., O'Sullivan, A., O'Sullivan, K. J. and O'Sullivan, L. W. (2022). Biocompatible 3D printing resins for medical applications: A review of marketed intended use, biocompatibility certification, and post-processing guidance. *Annals of 3D Printed Medicine*, 5: 100044.

Hafizović, L., Čaušević, A., Deumić, A., Bećirović, L. S., Pokvić, L. G. and Badnjević, A. (2021, October). The use of artificial intelligence in diagnostic medical imaging: systematic literature review. pp. 1–6. *In*: *2021 IEEE 21st International Conference on Bioinformatics and Bioengineering (BIBE)*. IEEE.

Hakala, P. (2015). *Direct Modeling in Global CAD Environment* (Master's thesis).

Hale, L., Linley, E. and Kalaskar, D. M. (2020). A digital workflow for design and fabrication of bespoke orthoses using 3D scanning and 3D printing, a patient-based case study. *Scientific Reports*, 10(1): 7028.

Haleem, A. and Javaid, M. (2019). 3D scanning applications in medical field: a literature-based review. *Clinical Epidemiology and Global Health*, 7(2): 199–210.

Hamrol, A., Górski, F., Grajewski, D. and Zawadzki, P. (2013). Virtual 3D atlas of a human body–development of an educational medical software application. *Procedia Computer Science*, 25: 302–314.

Harmonay, V. (2020). CT Scanner Components: What Does What? Atlantis Worldwide. Online: https://info.atlantisworldwide.com/blog/ct-scanner-components-what-does-what.

Harries, S., Abt, C. and Brenner, M. (2019). Upfront CAD—Parametric modeling techniques for shape optimization. *Advances in Evolutionary and Deterministic Methods for Design, Optimization and Control in Engineering and Sciences*, 191–211.

Hartcher-O'Brien, J., Evers, J. and Tempelman, E. (2019). Surface roughness of 3D printed materials: Comparing physical measurements and human perception. *Materials Today Communications*, 19: 300–305.

Hassan, I. and Selvaganapathy, P. R. (2022). A microfluidic printhead with integrated hybrid mixing by sequential injection for multimaterial 3D printing. *Additive Manufacturing*, 50: 102559.

Hau, M., Kim, J., Kattapuram, S., Hornicek, F. J., Rosenberg, A. E., Gebhardt, M. C. et al. (2002). Accuracy of CT-guided biopsies in 359 patients with musculoskeletal lesions. *Skeletal Radiology*, 31: 349–353.

Hendrixson, S. (2022). Understanding HP's Metal Jet: Beyond Part Geometry, Now It's About Modularity, Automation and Scale. *Additive Manufacturing*, Online: https://www.additivemanufacturing.media/articles/understanding-hps-metal-jet-beyond-part-geometry-now-its-about-modularity-automation-and-scale.

Hettesheimer, T., Hirzel, S. and Roß, H. B. (2018). Energy savings through additive manufacturing: An analysis of selective laser sintering for automotive and aircraft components. *Energy Efficiency*, 11: 1227–1245.

Hirschel, E. H. (2003). Towards the virtual product in aircraft design. pp. 453–464. *Periaux, J., Champion, M., Gagnepain, J.-J., Pironneau, O., Stoufflet, B., Thomas, P. (eds.). Fluid Dynamics and Aeronautics New Challenges. CIMNE Handbooks on Theory and Engineering Applications of Computational Methods, Barcelona, Spain.*

Hofmann, M., Burke, J., Pearlman, J., Fiedler, G., Hess, A., Schull, J. et al. (2016, October). Clinical and maker perspectives on the design of assistive technology with rapid prototyping technologies. pp. 251–256. *In*: *Proceedings of the 18th International ACM SIGACCESS Conference on Computers and Accessibility*.

Hohlweg-Majert, B., Pautke, C., Deppe, H., Metzger, M. C., Wagner, K. and Schulze, D. (2011). Qualitative and quantitative evaluation of bony structures based on DICOM dataset. *Journal of Oral and Maxillofacial Surgery*, 69(11): 2763–2770.

Hollywood Diagnostics Center. (2024). Innovative Ultrasound Diagnostics At Hollywood Diagnostics. Online: https://hdc1984.com/innovative-ultrasound-diagnostics-at-hollywood-diagnostics.

Horaud, R., Hansard, M., Evangelidis, G. and Ménier, C. (2016). An overview of depth cameras and range scanners based on time-of-flight technologies. *Machine Vision and Applications*, 27(7): 1005–1020.

Horien, C., Noble, S., Greene, A. S., Lee, K., Barron, D. S., Gao, S. et al. (2021). A hitchhiker's guide to working with large, open-source neuroimaging datasets. *Nature Human Behaviour*, 5(2): 185–193.

Hornick, J. (2017). 3D printing in Healthcare. *Journal of 3D Printing in Medicine*, 1(1): 13–17.

Hosseini, M. and Dixon, B. E. (2016). Syntactic interoperability and the role of standards. pp. 123–136. *In*: *Health Information Exchange*. Academic Press.

Howell, J. (2019). Principles and components of upper limb orthoses. pp. 134–145. *In*: *Atlas of Orthoses and Assistive Devices*. Elsevier.

Hozdić, E. (2015). Smart factory for industry 4.0: A review. *International Journal of Modern Manufacturing Technologies*, 7(1): 28–35.

Hsueh, M. H., Lai, C. J., Wang, S. H., Zeng, Y. S., Hsieh, C. H., Pan, C. Y. et al. (2021). Effect of printing parameters on the thermal and mechanical properties of 3d-printed pla and petg, using fused deposition modeling. *Polymers*, 13(11): 1758.

Hu, H. Z., Feng, X. B., Shao, Z. W., Xie, M., Xu, S., Wu, X. H. et al. (2019). Application and prospect of mixed reality technology in medical field. *Current Medical Science*, 39: 1–6.

Hu, W., Song, Y., Zhong, X., Feng, J., Wang, P. and Huang, C. (2016). improving doctor–patient communication: content validity examination of a novel urinary system-simulating physical model. *Patient Preference and Adherence*, 2519–2529.

Hu, Z., Zhu, H., Nie, X., Zhang, C., Zhang, H. and Zeng, X. (2018). On the role of atmospheric oxygen into mechanical properties and fracture behavior of selective laser melted AlCu5MnCdVA. *Materials & Design*, 150: 18–27.

Huamanchahua, D., Ortiz-Zacarias, J., Asto-Evangelista, J. and Quintanilla-Mosquera, I. (2021, December). Types of lower-limb orthoses for rehabilitation and assistance: a systematic review. pp. 0705–0711. *In*: *2021 IEEE 12th Annual Ubiquitous Computing, Electronics & Mobile Communication Conference (UEMCON)*. IEEE.

Huang, J., Qin, Q. and Wang, J. (2020). A review of stereolithography: Processes and systems. *Processes*, 8(9): 1138.

Hubert, K. F., Awa, K. N. and Zabelina, D. L. (2024). The current state of artificial intelligence generative language models is more creative than humans on divergent thinking tasks. *Scientific Reports*, 14(1): 3440.

Hudson, S. (2022). *The Metrology Handbook*. States Academic Press.

Hull, C. W. (2015). The birth of 3D printing. *Research-Technology Management*, 58(6): 25–30.

Hunde, B. R. and Woldeyohannes, A. D. (2022). Future prospects of computer-aided design (CAD)–A review from the perspective of artificial intelligence (AI), extended reality, and 3D printing. *Results in Engineering*, 14: 100478.

Hussaini, S. and Lahrman, G. (2012). Using modeling and simulation in the manufacturing of medical devices. *Advanced Materials Research*, 488: 1032–1038.

Iftekar, S. F., Aabid, A., Amir, A. and Baig, M. (2023). Advancements and limitations in 3D printing materials and technologies: a critical review. *Polymers*, 15(11): 2519.

Ilhami, M. A. and Masruroh, N. A. (2020). A mathematical model at the detailed design phase in the 3DCE new product development. *Computers & Industrial Engineering*, 146: 106617.

International Organization for Standardization. (2017). Additive manufacturing — General principles — Requirements for purchased AM parts (ISO/ASTM standard no. 52901:2017). Online: https://www.iso.org/standard/67288.html.

International Organization for Standardization. (2018). Additive manufacturing — Design — Requirements, guidelines and recommendations (ISO/ASTM standard no. 52910:2018). Online: https://www.iso.org/standard/67289.html.

International Organization for Standardization. (2021). Additive manufacturing — General principles — Fundamentals and vocabulary (ISO/ASTM standard no. 52900:2021). Online: https://www.iso.org/obp/ui/#iso:std:iso-astm:52900:ed-2:v1:en.

International Organization for Standardization. (2023). Additive manufacturing —General principles - Part positioning, coordinates and orientation (ISO standard no. 17295:2023). Online: https://www.iso.org/standard/76471.html.

Iouguina, A., Dawson, J. W., Hallgrimsson, B. and Smart, G. (2014). Biologically informed disciplines: A comparative analysis of bionics, biomimetics, biomimicry, and bio-inspiration among others. *International Journal of Design & Nature and Ecodynamics*, 9(3): 197–205.

Iqbal, N. V., Subramaniam, K. and P, S. A. (2018). A review on upper-limb myoelectric prosthetic control. *IETE Journal of Research*, 64(6): 740–752.

Izadi, M., Farzaneh, A., Mohammed, M., Gibson, I. and Rolfe, B. (2020). A review of laser engineered net shaping (LENS) build and process parameters of metallic parts. *Rapid Prototyping Journal*, 26(6): 1059–1078.

Jabehdar Maralani, P., Schieda, N., Hecht, E. M., Litt, H., Hindman, N., Heyn, C. et al. (2020). MRI safety and devices: An update and expert consensus. *Journal of Magnetic Resonance Imaging*, 51(3): 657–674.

Jackson, J. (2011). Specific treatment techniques. Physical Management for Neurological Conditions, 3rd ed. Edinburgh: Elsevier, 243–266.

Jaeschke, L., Steinbrecher, A. and Pischon, T. (2015). Measurement of waist and hip circumference with a body surface scanner: feasibility, validity, reliability, and correlations with markers of the metabolic syndrome. *PloS One*, 10(3): e0119430.

Jankowska, M. (2015). Prototyping of customized temporary lower limb prosthesis using additive manufacturing technologies. Diploma Thesis, Poznan University of Technology.

Javaid, M. and Haleem, A. (2018). Additive manufacturing applications in orthopaedics: a review. *Journal of Clinical Orthopaedics and Trauma*, 9(3): 202–206.

Javaid, M. and Haleem, A. (2020). Virtual reality applications toward medical field. *Clinical Epidemiology and Global Health*, 8(2): 600–605.

Javaid, M., Haleem, A., Singh, R. P., Suman, R. and Rab, S. (2021). Role of additive manufacturing applications towards environmental sustainability. *Advanced Industrial and Engineering Polymer Research*, 4(4): 312–322.

Jensen, L. and Konradsen, F. (2018). A review of the use of virtual reality head-mounted displays in education and training. *Education and Information Technologies*, 23: 1515–1529.

Jin, Y. A., Plott, J., Chen, R., Wensman, J. and Shih, A. (2015). Additive manufacturing of custom orthoses and prostheses—a review. *Procedia Cirp*, 36: 199–204.

Jin, Z., He, C., Fu, J., Han, Q. and He, Y. (2022). Balancing the customization and standardization: exploration and layout surrounding the regulation of the growing field of 3D-printed medical devices in China. *Bio-design and Manufacturing*, 5(3): 580–606.

Jin, Z., Li, Y., Yu, K., Liu, L., Fu, J., Yao, X. et al. (2021). 3D printing of physical organ models: recent developments and challenges. *Advanced Science*, 8(17): 2101394.

Johns Hopkins Medicine. (2024). Computed Tomography (CT or CAT) Scan of the Abdomen. Online: https://www.hopkinsmedicine.org/health/treatment-tests-and-therapies/computed-tomography-ct-or-cat-scan-of-the-abdomen.

Joshi, N. and Jain, S. (2016). Optimization of nonlocal means filtering technique for denoising magnetic resonance images: a review. pp. 1–15. In Proceedings of Fifth International Conference on Soft Computing for Problem Solving: SocProS 2015, Volume 1. Springer Singapore.

Jung, H. (2021). Basic physical principles and clinical applications of computed tomography. *Progress in Medical Physics*, 32(1): 1–17.

Junk, S. and Kuen, C. (2016). Review of open source and freeware CAD systems for use with 3D-printing. *Procedia CIRP*, 50: 430–435.

Junk, S. and Vögele, P. (2024, February). Investigation of bio-based and recycled printing materials for additive manufacturing. *In*: *Journal of Physics: Conference Series* (Vol. 2692, No. 1, p. 012042). IOP Publishing.

Kabasakal, İ., Keskin, F. D., Ventura, K. and Soyuer, H. (2017). From mass customization to product personalization in automotive industry: potentials of industry 4.0. *Journal of Management Marketing and Logistics*, 4(3): 244–250.

Kamada, H., Nakamura, M., Ota, H., Higuchi, S. and Takase, K. (2022). Blood flow analysis with computational fluid dynamics and 4D-flow MRI for vascular diseases. *Journal of Cardiology*, 80(5): 386–396.

Kara, L. B. and Shimada, K. (2008). Supporting early styling design of automobiles using sketch-based 3d shape construction. *Computer-Aided Design and Applications*, 5(6): 867–876.

Kargov, A., Pylatiuk, C., Oberle, R., Klosek, H., Werner, T., Roessler, W. et al. (2007, June). Development of a multifunctional cosmetic prosthetic hand. pp. 550–553. *In*: *2007 IEEE 10th International Conference on Rehabilitation Robotics*. IEEE.

Kaspar, J., Schneberger, J. H. and Vielhaber, M. (2020). LEAN Mobility–The spirit of a future Lightweight, Efficient, Application-oriented and Need-adapted road mobility concept. *Procedia Manufacturing*, 43: 64–71.

Kaule, S., Bock, A., Dierke, A., Siewert, S., Schmitz, K. P., Stiehm, M. et al. (2020). Medical Device Regulation and current challenges for the implementation of new technologies. *Current Directions in Biomedical Engineering*, 6(3): 334–337.

Kayğın Sel, S. (2013). *Industrial Design and the Mass Customization of Electronic Consumer Goods*. Ph.D. - Doctoral Program. Middle East Technical University.

Kazhdan, M. and Hoppe, H. (2013). Screened poisson surface reconstruction. *ACM Transactions on Graphics (ToG)*, 32(3): 1–13.

Keller, M., Guebeli, A., Thieringer, F. and Honigmann, P. (2021). In-hospital professional production of patient-specific 3D-printed devices for hand and wrist rehabilitation. *Hand Surgery and Rehabilitation*, 40(2): 126–133.

Kelly, S., Paterson, A. and Bibb, R. (2015). A review of wrist splint designs for additive manufacture.

Kendal, S.L. and Creen, M. (2007). *An Introduction to Knowledge Engineering.* London: Springer.

Khaleel, M., Ahmed, A. A. and Alsharif, A. (2023). Artificial Intelligence in Engineering. *Brilliance: Research of Artificial Intelligence*, 3(1): 32–42.

Khandagale, B. D. and Pise, U. V. (2022). Numerical and experimental investigation of hinged Ankle-Foot-Orthoses (AFO) using composite laminate material for Cerebral Palsy patient. *Materials Today: Proceedings*, 62: 2070–2080.

Kim, J. W., Wu, J., Shen, S. G., Xu, B., Shi, J. and Zhang, S. (2015). Interdisciplinary surgical management of multiple facial fractures with image-guided navigation. *Journal of Oral and Maxillofacial Surgery*, 73(9): 1767–1777.

Kim, J. Y. and Ko, S. J. (2017). Comparison of DICOM images and various types of images. *Journal of the Institute of Convergence Signal Processing*, 18(2): 76–83.

Kim, J., Hong, J. and Park, H. (2018). Prospects of deep learning for medical imaging. *Precision and Future Medicine*, 2(2): 37–52.

Kim, K. D., Yang, D. Y. and Jeong, J. H. (2006). Plaster casting process for prototyping of die casting based on rapid tooling. *The International Journal of Advanced Manufacturing Technology*, 28: 923–929.

Klein, K. and Stolk, P. (2018). Challenges and opportunities for the traceability of (biological) medicinal products. *Drug Safety*, 41: 911–918.

Kletetzka, I., Neitzel, F. and Schmid, H. J. (2023). Assessing the Impact of the Powder Production Method on Ceramic-filled Polyamide Composites made by Laser Sintering.

Kobayashi, D., Takahashi, O., Ueda, T., Deshpande, G. A., Arioka, H. and Fukui, T. (2013). Risk factors for adverse reactions from contrast agents for computed tomography. *BMC Medical Informatics and Decision Making*, 13: 1–6.

Komorowska, O. (2022). *Automation of Design of Modular Upper Limb Prostheses.* Master's thesis (supervision: Górski, F.), Poznan University of Technology.

Konieczny, R., Dudziak, A., Grajewski, D. and Górski, F. (2012). Techniques of optical measurements in reverse engineering (in Polish: Techniki pomiarów optycznych w inżynierii odwrotnej). *Comprint, Poznań.*

Koohi-Moghadam, M. and Bae, K. T. (2023). Generative AI in medical imaging: applications, challenges, and ethics. *Journal of Medical Systems*, 47(1): 94.

Körner, C. (2016). Additive manufacturing of metallic components by selective electron beam melting—a review. *International Materials Reviews*, 61(5): 361–377.

Kour, R. and Singh, H. (2023). Social and commercial trends of 3D printed orthoprosthesis. pp. 195–209. *In*: *3D Printing in Podiatric Medicine*. Academic Press.

Kowalski, M., Kowalski, M., Paszkiewicz, R., Kuczko, W., Wichniarek, R. and Zawadzki, P. (2011). Automation of the three-dimensional scanning process based on data obtained from the photogrammetric measurement. *Advances in Science and Technology*, 7(2011): 109–119.

Kowalski, M., Zawadzki, P. and Hamrol, A. (2021). Effectiveness of automatic CAM programming using machining templates for the manufacture of special production tooling. *Strojniski Vestnik-Journal of Mechanical Engineering*, 67(10): 475–489.

Kreuzahler, U. (2023). The Challenges of Creating Realistic 3D Objects using AI. CGI Backgrounds. Online: https://www.cgibackgrounds.com/blog/the-challenges-of-creating-realistic-3d-objects-using-ai.

Krishna, L. S. R., Reddy, M. B. and Venkatesh, S. (2011). Development of optimum preplanning for maxillofacial surgery using selective laser sintering. *International Journal of Engineering, Science and Technology*, 3(8): 185–196.

Kroes, P. and Kroes, P. (2012). *Engineering Design* (pp. 127–161). Springer Netherlands.

Kroma, A., Adamczak, O., Sika, R., Górski, F., Kuczko, W. and Grześkowiak, K. (2020). Modern reverse engineering methods used to modification of jewelry. *Advances in Science and Technology. Research Journal*, 14(4).

Kuang, X., Rong, Q., Belal, S., Vu, T., López López, A. M., Wang, N. et al. (2023). Self-enhancing sono-inks enable deep-penetration acoustic volumetric printing. *Science*, 382(6675): 1148–1155.

Kuczko, W., Wichniarek, R., Górski, F. and Banaszewski, J. (2018). Influence of sterilization of a product manufactured using FDM technology on its dimensional accuracy. *Advances in Science and Technology. Research Journal*, 12(1): 74–79.

Kuehnapfel, A., Ahnert, P., Loeffler, M., Broda, A. and Scholz, M. (2016). Reliability of 3D laser-based anthropometry and comparison with classical anthropometry. *Scientific Reports*, 6(1): 26672.

Kulon J., Mynors D.J. and Broomhead, P. (2006). A knowledge-based engineering design tool for metal forging, *Journal of Materials Processing Technology*, 177: 331–335.

Kumar, A., Jain, P. K. and Pathak, P. M. (2013). Reverse engineering in product manufacturing: an overview. *DAAAM International Scientific Book*, 39: 665–678.

Kumar, A., Kumar, P., Mittal, R. K. and Singh, H. (2023). Printing file formats for additive manufacturing technologies. *Advances in Additive Manufacturing*, 87–102.

Kumar, A., Mandal, S., Barui, S., Vasireddi, R., Gbureck, U., Gelinsky, M. et al. (2016). Low temperature additive manufacturing of three dimensional scaffolds for bone-tissue engineering applications: Processing related challenges and property assessment. *Materials Science and Engineering: R: Reports*, 103: 1–39.

Kumar, M. B., Sathiya, P. and Varatharajulu, M. (2021). Selective laser sintering. *Advances in Additive Manufacturing Processes; China Bentham Books: Beijing, China*, 28.

Kumar, P., Gupta, P. and Singh, I. (2023b). Performance analysis of acrylonitrile–butadiene–styrene–polycarbonate polymer blend filament for fused deposition modeling printing using hybrid artificial intelligence algorithms. *Journal of Materials Engineering and Performance*, 32(4): 1924–1937.

Kumar, R., Kumar, S., Kumar, M. and Luthra, G. (2022). Conventional and 3D printing technology for the manufacturing of metal-matrix composite: a study. pp. 197–223. *In*: *Metal Matrix Composites*. CRC Press.

Kumar, V. S., Aswath, S., Shashidhar, T. S. and Choudhary, R. K. (2017). A novel design of a full length prosthetic robotic arm for the disabled. pp. 271–287. *In*: *Robot Intelligence Technology and Applications 4: Results from the 4th International Conference on Robot Intelligence Technology and Applications*. Springer International Publishing.

Kumar, Y. R. (2011). Modelling and simulation of bio-medical components by selective laser sintering. *International Journal of Rapid Manufacturing*, 2(4): 234–257.

Kurowski, P. (2016). *Engineering Analysis with Solidworks Simulation 2016*. SDC Publications.

Kurzydło, W., Stach, B., Bober, A., Wodzińska, M. E. and Długosz, M. M. (2014). Structured-light 3D scanner in use to access the human body posture in physical therapy: a pilot study. *Folia Medica Cracoviensia*, 54(1).

Lacewell, D. (2021). Ray-tracing small voxel scenes. *Ray Tracing Gems II: Next Generation Real-Time Rendering with DXR, Vulkan, and OptiX*, 599–609.

Lacroix, D., Planell, J. A. and Prendergast, P. J. (2009). Computer-aided design and finite-element modelling of biomaterial scaffolds for bone tissue engineering. *Philosophical Transactions of the Royal Society A: Mathematical, Physical and Engineering Sciences*, 367(1895): 1993–2009.

Ladd, L. M. and Roth, T. D. (2017, October). Computed tomography and magnetic resonance imaging of bone tumors. pp. 209–226. *In*: *Seminars in Roentgenology* (Vol. 52, No. 4). WB Saunders.

Lagüela, S., Díaz-Vilariño, L. and Roca, D. (2016). Infrared thermography: Fundamentals and applications. pp. 113–138. *Non-Destructive Techniques for the Evaluation of Structures and Infrastructure*. CRC Press: Boca Raton, FL, USA.

Lalegani Dezaki, M., Mohd Ariffin, M. K. A. and Hatami, S. (2021). An overview of fused deposition modelling (FDM): Research, development and process optimisation. *Rapid Prototyping Journal*, 27(3): 562–582.

Lambert, J. W., Edic, P. M., FitzGerald, P. F., Torres, A. S. and Yeh, B. M. (2015). Complementary contrast media for metal artifact reduction in dual-energy computed tomography. *Journal of Medical Imaging*, 2(3): 033503–033503.

Land, K. M. (2023). Improving CAD Designs with Autodesk Fusion 360: A Project Based Guide to Modelling Effective Parametric Designs. *Packt Publishing*.

Lanzola, G., Quaglini, S. and Stefanelli, M. (1995). Knowledge-acquisition tools for medical knowledge-based systems. *Methods of Information in Medicine*, 34(01/02): 25–39.

Larobina, M. and Murino, L. (2014). Medical image file formats. *Journal of Digital Imaging*, 27: 200–206.

Lauer, M., Barker, J. P., Solano, M. and Dubin, J. (2017). FDA device regulation. *Missouri Medicine*, 114(4): 283.

Le Duigou, J., Bernard, A. and Perry, N. (2011). Framework for product lifecycle management integration in small and medium enterprises networks. *Computer-Aided Design and Applications*, 8(4): 531–544.

Leberfinger, A. N., Dinda, S., Wu, Y., Koduru, S. V., Ozbolat, V., Ravnic, D. J. et al. (2019). Bioprinting functional tissues. *Acta Biomaterialia*, 95: 32–49.

Lebude, B., Yadla, S., Albert, T., Anderson, D. G., Harrop, J. S., Hilibrand, A. et al. (2010). Defining "complications" in spine surgery: neurosurgery and orthopedic spine surgeons' survey. *Clinical Spine Surgery*, 23(8): 493–500.

Lee, K. (1999). *Principles of CAD/CAM/CAE Systems*. Addison-Wesley Longman Publishing Co., Inc..

Lee, S. H., Woo, S. H., Ryu, J. R. and Choo, S. Y. (2019). Automated building occupancy authorization using BIM and UAV-based spatial information: photogrammetric reverse engineering. *Journal of Asian Architecture and Building Engineering*, 18(2): 151–158.

Leering, R. (2017). 3D printing: a threat to global trade. ING Think. Online: https://think.ing.com/reports/3d-printing-a-threat-to-global-trade/.

Leirmo, T. L., Semeniuta, O. and Martinsen, K. (2020). Tolerancing from STL data: a legacy challenge. *Procedia Cirp*, 92: 218–223.

Leksowski, K., Rudzinska, M. and Rudzinski, J. (2011). Computed tomographic colonography in preoperative evaluation of colorectal tumors: a prospective study. *Surgical Endoscopy*, 25: 2344–2349.

Leordean, D., Dudescu, C., Marcu, T., Berce, P. and Balc, N. (2015). Customized implants with specific properties, made by selective laser melting. *Rapid Prototyping Journal*, 21(1): 98–104.

Li, C., Pisignano, D., Zhao, Y. and Xue, J. (2020). Advances in medical applications of additive manufacturing. *Engineering*, 6(11): 1222–1231.

Li, J., Ellis, D. G., Kodym, O., Rauschenbach, L., Rieß, C., Sure, U. et al. (2023). Towards clinical applicability and computational efficiency in automatic cranial implant design: An overview of the autoimplant 2021 cranial implant design challenge. *Medical Image Analysis*, 88: 102865.

Li, J., Gsaxner, C., Pepe, A., Morais, A., Alves, V., von Campe, G. et al. (2021). Synthetic skull bone defects for automatic patient-specific craniofacial implant design. *Scientific Data*, 8(1): 36.

Li, X., Morgan, P. S., Ashburner, J., Smith, J. and Rorden, C. (2016). The first step for neuroimaging data analysis: DICOM to NIfTI conversion. *Journal of Neuroscience Methods*, 264: 47–56.

Li, Y., Wang, J., Li, X. and Zhao, W. (2007). Design creativity in product innovation. *The International Journal of Advanced Manufacturing Technology*, 33: 213–222.

Li, Z., Yang, H. and Xu, J. (2022). How to adopt mass customization strategy: Understanding the role of consumers' perceived brand value. *Computers & Industrial Engineering*, 173: 108666.

Liacouras, P. C., Sahajwalla, D., Beachler, M. D., Sleeman, T., Ho, V. B. and Lichtenberger, J. P. (2017). Using computed tomography and 3D printing to construct custom prosthetics attachments and devices. *3D Printing in Medicine*, 3: 1–7.

Liao, S. H. (2005). Expert system methodologies and applications—a decade review from 1995 to 2004. *Expert Systems with Applications*, 28(1): 93–103.

Lie, S. S., Helle, N., Sletteland, N. V., Vikman, M. D. and Bonsaksen, T. (2023). Implementation of virtual reality in health professions education: scoping review. *JMIR Medical Education*, 9: e41589.

Lim, K. H. A., Loo, Z. Y., Goldie, S. J., Adams, J. W. and McMenamin, P. G. (2016). Use of 3D printed models in medical education: A randomized control trial comparing 3D prints versus cadaveric materials for learning external cardiac anatomy. *Anatomical Sciences Education*, 9(3): 213–221.

Lin, B. T. and Hsu, S. H. (2008). Automated design system for drawing dies. *Expert Systems with Applications*, 34(3): 1586–1598.

Linder, W. (2009). *Digital Photogrammetry* (Vol. 1). Berlin/Heidelberg, Germany: Springer.

Liravi, F. and Toyserkani, E. (2018). Additive manufacturing of silicone structures: A review and prospective. *Additive Manufacturing*, 24: 232–242.

Liu, K., Tang, T., Wang, A. and Cui, S. (2015). Surgical revision for stump problems after traumatic above-ankle amputations of the lower extremity. *BMC Musculoskeletal Disorders*, 16: 1–7.

Liu, L. and Ciftci, O. N. (2021). Effects of high oil compositions and printing parameters on food paste properties and printability in a 3D printing food processing model. *Journal of Food Engineering*, 288: 110135.

Lojo, J. M., Pérez-Barquero, J. A., Bonmatí, F. G. S., Agustín-Panadero, R., Yilmaz, B. and Revilla-León, M. (2023). Influence of print orientation on the accuracy (trueness and precision) of diagnostic casts manufactured with a daylight polymer printer. *The Journal of Prosthetic Dentistry*.

Lopes da Silva, J. V. (2013, October). 3D technologies and the new digital ecosystem: a Brazilian experience. pp. 278–284. *In*: *Proceedings of the Fifth International Conference on Management of Emergent Digital EcoSystems*.

Lorenz, R., Benninghaus, C., Friedli, T. and Netland, T. H. (2020). Digitization of manufacturing: the role of external search. *International Journal of Operations & Production Management*, 40(7/8): 1129–1152.

Luhmann, T. (2010). Close range photogrammetry for industrial applications. *ISPRS Journal of Photogrammetry and Remote Sensing*, 65(6): 558–569.

Lumbroso, B. and Rispoli, M. (2014). *Practical Retinal OCT*. Jaypee Brothers Medical Publishers.

Lusic, H. and Grinstaff, M. W. (2013). X-ray-computed tomography contrast agents. *Chemical Reviews*, 113(3): 1641–1666.

Maas, H. G. (2008, July). Close range photogrammetry sensors. pp. 81–90. *In*: *Advances in Photogrammetry, Remote Sensing and Spatial Information Sciences: 2008 ISPRS Congress Book*. CRC Press.

Magomedov, I. A. and Sebaeva, Z. S. (2020, April). Comparative study of finite element analysis software packages. *In*: *Journal of Physics: Conference Series* (Vol. 1515, No. 3, p. 032073). IOP Publishing.

Maik, M., Sobociński, P., Walczak, K., Strugała, D., Górski, F. and Zawadzki, P. (2022, November). Flexible photorealistic VR training system for electrical operators. pp. 1–9. *In*: *Proceedings of the 27th International Conference on 3D Web Technology*.

Makris, S. (2021). *Cooperating Robots for Flexible Manufacturing* (pp. 123–132). Cham: Springer.

Malik, A., Haq, M. I. U., Raina, A. and Gupta, K. (2022). 3D printing towards implementing Industry 4.0: sustainability aspects, barriers and challenges. *Industrial Robot: The International Journal of Robotics Research and Application*, 49(3): 491–511.

Manero, A., Smith, P., Sparkman, J., Dombrowski, M., Courbin, D., Kester, A. et al. (2019). Implementation of 3D printing technology in the field of prosthetics: Past, present, and future. *International Journal of Environmental Research and Public Health*, 16(9): 1641.

Manshoori Yeganeh, A., Movahhedy, M. R. and Khodaygan, S. (2019). An efficient scanning algorithm for improving accuracy based on minimising part warping in selected laser sintering process. *Virtual and Physical Prototyping*, 14(1): 59–78.

Manuel de A. Monteiro Ramos, A. and António Simões, J. (2009). CAD-CAM-RTV–lost-wax casting technology for medical implants. *Rapid Prototyping Journal*, 15(3): 211–215.

Marcus, H. J., Payne, C. J., Hughes-Hallett, A., Marcus, A. P., Yang, G. Z., Darzi, A. et al. (2016). Regulatory approval of new medical devices: cross sectional study. *bmj*, 353.

Markatos, K., Tsoucalas, G. and Sgantzos, M. (2016). Hallmarks in the history of orthopaedic implants for trauma and joint replacement. *Acta medico-historica Adriatica: AMHA*, 14(1): 161–176.

Markiewicz, M. R. and Bell, R. B. (2011). The use of 3D imaging tools in facial plastic surgery. *Facial Plastic Surgery Clinics*, 19(4): 655–682.

Maroti, P., Varga, P., Abraham, H., Falk, G., Zsebe, T., Meiszterics, Z. et al. (2018). Printing orientation defines anisotropic mechanical properties in additive manufacturing of upper limb prosthetics. *Materials Research Express*, 6(3): 035403.

Martínez Gómez, J. M., López Gualdrón, C. I., Murillo Bohórquez, A. P. and Garnica Bohórquez, I. (2019). PLM strategy for developing specific medical devices and lower limb prosthesis at healthcare sector: case reports from the academia. *Product Lifecycle Management (Volume 4): The Case Studies*, 201–221.

Martínez-García, A., Monzón, M. and Paz, R. (2021). Standards for additive manufacturing technologies: Structure and impact. pp. 395–408. *In*: *Additive Manufacturing*. Elsevier.

Martinez-Marquez, D., Jokymaityte, M., Mirnajafizadeh, A., Carty, C. P., Lloyd, D. and Stewart, R. A. (2019). Development of 18 quality control gates for additive manufacturing of error free patient-specific implants. *Materials*, 12(19): 3110.

Martins, N. C., Marques, B., Alves, J., Araújo, T., Dias, P. and Santos, B. S. (2022). Augmented reality situated visualization in decision-making. *Multimedia Tools and Applications*, 81(11): 14749–14772.

Mathieu, V., Vayron, R., Richard, G., Lambert, G., Naili, S., Meningaud, J. P. et al. (2014). Biomechanical determinants of the stability of dental implants: Influence of the bone–implant interface properties. *Journal of Biomechanics*, 47(1): 3–13.

Matsiushevich, K., Belvedere, C., Leardini, A. and Durante, S. (2019). Quantitative comparison of freeware software for bone mesh from DICOM files. *Journal of Biomechanics*, 84: 247–251.

Mavroidis, C., Ranky, R. G., Sivak, M. L., Patritti, B. L., DiPisa, J., Caddle, A. et al. (2011). Patient specific ankle-foot orthoses using rapid prototyping. *Journal of Neuroengineering and Rehabilitation*, 8: 1–11.

Mazzoli, A. (2013). Selective laser sintering in biomedical engineering. *Medical & Biological Engineering & Computing*, 51: 245–256.

McGill University. (2024). Mechanical Design. Online: https://www.mcgill.ca/engineeringdesign/step-step-design-process/design-software/mechanical-design.

Meglioli, M., Naveau, A., Macaluso, G. M. and Catros, S. (2020). 3D printed bone models in oral and cranio-maxillofacial surgery: a systematic review. *3D Printing in Medicine*, 6: 1–19.

Memon, A. R., Wang, E., Hu, J., Egger, J. and Chen, X. (2020). A review on computer-aided design and manufacturing of patient-specific maxillofacial implants. *Expert Review of Medical Devices*, 17(4): 345–356.

Memon, M. and Rizzo III, J. F. (2009). The development of visual prosthetic devices to restore vision to the blind. pp. 723–742. *In*: *Neuromodulation*. Academic Press.

Mendaza-DeCal, R., Peso-Fernandez, S. and Rodriguez-Quiros, J. (2023). Orthotics and prosthetics by 3D-printing: Accelerating its fabrication flow. *Research in Veterinary Science*, 162: 104960.

Merlo, A., Sánchez Belenguer, C., Vendrell Vidal, E., Fantini, F. and Aliperta, A. (2013). 3D model visualization enhancements in real-time game engines. *The International Archives of the Photogrammetry, Remote Sensing and Spatial Information Sciences*, 40: 181–188.

Milgram, P., Takemura, H., Utsumi, A. and Kishino, F. (1995, December). Augmented reality: A class of displays on the reality-virtuality continuum. pp. 282–292. *In*: Telemanipulator and Telepresence Technologies (Vol. 2351). Spie Digital Library.

Minoufekr, M. and Plapper, P. (2020). Holo Study–Mixed Reality Framework for Industrial Engineering Education and Training. *Trends in Technical & Scientific Research*, 4(5).

Mintz, Y. and Brodie, R. (2019). Introduction to artificial intelligence in medicine. *Minimally Invasive Therapy & Allied Technologies*, 28(2): 73–81.

Mohamed, O. A., Masood, S. H. and Bhowmik, J. L. (2015). Optimization of fused deposition modeling process parameters: a review of current research and future prospects. *Advances in Manufacturing*, 3: 42–53.

Mohammed, M. I., Fitzpatrick, A. P. and Gibson, I. (2017). Customised design of a patient specific 3D printed whole mandible implant. *KnE Engineering*, 104–111.

Molotla, O., Peña-Cabrera, J. M. and Lomas-Barrie, V. (2023). Configurable hybrid integral manufacturing platform: Subtractive-additive process with industrial robot arm, proof of concept results. *IEEE Latin America Transactions*, 21(11): 1227–1235.

Mostafaei, A., Elliott, A. M., Barnes, J. E., Li, F., Tan, W., Cramer, C. L. et al. (2021). Binder jet 3D printing—Process parameters, materials, properties, modeling, and challenges. *Progress in Materials Science*, 119: 100707.

Mowry, S. E., Jammal, H., Myer IV, C., Solares, C. A. and Weinberger, P. (2015). A novel temporal bone simulation model using 3D printing techniques. *Otology & Neurotology*, 36(9): 1562–1565.

Muhanna, M. A. (2015). Virtual reality and the CAVE: Taxonomy, interaction challenges and research directions. *Journal of King Saud University-Computer and Information Sciences*, 27(3): 344–361.

Munsch, M. (2017). Laser additive manufacturing of customized prosthetics and implants for biomedical applications. pp. 399–420. *In*: *Laser Additive Manufacturing*. Woodhead Publishing.

Muramoto, Y. (2020). *Study of Drilling Properties in Bone Biomodel* (Doctoral dissertation, Tohoku University).

Murphy, S. V., De Coppi, P. and Atala, A. (2020). Opportunities and challenges of translational 3D bioprinting. *Nature Biomedical Engineering*, 4(4): 370–380.

Musso, G., Lentini, G., Enrietti, L., Volpe, C., Ambrosio, E. P., Lorusso, M. et al. (2016). Portable on orbit printer 3D: 1st European additive manufacturing machine on international space station. pp. 643–655. *In*: *Advances in Physical Ergonomics and Human Factors: Proceedings of the AHFE 2016 International Conference on Physical Ergonomics and Human Factors, July 27–31, 2016, Walt Disney World®, Florida, USA*. Springer International Publishing.

Mustafa, N. N. M., Kadir, A. Z. A., Ngadiman, N. A., Ma'aram, A. and Zakaria, K. (2020). Comparison of different additive manufacturing patterns on the performance of rapid vacuum casting for mating parts via the Taguchi method. *Journal of Mechanical Engineering and Sciences*, 14(1): 6417–6429.

Nachtigall, W. and Wisser, A. (2014). *Bionics by Examples*. Heidelberg: Springer.
Najjar, R. (2023). Redefining radiology: a review of artificial intelligence integration in medical imaging. *Diagnostics*, 13(17): 2760.
Nakajima, K. and Schwarz, O. (2014). How to use the ovipositor drilling mechanism of hymenoptera for developing a surgical instrument in biomimetic design. *International Journal of Design & Nature and Ecodynamics*, 9(3): 177–189.
Nassar, A. H., Maselli, A. M., Manstein, S., Shiah, E., Slatnick, B. L., Dowlatshahi, A. S. et al. (2022). Comparison of various modalities utilized for preoperative planning in microsurgical reconstructive surgery. *Journal of Reconstructive Microsurgery*, 38(03): 170–180.
Natarajan, P., Krishnan, N., Kenkre, N. S., Nancy, S. and Singh, B. P. (2012, December). Tumor detection using threshold operation in MRI brain images. pp. 1–4. *In*: *2012 IEEE International Conference on Computational Intelligence and Computing Research*. IEEE.
Neethan, R., Nidershan, S., Mugilgeethan, V., Tharsika, T. and Anburuvel, A. (2020). A study of three-dimensional (3-D) printed prosthetic upper limb models in local context. *Materials Today: Proceedings*, 23: 8–11.
NEMA. (2024). *DICOM PS3.1 2024c – Introduction and Overview*. Online: https://dicom.nema.org/medical/dicom/current/output/chtml/part01/chapter_1.html#sect_1.1.
Nikitichev, D. I., Patel, P., Avery, J., Robertson, L. J., Bucking, T. M., Aristovich, K. Y. et al. (2018). Patient-specific 3D printed models for education, research and surgical simulation. *3D Printing*, 115–134.
Nof, S. Y. (2009). Automation: What it means to us around the world. *Springer Handbook of Automation*, 13–52.
Noor, N., Shapira, A., Edri, R., Gal, I., Wertheim, L. and Dvir, T. (2019). 3D printing of personalized thick and perfusable cardiac patches and hearts. *Advanced Science*, 6(11): 1900344.
Nowik, K., Langwińska-Wośko, E., Skopiński, P., Nowik, K. E. and Szaflik, J. P. (2020). Bionic eye review–An update. *Journal of Clinical Neuroscience*, 78: 8–19.
O'Connor, H. J., Dickson, A. N. and Dowling, D. P. (2018). Evaluation of the mechanical performance of polymer parts fabricated using a production scale multi jet fusion printing process. *Additive Manufacturing*, 22: 381–387.
Oldham, K., Kneebone, S., Callot, M., Murton, A., Brimble, R. (1998). MOKA–A methodology and tools oriented to knowledge-based engineering applications. pp. 198–207. *In*: Mårtensson, N., Mackay, R. and Björgvinsson, S. (eds.). Changing the Ways We Work, Amsterdam, IOS Press.
Opie, N. L., Ayton, L. N., Apollo, N. V., Ganesan, K., Guymer, R. H. and Luu, C. D. (2014). Optical coherence tomography-guided retinal prosthesis design: model of degenerated retinal curvature and thickness for patient-specific devices. *Artificial Organs*, 38(6): E82–E94.
Ortega, B. P. and Olmedo, J. M. J. (2017). Application of motion capture technology for sport performance analysis. *Retos: nuevas tendencias en educación física, deporte y recreación*, (32): 241–247.
Otero, J. J., Vijverman, A. and Mommaerts, M. Y. (2017). Use of fused deposit modeling for additive manufacturing in hospital facilities: European certification directives. *Journal of Cranio-Maxillofacial Surgery*, 45(9): 1542–1546.
Ozsahin, I., Chen, L., Könik, A., King, M. A., Beekman, F. J. and Mok, G. S. (2020). The clinical utilities of multi-pinhole single photon emission computed tomography. *Quantitative Imaging in Medicine and Surgery*, 10(10): 2006.
Pabiszczak, M., Banaszewski, J., Pastusiak, T., Buczkowska, A., Kuczko, W., Wichniarek, R. et al. (2016). Mandibular reconstruction with osteo-cutaneous free flaps in a patient after extensive surgery supported with 3d printed models. *J. Integr. Oncol.*, 5: 177.
Pacurar, R. and Pacurar, A. (2016). Applications of the selective laser melting technology in the industrial and medical fields. *New Trends in 3D Printing*, 175–183.
Păcurar, R. I., Sanfilippo, F., Økter, M. B., Băilă, D. I., Zaharia, C., Nicoară, A. I. et al. (2024). Use of high-performance polymeric materials in customized low-cost robotic grippers for biomechatronic applications: experimental and analytical research. *Frontiers in Materials*, 11: 1304339.
Pająk, E., Dudziak, A., Górski, F. and Wichniarek, R., (2011). Additive manufacturing technologies and virtual reality in processes of production preparation (in Polish: Techniki przyrostowe i wirtualna rzeczywistość w procesach przygotowania produkcji), ARP Promocja 21.

Pallari, J. H., Dalgarno, K. W. and Woodburn, J. (2010). Mass customization of foot orthoses for rheumatoid arthritis using selective laser sintering. *IEEE Transactions on Biomedical Engineering*, 57(7): 1750–1756.

Palousek, D., Rosicky, J., Koutny, D., Stoklásek, P. and Navrat, T. (2014). Pilot study of the wrist orthosis design process. *Rapid Prototyping Journal*, 20(1): 27–32.

Palumbo, A. (2022). Microsoft HoloLens 2 in medical and healthcare context: state of the art and future prospects. *Sensors*, 22(20): 7709.

Pan, R., Yang, L., Zheng, L., Hao, L. and Li, Y. (2020). Microscopic morphology, thermodynamic and mechanical properties of thermoplastic polyurethane fabricated by selective laser sintering. *Materials Research Express*, 7(5): 055301.

Park, J. B. (2012). *Biomaterials Science and Engineering*. Springer Science & Business Media.

Parmar, A. J., Tyagi, S. K., Dabas, V. S., Mistry, J. N., Jhala, S. K., Suthar, D. N. et al. (2014). Assessment of the physical and mechanical properties of plaster of Paris bandage cast used as a splinting and casting materials. *Veterinary World*, 7(12).

Parras, D., Cavas-Martínez, F., Nieto, J., Cañavate, F. J. and Fernández-Pacheco, D. G. (2018). Reconstruction by low cost software based on photogrammetry as a reverse engineering process. pp. 145–154. *In*: *Virtual, Augmented and Mixed Reality: Interaction, Navigation, Visualization, Embodiment, and Simulation: 10th International Conference, VAMR 2018, Held as Part of HCI International 2018, Las Vegas, NV, USA, July 15–20, 2018, Proceedings, Part I 10*. Springer International Publishing.

Pasierbek, A. (2021). *Automation of Design of Pre-surgery Supplies in Maxillofacial Surgery*. Master's thesis (supervision: Górski, F.), Poznan University of Technology.

Patalas-Maliszewska, J. and Kłos, S. (2018). Knowledge network for the development of software projects (KnowNetSoft). *IFAC-PapersOnLine*, 51(11): 776–781.

Patel, J., Gondaliya, B., Patel, N. and Patel, A. M. (2019). A novel integration of laser engraver and 3D printer.

Paternò, L., Ibrahimi, M., Gruppioni, E., Menciassi, A. and Ricotti, L. (2018). Sockets for limb prostheses: a review of existing technologies and open challenges. *IEEE Transactions on Biomedical Engineering*, 65(9): 1996–2010.

Paterson, A. M., Bibb, R., Campbell, R. I. and Bingham, G. (2015). Comparing additive manufacturing technologies for customised wrist splints. *Rapid Prototyping Journal*, 21(3): 230–243.

Pauzon, C. N. G. (2021). *Tailored Process Gases for Laser Powder Bed Fusion*. Chalmers Tekniska Hogskola (Sweden).

Paxton, N. C., Nightingale, R. C. and Woodruff, M. A. (2022). Capturing patient anatomy for designing and manufacturing personalized prostheses. *Current Opinion in Biotechnology*, 73: 282–289.

Pejić, P., Rizov, T., Krasić, S. and Stajić, B. (2014, October). Augmented reality application in engineering. *In*: *3rd International Congress, SMAT* (No. s 39, p. 44).

Pell, M. (2017). *Envisioning Holograms: Design Breakthrough Experiences for Mixed Reality*. Apress.

Pelletier-Galarneau, M. and Ruddy, T. D. (2021). The potential of regional myocardial blood flow measurement with SPECT. *Journal of Nuclear Cardiology*, 28: 260–262.

Penumakala, P. K., Santo, J. and Thomas, A. (2020). A critical review on the fused deposition modeling of thermoplastic polymer composites. *Composites Part B: Engineering*, 201: 108336.

Pereverzyev, V. (2023), Best Software for Dental Implant Planning: A Comprehensive Review. Online: https://www.vladpereverzyev.com/the-best-software-for-dental-implant-planning-a-comprehensive-review/.

Peter, S. C., Dhanjal, J. K., Malik, V., Radhakrishnan, N., Jayakanthan, M., Sundar, D. et al. (2019). Encyclopedia of bioinformatics and computational biology. *Ranganathan, S., Grib-skov, M., Nakai, K., Schönbach, C., Eds*, 661–676.

Petralia, G., Preda, L., Giugliano, G., Jereczek-Fossa, B. A., Rocca, A., D'Andrea, G. et al. (2009). Perfusion computed tomography for monitoring induction chemotherapy in patients with squamous cell carcinoma of the upper aerodigestive tract: correlation between changes in tumor perfusion and tumor volume. *Journal of Computer Assisted Tomography*, 33(4): 552–559.

Pham, D. and Dimov, S. S. (2012). *Rapid Manufacturing: The Technologies and Applications of Rapid Prototyping and Rapid Tooling*. Springer Science & Business Media.

Pine, B. J. (1999). *Mass Customization: The New Frontier in Business Competition*. Harvard Business Press.

Pine, B. J. (2018). Customering: The next stage in the shift to mass customization. pp. 13–28. *In*: *Mass Customization and Design Democratization*. Routledge.

Polish Science. (2023). Innovative prosthesis for children wins in the "EUREKA! DGP – we discover Polish inventions". Online: https://polishscience.pl/en/innovative-prosthesis-for-children-wins-in-the-eureka-gpd-we-discover-polish-inventions/.

Poly Haven .(2024). Poly Haven. Online: https://polyhaven.com/about-contact.

Popescu, D., Lăptoiu, D. and Căruțașu, N. L. (2024). Considerations on the Design, Printability and Usability of Customized 3D-Printed Upper Limb Orthoses. *Applied Sciences*, 14(14): 6157.

Post, D. E. and Kendall, R. P. (2021). *Creating and Using Virtual Prototyping Software: Principles and Practices*. Addison-Wesley Professional.

Prashar, G., Vasudev, H. and Bhuddhi, D. (2023). Additive manufacturing: expanding 3D printing horizon in industry 4.0. *International Journal on Interactive Design and Manufacturing (IJIDeM)*, 17(5): 2221–2235.

Pyteraf, J., Pacławski, A., Jamróz, W., Mendyk, A., Paluch, M. and Jachowicz, R. (2022). Application and multi-stage optimization of daylight polymer 3D printing of personalized medicine products. *Pharmaceutics*, 14(4): 843.

Qi, Y., Mao, Z., Zhang, M. and Guo, H. (2020). Manufacturing practices and servitization: The role of mass customization and product innovation capabilities. *International Journal of Production Economics*, 228: 107747.

Qu, H. (2020). Additive manufacturing for bone tissue engineering scaffolds. *Materials Today Communications*, 24: 101024.

Raja, V. and Fernandes, K. J. (eds.). (2007). *Reverse Engineering: An Industrial Perspective*. Springer Science & Business Media.

Rangarajan, K., Davis, H. and Pucher, P. H. (2020). Systematic review of virtual haptics in surgical simulation: a valid educational tool?. *Journal of Surgical Education*, 77(2): 337–347.

Raphael, B. and Smith, I. F. (2003). *Fundamentals of Computer-Aided Engineering*. John Wiley & sons.

Rausch, M., Hiestand, P., Baumann, D., Cannet, C. and Rudin, M. (2003). MRI-based monitoring of inflammation and tissue damage in acute and chronic relapsing EAE. *Magnetic Resonance in Medicine: An Official Journal of the International Society for Magnetic Resonance in Medicine*, 50(2): 309–314.

Redmann, A., Oehlmann, P., Scheffler, T., Kagermeier, L. and Osswald, T. A. (2020). Thermal curing kinetics optimization of epoxy resin in Digital Light Synthesis. *Additive Manufacturing*, 32: 101018.

Reguia, C. (2014). Product innovation and the competitive advantage. *European Scientific Journal*, 1(1): 140–157.

Revilla-León, M., Gómez-Polo, M., Vyas, S., Barmak, B. A., Galluci, G. O., Att, W. et al. (2023). Artificial intelligence applications in implant dentistry: A systematic review. *The Journal of Prosthetic Dentistry*, 129(2): 293–300.

Rieber, A. B. H. G. A., Brambs, H. J., Gabelmann, A., Heilmann, V., Kreienberg, R. and Kühn, T. (2002). Breast MRI for monitoring response of primary breast cancer to neo-adjuvant chemotherapy. *European Radiology*, 12: 1711–1719.

Rigger, E., Shea, K. and Stankovic, T. (2018). Task categorisation for identification of design automation opportunities. *Journal of Engineering Design*, 29(3): 131–159.

Rimini, S., Vincent, J. and Wake, N. (2022). Computer-Aided Design Principles for Anatomic Modeling. pp. 45–59. *In*: *3D Printing for the Radiologist*. Elsevier.

Rizzo, A., Reger, G., Gahm, G., Difede, J. and Rothbaum, B. O. (2009). Virtual reality exposure therapy for combat-related PTSD. *Post-traumatic Stress Disorder: Basic Science and Clinical Practice*, 375–399.

Robinson, P. M. and O'meara, M. J. (2009). The Thomas splint: its origins and use in trauma. *The Journal of Bone & Joint Surgery British Volume*, 91(4): 540–544.

Robiony, M., Salvo, I., Costa, F., Zerman, N., Bazzocchi, M., Toso, F. et al. (2007). Virtual reality surgical planning for maxillofacial distraction osteogenesis: the role of reverse engineering rapid prototyping and cooperative work. *Journal of Oral and Maxillofacial Surgery*, 65(6): 1198–1208.

Rojek, I., Mikołajewski, D., Kotlarz, P., Macko, M. and Kopowski, J. (2021). Intelligent system supporting technological process planning for machining and 3D printing. *Bulletin of the Polish Academy of Sciences: Technical Sciences*.

Rolland, J. P. (2016). Functional Materials for 3D Manufacturing using Carbon's CLIP Technology. *Journal of Photopolymer Science and Technology*, 29(3): 451–452.

Rouhana, C. M., Aoun, M. S., Faek, F. S., Eljazzar, M. S. and Hamzeh, F. R. (2014, June). The reduction of construction duration by implementing contourontour crafting (3D printing). pp. 1031–1042. *In*: *Proceedings of the 22nd Annual Conference of the International Group for Lean Construction: Understanding and Improving Project Based Production, IGLC*.

Ruggieri, P., Mavrogenis, A. F., Pala, E., Romantini, M., Manfrini, M. and Mercuri, M. (2013). Outcome of expandable prostheses in children. *Journal of Pediatric Orthopaedics*, 33(3): 244–253.

Rumbo-Rodríguez, L., Sánchez-SanSegundo, M., Ferrer-Cascales, R., García-D'Urso, N., Hurtado-Sánchez, J. A. and Zaragoza-Martí, A. (2021). Comparison of body scanner and manual anthropometric measurements of body shape: a systematic review. *International Journal of Environmental Research and Public Health*, 18(12): 6213.

Runge, V. M. and Heverhagen, J. T. (2022). The clinical utility of magnetic resonance imaging according to field strength, specifically addressing the breadth of current state-of-the-art systems, which include 0.55 T, 1.5 T, 3 T, and 7 T. *Investigative Radiology*, 57(1): 1–12.

Rupp, A. I. and Gruber, P. (2022). A case study in biomimetic roofing: Moisture dissipation from leaf-shaped shingles. pp. 467–487. *In*: *Biomimicry for Materials, Design and Habitats*. Elsevier.

Rybarczyk, J., Dorna, P., Górski, F., Kuczko, W. and Żukowska, M. (2024b, June). Personalized Mechatronic Lower Limb Orthosis. pp. 325–336. *In*: *International Conference Innovation in Engineering*. Cham: Springer Nature Switzerland.

Rybarczyk, J., Kuczko, W., Wichniarek, R., Górski, F. and Żukowska, M. (2024a). Study of individualized leg orthoses made by 3D printing of composite material. pp. 59–67. *In*: *Innovations in Biomedical Engineering 2023*. Cham: Springer Nature Switzerland.

Ryu, C. W., Kim, J. K., Kim, S. J., Lee, J. H., Kim, J. H., Ha, H. I. et al. (2009). Head and neck vascular lesions: characterization of the flow pattern by the use of three-phase CT. *Korean Journal of Radiology*, 10(4): 323–332.

Saaksvuori, A. and Immonen, A. (2008). *Product Lifecycle Management Systems* (pp. 27–44). Springer Berlin Heidelberg.

Sacco, R., Munoz, M. A., Billuart, F., Lalevée, M. and Beldame, J. (2023). Validation of an automated optical scanner for a comprehensive anthropometric analysis of the foot and ankle. *Bioengineering*, 10(8): 968.

Saito, Y., Sugimoto, M., Imura, S., Morine, Y., Ikemoto, T., Iwahashi, S. et al. (2020). Intraoperative 3D hologram support with mixed reality techniques in liver surgery. *Annals of Surgery*, 271(1): e4–e7.

Salehi, V. and Wang, S. (2019). Web-based visualization of 3D factory layout from hybrid Modeling of CAD and point cloud on virtual globe DTX solution. *Computer-Aided Design and Applications*, 16(2): 243–255.

Salmi, M. (2016). Possibilities of preoperative medical models made by 3D printing or additive manufacturing. *Journal of Medical Engineering*, 2016(1): 6191526.

Salmi, M. (2021). Additive manufacturing processes in medical applications. *Materials*, 14(1): 191.

Saniuk, S., Grabowska, S. and Gajdzik, B. (2020). Personalization of products in the industry 4.0 concept and its impact on achieving a higher level of sustainable consumption. *Energies*, 13(22): 5895.

Sarmah, S. S. (2019). Concept of artificial intelligence, its impact and emerging trends. *Int. Res. J. Eng. Technol.*, 6(11): 2164–2168.

Schaub, P. B. (2004). Strategies of experts in engineering design: between innovation and routine behaviour. *Journal of Design Research*, 4(2): 125–143.

Schlarp, J., Csencsics, E. and Schitter, G. (2019). Optical scanning of a laser triangulation sensor for 3-D imaging. *IEEE Transactions on Instrumentation and Measurement*, 69(6): 3606–3613.

Schlégl, Á. T., Told, R., Kardos, K., Szőke, A., Ujfalusi, Z. and Maróti, P. (2022). Evaluation and comparison of traditional plaster and fiberglass casts with 3D-printed PLA and PLA–CaCO3 composite splints for bone-fracture management. *Polymers*, 14(17): 3571.

Schmidt, C., Finsterwalder, F., Griesbaum, R. and Sehrt, J. T. (2022, July). Cost analysis of automated additive printer farms. *In*: *IOP Conference Series: Earth and Environmental Science* (Vol. 1048, No. 1, p. 012008). IOP Publishing.

Schreiber, G. (2007). Knowledge Engineering: Handbook of Knowledge Representation, Elsevier, 929–946.

Schreiber, G. (2012). Knowledge acquisition and the web. *International Journal of Human-Computer Studies*. Elsevier, 71: 206–210.

Schreiber, G., Akkermans, H., Anjewierden, A., de Hoog, R., Shadbolt, N., Van de Velde, W. et al. (1999). Knowledge engineering and man-agement: The CommonKADS Methodology. Cambridge, MA, MIT Press.

Schwartz, D. A. and Schofield, K. A. (2023). Utilization of 3D printed orthoses for musculoskeletal conditions of the upper extremity: A systematic review. *Journal of Hand Therapy*, 36(1): 166–178.

Scott, L. P. (2018). Producer Management Systems—The Need for Automation and Integration. pp. 4–1. *In*: *Insurance Technology Handbook*. CRC Press.

Sefene, E. M. (2022). State-of-the-art of selective laser melting process: A comprehensive review. *Journal of Manufacturing Systems*, 63: 250–274.

Semba, J. A., Mieloch, A. A. and Rybka, J. D. (2020). Introduction to the state-of-the-art 3D bioprinting methods, design, and applications in orthopedics. *Bioprinting*, 18: e00070.

Semba, J. A., Mieloch, A. A., Tomaszewska, E., Cywoniuk, P. and Rybka, J. D. (2023). Formulation and evaluation of a bioink composed of alginate, gelatin, and nanocellulose for meniscal tissue engineering. *International Journal of Bioprinting*, 9(1).

Shahar, F. S., Sultan, M. H., Shah, A. U. M. and Safri, S. A. (2020). A comparative analysis between conventional manufacturing and additive manufacturing of ankle-foot orthosis. *Applied Science and Engineering Progress*, 13(2): 96–103.

Shanmugam, V., Pavan, M. V., Babu, K. and Karnan, B. (2021). Fused deposition modeling based polymeric materials and their performance: A review. *Polymer Composites*, 42(11): 5656–5677.

Shaoki, A., Xu, J. Y., Sun, H., Chen, X. S., Ouyang, J., Zhuang, X. M. et al. (2016). Osseointegration of three-dimensional designed titanium implants manufactured by selective laser melting. *Biofabrication*, 8(4): 045014.

Sharma, A., Sharma, Y., Singh, K. R., Vardhan, H., Bhatt, H. and Agrahari, R. (2024). Bionics in medicine: the future of organ replacement. *Acta Pharma Reports*.

Sheoran, A. J., Kumar, H., Arora, P. K. and Moona, G. (2020). Bio-medical applications of additive manufacturing: a review. *Procedia Manufacturing*, 51: 663–670.

Siiskonen, M., Malmqvist, J. and Folestad, S. (2020). Integrated product and manufacturing system platforms supporting the design of personalized medicines. *Journal of Manufacturing Systems*, 56: 281–295.

Silesian University of Technology. (2007), MOKA KA Application, online: https://kpkm.polsl.pl/PROJEKTY/KBE/moka_ka/moka_ka_en.html.

Silveira, A. T., Souza, M. A. D., Fernandes, B. L. and Nohama, P. (2018). From the past to the future of therapeutic orthoses for upper limbs rehabilitation. *Research on Biomedical Engineering*, 34: 368–380.

Sing, S. L., Yeong, W. Y., Wiria, F. E., Tay, B. Y., Zhao, Z., Zhao, L. et al. (2017). Direct selective laser sintering and melting of ceramics: a review. *Rapid Prototyping Journal*, 23(3): 611–623.

Skarbez, R., Smith, M. and Whitton, M. C. (2021). Revisiting Milgram and Kishino's reality-virtuality continuum. *Frontiers in Virtual Reality*, 2: 647997.

Skarka, W. (2007). Application of MOKA methodology in generative model creation using CATIA. *Engineering Applications of Artificial Intelligence*, 20(5): 677–690.

Skarka, W., Nalepa, R. and Musik, R. (2023). Integrated aircraft design system based on generative modelling. *Aerospace*, 10(8): 677.

Smith, C. F., Tollemache, N., Covill, D. and Johnston, M. (2018). Take away body parts! An investigation into the use of 3D-printed anatomical models in undergraduate anatomy education. *Anatomical Sciences Education*, 11(1): 44–53.

Sofronov, Y., Zagorski, M., Todorov, G. and Gavrailov, T. (2019). Approach for reverse engineering of complex geometry components. *BulTrans, Sozopol, Bulgaria*.

Song, C., Xia, S., Zhang, H., Zhang, L., Li, X., Wang, K. et al. (2022). Novel endovascular interventional surgical robotic system based on biomimetic manipulation. *Micromachines*, 13(10): 1587.

Souto, A. L. M., Souto, R. M., Teixeira, I. C. R. and Nacif, M. S. (2017). Myocardial viability on cardiac magnetic resonance. *Arquivos brasileiros de cardiologia*, 108(5): 458–469.

Spahiu, T., Canaj, E. and Shehi, E. (2020). 3D printing for clothing production. *Journal of Engineered Fibers and Fabrics*, 15: 1558925020948216.

Spierings, A. B. and Klahn, C. (2023). Benefits using additive manufacturing. pp. 115–124. *In*: *Springer Handbook of Additive Manufacturing*. Cham: Springer International Publishing.

Sprawls, P. (2000). *Magnetic Resonance Imaging: Principles, Methods, and Techniques*. Madison: Medical Physics Publishing.

Sree, K. A., Yagnik, A., Chordiya, P. K., Choudhury, K. and Kumar, P. (2023). Nature-inspired vascularised materials and devices for biomedical engineering. *Current Opinion in Biomedical Engineering*, 25: 100444.

Sriramakrishnan, P., Kalaiselvi, T., Padmapriya, S. T., Shanthi, N., Ramkumar, S. and Kalaichelvi, N. (2019). An medical image file formats and digital image conversion. *Int. J. Eng. Adv. Technol*, 9(1S3): 74–78.

Stamenković, D., Obradović-Đuričić, K., Rudolf, R., Bobovnik, R. and Stamenković, D. (2019). Selective laser melting and sintering technique of the cobalt-chromium dental alloy. *Srpski arhiv za celokupno lekarstvo*, 147(11-12): 664–669.

Standaert, W. (2022). Product digitalization at Nike: The future is now. *Journal of Information Technology Teaching Cases*, 12(1): 28–34.

Stark, E., Haffner, O. and Kučera, E. (2022). Low-cost method for 3D body measurement based on photogrammetry using smartphone. *Electronics*, 11(7): 1048.

State, G. (2023). Rapidly Generate 3D Assets for Virtual Worlds with Generative AI. NVidia Developer Blog. Online: https://developer.nvidia.com/blog/rapidly-generate-3d-assets-for-virtual-worlds-with-generative-ai/.

Steier, G., Peukert, S. and Lanza, G. (2022). An integrated approach for developing production strategy and digital business model in global production. pp. 666–674. *In*: *Production at the Leading Edge of Technology: Proceedings of the 11th Congress of the German Academic Association for Production Technology (WGP), Dresden, September 2021*. Springer International Publishing.

Stokes, M. (2001). Managing Engineering Knowledge; MOKA: Methodology for Knowledge Based Engineering Applications, Professional Engineering Publishing, London 2001.

Storrs, C. (2013). Do CT scans cause cancer?. *Scientific American*, 309(1): 30–33.

Strong, D., Kay, M., Conner, B., Wakefield, T. and Manogharan, G. (2018). Hybrid manufacturing–integrating traditional manufacturers with additive manufacturing (AM) supply chain. *Additive Manufacturing*, 21: 159–173.

Study in Poland. (2020). Poznan University of Technology has joined #printfordoctor #drukujdlalekarza campaign. Online: https://study.gov.pl/news/poznan-university-technology-has-joined-printfordoctor-drukujdlalekarza-campaign.

Su, A. and Al'Aref, S. J. (2018). History of 3D printing. pp. 1–10. *In*: *3D Printing Applications in Cardiovascular Medicine*. Academic Press.

Subramani, R., Mustafa, M. A., Ghadir, G. K., Al-Tmimi, H. M., Alani, Z. K., Rusho, M. A. et al. (2024). Advancements in 3D printing materials: A comparative analysis of performance and applications. *Applied Chemical Engineering*, 3867–3867.

Suetens, P. (2017). *Fundamentals of Medical Imaging*. Cambridge University Press.

Swinnen, E. and Kerckhofs, E. (2015). Compliance of patients wearing an orthotic device or orthopedic shoes: a systematic review. *Journal of Bodywork and Movement Therapies*, 19(4): 759–770.

Tai, Y. M. (2017). Effects of product lifecycle management systems on new product development performance. *Journal of Engineering and Technology Management*, 46: 67–83.

Targońska, S., Dobrzyńska-Mizera, M., Di Lorenzo, M. L., Knitter, M., Longo, A., Dobrzyński, M. et al. (2024). Design, clinical applications and post-surgical assessment of bioresorbable 3D-printed craniofacial composite implants. *Biomaterials Science*.

Ten Kate, J., Smit, G. and Breedveld, P. (2017). 3D-printed upper limb prostheses: a review. *Disability and Rehabilitation: Assistive Technology*, 12(3): 300–314.

Tettey, F., Parupelli, S. K. and Desai, S. (2024). A review of biomedical devices: classification, regulatory guidelines, human factors, software as a medical device, and cybersecurity. *Biomedical Materials & Devices*, 2(1): 316–341.

Thirunavukarasu, A. J., Ting, D. S. J., Elangovan, K., Gutierrez, L., Tan, T. F. and Ting, D. S. W. (2023). Large language models in medicine. *Nature Medicine*, 29(8): 1930–1940.

Thomas, D. J., Azmi, M. M. and Tehrani, Z. (2014). 3D additive manufacture of oral and maxillofacial surgical models for preoperative planning. *The International Journal of Advanced Manufacturing Technology*, 71: 1643–1651.

Tickoo, S. (2018). *CATIA V5-6R2018 for Designers*. Cadcim Technologies.

Tissir, S., El Fezazi, S. and Cherrafi, A. (2020, December). Industry 4.0 impact on lean manufacturing: literature review. pp. 1–5. *In*: *2020 IEEE 13th International Colloquium of Logistics and Supply Chain Management (LOGISTIQUA)*. IEEE.

Tiwari, S. K., Pande, S., Agrawal, S. and Bobade, S. M. (2015). Selection of selective laser sintering materials for different applications. *Rapid Prototyping Journal*, 21(6): 630–648.

Tobin, J. J. and Walsh, G. (2023). *Medical Product Regulatory Affairs: Pharmaceuticals, Diagnostics, Medical Devices*. John Wiley & Sons.

Tomić, I., Juric, I., Dedijer, S. and Adamović, S. (2023, September). Artificial intelligence in graphic design. pp. 85–93. *In*: *Proceedings of the 54th Annual Scientific Conference of the International Circle of Educational Institutes of Graphic-Media Technology and Management, The Hellenic Union of Graphic Arts and Media Technology Engineers, Greece*.

Török iD, J., Vodilka, A., Kaščak, J. and Kočiško, M. (2024). Check for design of personalized orthoses with support of PTC Creo and FDM technology. *Advances in Manufacturing IV: Volume 5-Biomedical Engineering: Digitalization, Sustainability and Industry Applications*, 5: 149.

Toth, T., Hudak, R. and Zivcak, J. (2015). Dimensional verification and quality control of implants produced by additive manufacturing. *Quality Innovation Prosperity*, 19(1): 9–21.

Tripathi, K. P. (2011). A review on knowledge-based expert system: concept and architecture. *IJCA Special Issue on Artificial Intelligence Techniques-Novel Approaches & Practical Applications*, 4: 19–23.

Triwiyanto, T., Maghfiroh, A. M., Musvika, S. D., Amrinsani, F., Syaifudin, Mak'ruf, R. and Sulowicz, M. (2023, April). State of the art methods of machine learning for prosthetic hand development: a review. pp. 555–574. *In*: *Proceeding of the 3rd International Conference on Electronics, Biomedical Engineering, and Health Informatics: ICEBEHI 2022, 5–6 October, Surabaya, Indonesia*. Singapore: Springer Nature Singapore.

Tseng, M. M. and Jiao, J. (2001). Mass customization. *Handbook of Industrial Engineering*, 3: 684–709.

Tsioukas, V., Pikridas, C. and Karolos, I. A. (2020). Challenges, opportunities, and limitations in 3D printing. *3D Printing: Applications in Medicine and Surgery*, 151–155.

Tsoulfas, G., Bangeas, P. I. and Suri, J. (eds.). (2019). *3D Printing: Application in Medical Surgery E-Book*. Elsevier Health Sciences.

Tuazon, B. J., Custodio, N. A. V., Basuel, R. B., Delos Reyes, L. A. and Dizon, J. R. C. (2022). 3D printing technology and materials for automotive application: a mini-review. *Key Engineering Materials*, 913: 3–16.

Turner, J. S. and Soar, R. C. (2008, May). Beyond biomimicry: What termites can tell us about realizing the living building. pp. 1–18. *In*: *First International Conference on Industrialized, Intelligent Construction at Loughborough University*.

Turostowska, K. (2016), *Methodology of Data Preparation for Rapid Manufacturing of Surgery Supporting Models*. Master's Thesis (advisor: Górski F.), Poznan University of Technology.

Tyhaar, J. L. (2021). *The Continuous Exploration of Needs: Sentiment Feedback from Twitter for Product Innovation and Improvement (At the Example of the Apple iPhone)* (Bachelor's thesis, University of Twente).

Uriondo, A., Esperon-Miguez, M. and Perinpanayagam, S. (2015). The present and future of additive manufacturing in the aerospace sector: A review of important aspects. *Proceedings of the Institution of Mechanical Engineers, Part G: Journal of Aerospace Engineering*, 229(11): 2132–2147.

Vadiraj, A., Abraham, M. and Bharadwaj, A. S. (2019). Trends in automotive light weighting. pp. 89–102. *In*: *Light Weighting for Defense, Aerospace, and Transportation*. Singapore: Springer Singapore.

Valls, A. E. (2023). *Comparison of Intraoral Scans and Desktop Scans for Assessment of Dimensional Changes of Invisalign® Attachments* (Master's thesis, The University of Texas School of Dentistry at Houston).

Van Alsenoy, K., Ryu, J. H. and Girard, O. (2019). The effect of EVA and TPU custom foot orthoses on running economy, running mechanics, and comfort. *Frontiers in Sports and Active Living*, 1: 34.

VBionic. (2020). TOLKA: Functional hand prosthesis. Online: https://vbionic.com/en/tolka-bionic-prosthesis/.

Vendittoli, P. A., Riviere, C., Hirschmann, M. T. and Bini, S. (2023). Why personalized surgery is the future of hip and knee arthroplasty: a statement from the Personalized Arthroplasty Society. *EFORT Open Reviews*, 8(12): 874–882.

Verlinden, J. C., de Smit, A., Horváth, I., Epema, E. and de Jong, M. (2003). Time compression characteristics of the Augmented Prototyping Pipeline. *In*: *Proceedings of the Euro-uRapid Conference*.

Verma, G. (2018). *Autodesk Fusion 360 Black Book*. BPB Publications.

Vidakis, N., Petousis, M. and Maniadi, A. (2021). Sustainable additive manufacturing: Mechanical response of high-density polyethylene over multiple recycling processes. *Recycling*, 6(1): 4.

Vitković, N., Stojković, J. R., Korunović, N., Teuţan, E., Pleşa, A., Ianoşi-Andreeva-Dimitrova, A. et al. (2023a). Extra-articular distal humerus plate 3D model creation by using the method of anatomical features. *Materials*, 16(15): 5409.

Vitković, N., Stojković, M., Trajanović, M. and Górski, F. (2023b). Creation of personalized 3D model of bone scaffold by using curve and pattern based methodologies. *ACTA TECHNICA NAPOCENSIS-Series: APPLIED MATHEMATICS, MECHANICS, and ENGINEERING*, 66(3).

Vora, H. D. and Sanyal, S. (2020). A comprehensive review: metrology in additive manufacturing and 3D printing technology. *Progress in Additive Manufacturing*, 5(4): 319–353.

Waguespack, C. (2013). *Mastering Autodesk Inventor 2014 and Autodesk Inventor LT 2014: Autodesk Official Press*. John Wiley & Sons.

Waisberg, E., Ong, J., Masalkhi, M., Kamran, S. A., Zaman, N., Sarker, P. et al. (2023). GPT-4: a new era of artificial intelligence in medicine. *Irish Journal of Medical Science (1971-)*, 192(6): 3197–3200.

Wald, L. L., McDaniel, P. C., Witzel, T., Stockmann, J. P. and Cooley, C. Z. (2020). Low-cost and portable MRI. *Journal of Magnetic Resonance Imaging*, 52(3): 686–696.

Walloch, M. (2023). DraftSight 3D Modeling Part 2 – Creating Solid Bodies. Javelin Tech. Online: https://www.javelin-tech.com/blog/2023/06/draftsight-3d-modeling-part-2-creating-solid-bodies/.

Wang, D., Ma, D., Wong, M. L. and Wáng, Y. X. J. (2015). Recent advances in surgical planning & navigation for tumor biopsy and resection. *Quantitative Imaging in Medicine and Surgery*, 5(5): 640.

Wang, G. G. (2002). Definition and review of virtual prototyping. *J. Comput. Inf. Sci. Eng.*, 2(3): 232–236.

Wang, M., Nong, Q., Liu, Y. and Yu, H. (2022). Design of lower limb prosthetic sockets: a review. *Expert Review of Medical Devices*, 19(1): 63–73.

Wang, Y. M., Yu, S. Y., Ren, S., Cheng, S. and Liu, J. Z. (2020). Close-range industrial photogrammetry and application: review and outlook. *AOPC 2020: Optics Ultra Precision Manufacturing and Testing*, 11568: 152–162.

Wartofsky, M. W. (2012). *Models: Representation and the Scientific Understanding* (Vol. 48). Springer Science & Business Media.

Webster, J. G. and Eren, H. (2018). *Measurement, Instrumentation, and Sensors Handbook: Two-Volume Set*. CRC press.

Weitz, J., Deppe, H., Stopp, S., Lueth, T., Mueller, S. and Hohlweg-Majert, B. (2011). Accuracy of templates for navigated implantation made by rapid prototyping with DICOM datasets of cone beam computer tomography (CBCT). *Clinical Oral Investigations*, 15: 1001–1006.

Wernke, M. M., Schroeder, R. M., Haynes, M. L., Nolt, L. L., Albury, A. W. and Colvin, J. M. (2017). Progress toward optimizing prosthetic socket fit and suspension using elevated vacuum to promote residual limb health. *Advances in Wound Care*, 6(7): 233–239.

Wichniarek, R., Górski, F. and Kuczko, W. (2014). Application of additive technology of Fused Deposition Modeling for manufacturing small series of elastic models. *Applied Computer Science*, 10(2): 20–31.

Wichniarek, R., Górski, F., Kuczko, W. and Żukowska, M. (2020). Accuracy and Repeatability of Limb Scans Obtained on the Semi-Automatic Measuring Station. *Advances in Science and Technology. Research Journal*, 14(4).

Wodzinski, M., Daniol, M., Socha, M., Hemmerling, D., Stanuch, M. and Skalski, A. (2022). Deep learning-based framework for automatic cranial defect reconstruction and implant modeling. *Computer Methods and Programs in Biomedicine*, 226: 107173.

Wolszczak, P., Lygas, K., Paszko, M. and Wach, R. A. (2018). Heat distribution in material during fused deposition modelling. *Rapid Prototyping Journal*, 24(3): 615–622.

Woo, E. J. (2014). Postmarketing safety of biologics and biological devices. *The Spine Journal*, 14(3): 560–565.

World Health Organization. (2017). WHO global model regulatory framework for medical devices including *in vitro* diagnostic medical devices.

Wortmann, M. and Frese, N. (2022). Industrial-scale vacuum casting with silicone molds: A review. *Applied Research*, 1(1-2): e202100012.

Wu, D., Rosen, D. W., Wang, L. and Schaefer, D. (2015). Cloud-based design and manufacturing: A new paradigm in digital manufacturing and design innovation. *Computer-aided Design*, 59: 1–14.

Wu, T., Yuan, Y. J., Zhang, L. X., Yang, J., Cao, Y. P., Yan, L. Q. et al. (2024). Recent advances in 3d gaussian splatting. *Computational Visual Media*, 1–30.

Xu, K. and Qin, S. (2023, May). An Interdisciplinary approach and advanced techniques for enhanced 3D-printed upper limb prosthetic socket design: A literature review. *In*: *Actuators* (Vol. 12, No. 6, p. 223). MDPI.

Xue, S. (2019). Intelligent system for products personalization and design using genetic algorithm. *Journal of Intelligent & Fuzzy Systems*, 37(1): 63–70.

Yamashita, K., Higashino, K., Wada, K., Morimoto, M., Abe, M., Takata, Y. et al. (2016). Radiation exposure to the surgeon and patient during a fluoroscopic procedure: how high is the exposure dose? A cadaveric study. *Spine*, 41(15): 1254–1260.

Yan, C., Kleiner, C., Tabigue, A., Shah, V., Sacks, G., Shah, D. et al. (2023). PETG: applications in modern medicine. *Engineered Regeneration*.

Yang, J., Shi, Y., Shen, Q. and Yan, C. (2009). Selective laser sintering of HIPS and investment casting technology. *Journal of Materials Processing Technology*, 209(4): 1901–1908.

Yap, C. Y., Chua, C. K., Dong, Z. L., Liu, Z. H., Zhang, D. Q., Loh, L. E. et al. (2015). Review of selective laser melting: Materials and applications. *Applied Physics Reviews*, 2(4).

Ye, X., Wang, Z., Zhang, X., Zhou, M. and Cai, L. (2014). Hemocompatibility research on the micro-structure surface of a bionic heart valve. *Bio-Medical Materials and Engineering*, 24(6): 2361–2369.

Yeung, A. W. K., Tosevska, A., Klager, E., Eibensteiner, F., Laxar, D., Stoyanov, J. et al. (2021). Virtual and augmented reality applications in medicine: analysis of the scientific literature. *Journal of Medical Internet Research*, 23(2): e25499.

Young, B. H. (2022). The bionic-hand arms race: high-tech hands are complicated, costly, and often impractical. *IEEE Spectrum*, 59(10): 24–30.

Yount, Z. F., Kass, S. J. and Arruda, J. E. (2022). Route learning with augmented reality navigation aids. *Transportation Research Part F: Traffic Psychology and Behaviour*, 88: 132–140.

Yu, J., Tan, M., Wang, S. and Chen, E. (2004). Development of a biomimetic robotic fish and its control algorithm. *IEEE Transactions on Systems, Man, and Cybernetics, Part B (Cybernetics)*, 34(4): 1798–1810.

Yusupbekov, N. R., Gulyamov, S. M., Kasimov, S. S. and Usmanova, N. B. (2019). Knowledge-based planning for industrial automation systems: The way to support decision making. pp. 873–879. *In*: *13th International Conference on Theory and Application of Fuzzy Systems and Soft Computing—ICAFS-2018 13*. Springer International Publishing.

Zadpoor, A. A. and Malda, J. (2017). Additive manufacturing of biomaterials, tissues, and organs. *Annals of Biomedical Engineering*, 45: 1–11.

Zarek, M., Mansour, N., Shapira, S. and Cohn, D. (2017). 4D printing of shape memory-based personalized endoluminal medical devices. *Macromolecular Rapid Communications*, 38(2): 1600628.

Zawadzki, P. (2016). Methodology of building an automated system for design of configurable products using knowledge based engineering. PhD Thesis, Poznan University of Technology. Online: https://sin.put.poznan.pl/files/download/5660.

Zawadzki, P. (2018). Methodology of KBE system development for automated design of multivariant products. pp. 239–248. *In*: *Advances in Manufacturing*. Springer International Publishing.

Zawadzki, P. and Żywicki, K. (2016). Smart product design and production control for effective mass customization in the Industry 4.0 concept. *Management and Production Engineering Review*.

Zawadzki, P., Wichniarek, R., Kuczko, W., Slupińska, S. and Żukowska, M. (2022, May). Automated design and rapid manufacturing of low-cost customized upper limb prostheses. *In*: *Journal of Physics: Conference Series* (Vol. 2198, No. 1, p. 012040). IOP Publishing.

Zema, L., Melocchi, A., Maroni, A. and Gazzaniga, A. (2017). Three-dimensional printing of medicinal products and the challenge of personalized therapy. *Journal of Pharmaceutical Sciences*, 106(7): 1697–1705.

Zenn, M. and Jones, G. (2012). *Reconstructive Surgery: Anatomy, Technique, and Clinical Application*. CRC Press.

Zhan, H., Jiang, H. and Jiang, R. (2020). Three-dimensional images generated from diffuse ultrasound wave: Detections of multiple cracks in concrete structures. *Structural Health Monitoring*, 19(1): 12–25.

Zhang, H., Huang, L., Tan, M., Zhao, S., Liu, H., Lu, Z. et al. (2022). Overview of 3D-printed silica glass. *Micromachines*, 13(1): 81.

Zhang, Y. J. (2018). *Geometric Modeling and Mesh Generation from Scanned Images*. Chapman and Hall/CRC.

Zhang, Y., Feng, H., Zhao, Y. and Zhang, S. (2024). Exploring the application of the artificial-intelligence-integrated platform 3D Slicer in medical imaging education. *Diagnostics*, 14(2): 146.

Zhang, Y., Jedeck, S., Yang, L. and Bai, L. (2019). Modeling and analysis of the on-demand spare parts supply using additive manufacturing. *Rapid Prototyping Journal*, 25(3): 473–487.

Zhang, Y., Wang, J., Ahmad, R. and Li, X. (2022). Integrating lean production strategies, virtual reality technique and building information modeling method for mass customization in cabinet manufacturing. *Engineering, Construction and Architectural Management*, 29(10): 3970–3996.

Zhao, J., Yang, Y. and Li, L. (2020a). A comprehensive evaluation for different post-curing methods used in stereolithography additive manufacturing. *Journal of Manufacturing Processes*, 56: 867–877.

Zhao, K., Kim, K., Craig, J. R. and Palmer, J. N. (2020b). Using 3D printed sinonasal models to visualize and optimize personalized sinonasal sinus irrigation strategies. *Rhinology*, 58(3): 266.

Zhou, X., Ren, L., Song, Z., Li, G., Zhang, J., Li, B. et al. (2023). Advances in 3D/4D printing of mechanical metamaterials: From manufacturing to applications. *Composites Part B: Engineering*, 254: 110585.

Zou, Q., Feng, H. Y. and Gao, S. (2023). Variational direct modeling: A framework towards integration of parametric modeling and direct modeling in CAD. *Computer-Aided Design*, 157: 103465.

Żukowska, M., Boberska, D., Górski, F. and Okła, M. (2024a). Rapid Manufacturing of Individualized Supplies for Eye Socket Reconstructive Surgery. pp. 249–256. *In*: *Innovations in Biomedical Engineering 2023*. Cham: Springer Nature Switzerland.

Żukowska, M., Górski, F. and Bromiński, G. (2019a). Rapid manufacturing and virtual prototyping of pre-surgery aids. pp. 399–403. *In*: *World Congress on Medical Physics and Biomedical Engineering 2018: June 3–8, 2018, Prague, Czech Republic (Vol. 3)*. Springer Singapore.

Żukowska, M., Górski, F., Wichniarek, R. and Kuczko, W. (2019b). Methodology of low cost rapid manufacturing of anatomical models with material imitation of soft tissues. *Advances in Science and Technology. Research Journal*, 13(4): 120–128.

Żukowska, M., Górski, F., Wichniarek, R., Kuczko, W., Buczkowska-Andruszko, A. and Banaszewski, J. (2024b, March). Assessment of the usefulness of additively manufactured anatomical models in the process of preoperative support and education. pp. 121–129. *In*: *International Scientific-Technical Conference MANUFACTURING*. Cham: Springer Nature Switzerland.

Żukowska, M., Jezińska, R., Górski, F., Kuczko, W., Wichniarek, R., Banaszewski, J. and Buczkowska-Andruszko, A. (2022, April). Methodology of the rapid manufacturing of an individualized anatomical model of the tongue with a tumor for the preparation of an organ reconstruction operation. pp. 45–58. *In*: *International Scientific-Technical Conference MANUFACTURING*. Cham: Springer International Publishing.

Żukowska, M., Rad, M. A. and Górski, F. (2023). Additive manufacturing of 3D anatomical models—review of processes, materials and applications. *Materials*, 16(2): 880.

Zupan, B., Pakrac, V. and Nabergoj, A. S. (2022). Using design thinking to develop a new product: case study of developing a novel medical device. pp. 161–185. *In*: *Design Thinking*. CRC Press.

Index

For Product Safety Concerns and Information please contact our
EU representative GPSR@taylorandfrancis.com Taylor & Francis
Verlag GmbH, Kaufingerstraße 24, 80331 München, Germany